INTEGRATED APPROACHES TO WATER POLLUTION PROBLEMS

Proceedings of the
International Symposium (SISIPPA)
(Lisbon, Portugal 19-23 June 1989)

ACKNOWLEDGEMENTS

SISIPPA 89 was jointly organised by the Laboratório Nacional de Engenharia Civil (LNEC), the Portuguese Water Resources Association (APRH) and the International Water Resources Association (IWRA), with the cooperation of the United Nations (UN) through its specialised agencies UN/DTCD and UNESCO. Approximately 450 participants from 40 countries attended the Symposium.

The SISIPPA 89 Organising Committee wishes to express deep appreciation to the rapporteurs of the Symposium, Vladimir Novozny, Miguel Solanes and Zbignew Jan Brzezinsky, who have written the thematic papers and have drawn the conclusions of the Symposium presented hereinafter.

The SISIPPA 89 Organising Committee also acknowledges the subsidy conceded by the Calouste Gulbenkian Foundation for editing the papers.

INTEGRATED APPROACHES TO WATER POLLUTION PROBLEMS

EDITORS

João Bau
J. P. Lobo Ferreira
José Duarte Henriques
José de Oliveira Raposo

Published by
ELSEVIER APPLIED SCIENCE
LONDON and NEW YORK

Organised by the International Symposium on
Integrated Approaches to Water Pollution Problems (SISIPPA)

ELSEVIER SCIENCE PUBLISHERS LTD
Crown House, Linton Road, Barking, Essex IG11 8JU, England

Sole Distributor in the USA and Canada
ELSEVIER SCIENCE PUBLISHING CO., INC.
655 Avenue of the Americas, New York, NY 10010, USA

WITH 53 TABLES AND 118 ILLUSTRATIONS

© 1991 THE ORGANISING COMMITTEE, SISIPPA '89

British Library Cataloguing in Publication Data

SISIPPA '89; Lisbon Portugal
Integrated approaches to water pollution problems.
1. Water. Pollution
I. Title   II. Bau, J.
363.7394

ISBN 1-85166-659-1

Library of Congress CIP data applied for

\

Printed in Northern Ireland by The Universities Press (Belfast) Ltd.

CONTENTS

INTRODUCTION AND CONCLUSIONS OF THE SYMPOSIUM

In the last decades, social and economic development has brought about considerable changes in natural and human environmental conditions, particularly in population structure, life pattern, power sources, industrial and agricultural production, transport and tourism.

All these factors, either by themselves or in interaction, place an increasing pressure on available water resources, so producing serious disfunctions in the water ecosystem and harmful effects on human health as well.

Thus, it is important to ensure that social and economic development takes place with due respect for essential ecological balances, and safeguards public health and welfare. With this aim, the scientific and technical community throughout the world has been devoting growing attention to the study of the technical, economic, legal and institutional aspects of water resources management. In June 1989, the 'International Symposium of Integrated Approaches to Water Pollution Problems (SISIPPA 89)' was held at Laboratório Nacional de Engenharia Civil, in Lisbon, Portugal. The purpose of the Symposium was to discuss problems related to point and non-point pollution, bringing together professionals working in a broad range of fields such as scientists, engineers, industrialists, agricultural and land use planners and decision-makers, public health technicians and managers.

The papers presented at the Symposium cover three main areas: technical solutions to water pollution problems; legal economic and institutional aspects; and public health and welfare. English, French and Portuguese were the working languages of the Symposium. This collection presents the papers written in English and for ease of reference they have been divided up into six parts: integrated environmental management strategies; policies for pollution control; groundwater and river contamination; industrial and urban pollution; agricultural pollution; and measurement and data.

CONCLUSIONS

The group of participants, as a whole, realised that pollution was not a nuisance that had to be tolerated, but a serious threat to the very existence of the human population in many parts of this planet. Over the last fifty years the pollution problem had changed from localised pollution incidents in industrialised countries to a global problem. The question is not whether we can afford to combat pollution but how to solve the problem in an efficient, economic and equitable way. Past experience had shown that pollution problems must be attacked in a comprehensive, integrated manner. Solving only a portion of the problem in one area may create new problems and/or shift the pollution to other areas.

The papers and discussions presented to the Symposium indicated that integrated approaches are required and are the only way to achieve the objective of a cleaner environment. Solutions to the problem of pollution covered and suggested by the Symposium were structural (and expensive) such as sewers and treatment plants, and non-structural (less expensive or no cost), such as better methods of using the land, better management of agricultural wastes and reduction of erosion. Such integrated approaches and plans of action have four dimensions:

1. The plan must be multi-disciplinary, involving not only scientists and engineers, but also economists and lawyers, who can advise on who should pay and design regulations and incentive schemes, to overcome the externality problems. Moreover, policy-makers and political scientists must be involved to design and implement the necessary political and institutional infrastructures.

2. In order to take advantage of economies of scale the plan must be regional, meaning encompassing either a drainage basin or, in some cases, a political or cultural unit, ranging from a tribal territory to a state or province. Individual local efforts are inefficient. Effective approaches to pollution abatement must be regional, national or international.

3. The plan must be comprehensive, addressing all pollution loads from all sources and comparing them to the waste assimilative capacity of receiving surface and groundwaters. The plans of action should develop the most economical ways to reduce the loads below this capacity in order to implement the plan.
   Many pollutants, such as biodegradable organics or metals, become pollutants only when the loads to the environment exceed its capacity to assimilate waste. In many cases these limits can be established by modelling, and several excellent papers were presented on modelling to determine waste assimilative capacity. Papers were also presented on determining the waste loads from various sources and on possible treatment.

Both point and non-point sources of pollution must be considered and, once the waste assimilative capacities are established, trade-offs between the allowable source loads are possible and should be considered.

4. Efficiency, equity and fairness must be considered in the given institutional, economic and legal framework, and the technical approaches must fit within that framework. The technical solutions must be designed in a way that would be fair to all involved groups and would not put excessive burdens on one group of users, at the expense of others. In many countries political and institutional structures will have to be modified in order to carry out the technical pollution abatement efforts.

From a technical point-of-view, the following conclusions were drawn:

1. In general, pollution abatement planning today must be an integrated effort that will address: the pollution abatement of dry weather and wet weather point sources; reduction and mitigation of non-point pollution; land use and its pollution impact; the impact of drainage on pollution reduction; and the waste assimilative capacity of the receiving water bodies. Although in some parts of the world raw sewage discharges are still unabated, emphasis may be shifting from removal of organic, oxygen demanding pollution from urban and industrial waste-water sources, to controlling nutrients and toxics, both from point and non-point sources.

2. Traditionally, structural solutions are more common for urban and industrial point sources while non-structural management is applied to rural, primarily agricultural non-point sources. In many cases, enhancement of waste assimilative capacity and reduction of delivery of pollutants are feasible. The efficiency of the entire integrated approach will be increased if such measures are considered and included. It is necessary for political and legislative bodies to address the possible economic externalities associated with pollution abatement programmes if equitable and efficient solutions are to be achieved.

3. It is more important that the 1990s and the beginning of the next millennium will see a new, increased emphasis on solutions for the severe and large-scale pollution problems, such as: the pollution of the northern Adriatic Sea, the North Sea, and many large European and US rivers; contamination of the US – Canada Great Lakes; adverse effects of deforestation; severe degradation of surface and sub-surface water resources by agricultural, industrial and solid waste deposition practices; and pollution from urban stormwater runoff. These large-scale problems will require co-ordinated, integrated approaches and a change in the way the land is used. Losses of chemicals into water bodies in agricultural areas can be reduced by practicing sustainable agriculture. At the same time, attention at the local levels is and will be on protecting water resources and providing safe water for various beneficial uses. New, integrated approaches will emerge under a favourable political and economical climate.

On the policy, regulation and economic aspects a number of conclusions were reached:

1. Prevention, control, management, and elimination of surface and groundwater pollution can be achieved through a variety of planning, legal, economic, financial and technical measures.

2. Legal measures for water pollution control must include standards for effluents or emissions as well as for the receiving bodies, established according to intended water uses or/and ecological goals.

3. The legal system must provide: procedural alternatives for swift enforcement of legislation against public and private pollution entities; for public participation in policy making, water management and legal proceedings; and for flexible alternatives to resolve differences and conflicts related to water pollution.

4. Appropriate planning is crucial to the prevention, and elimination of pollution, resulting from non-point sources. Appropriate planning measures can include the regulation and eventually the prohibition of certain products, such as fertilizers and pesticides.

The implementation of the 'polluter pays principle' is a key element in achieving desired water quality objectives. Yet the application of the 'polluter pays principle' must not obscure the eventual need for affirmative public action.

5. Toxic pollutants and effluents shall be subject to stringent regulation measures.

6. The ultimate success of pollution control policies lies in, and is related to:
   (a) The development, adoption and implementation of prevention measures, such as the clean technologies;
   (b) The development of public awareness;
   (c) The concentration of pollution control power in a single, identified administrative unit;
   (d) The selection of appropriate territorial, regional or physical units, such as the river basin or other geographic singularity, for the implementation of programmes, plans and measures;

(e) The granting of appropriate legal authorities (issuance of permits, access to records, requirements for information, entrance and inspection, issuance of cease and desist orders) to administrative agencies entrusted with pollution prevention, control, and elimination;

(f) The development and dissemination of remedies for civil responsibilities based on the principle of strict liability.

7. Pollution of international natural resources shall be controlled by the principles of international law.

8. The implementation of environmental measures in general, and the protection of water resources in particular, is seriously and irreversibly hampered by the paucity of resources and the economic crisis faced by developing countries.

The following conclusions, relating to public health and welfare were drawn:

1. Social, economic and industrial developments are associated with an increase of known health risks originating from water pollution. However, at the same time, scientific and technological progress offers efficient ways and means of solving the existing problems, creating new challenges.

2. To take full advantage of the available knowledge and technology, close multidisciplinary co-operation is required. This applies equally to operational programmes and research.

3. There is a need for studies into the health risks and impact of water supply and sanitation. These may be best undertaken by the combined efforts of engineers, social scientists and epidemiologists. Such studies should pay special attention to appropriate measures of exposure to water-related health risks and to more precise measures of the health outcome.

João Bau (Chairman),
J P Lobo Ferreira,
José Duarte Henriques,
José de Oliveira Raposo,
Organising Committee of SISIPPA 89,
Lisbon, June 1989.

PART I

Integrated environmental management strategies

Chapter 1

SUSTAINABLE DEVELOPMENT OF WATER RESOURCES

L V Da Cunha (European Institute for Water)

ABSTRACT

The concept of sustainable development is explained, with particular reference to the Brundtland Report, pleading for the need to promote a sustainable development of water resources along the lines put forward by this report. A broad description of the world water resources situation is presented, and the main problems threatening the sustainable development of water resources are identified, with special reference to global climate change impacts. The paper concludes by clarifying the main actions required to ensure a sustainable development of water resources.

Key words: sustainable development, water resources, world water assessment, global climate change

1 - THE CONCEPT OF SUSTAINABLE DEVELOPMENT AND THE BRUNDTLAND REPORT

The concept of sustainable development has become a very popular one in the environment arena following publication of the report "Our Common Future" in 1987 (WCED 1987), in response to an initiative of the General Assembly of the United Nations. This report was prepared by an international commission of 22 leading personalities, the World Commission on Environment and Development, and is currently known as the "Brundtland Report" named after its Chairman, the Prime Minister of Norway who is a previous Norwegian Minister of the Environment. The aims of the Brundtland Report are, according to the Scope Statement of the Commission, as follows:

- "to propose long-term environmental strategies for achieving sustainable development by the year 2000 and beyond;
- to recommend ways concern for the environment may be translated into greater co-operation among developing countries and between countries at different stages of economic and social development and lead to the achievement of common and mutually supportive objectives that take account of the interrelationships between people, resources, environment, and development;
- to consider ways and means by which the international community can deal more effectively with environmental concerns;
- to help define shared perceptions of long-term environmental issues and the appropriate efforts needed to deal successfully with the problems of protecting and enhancing the environment, a long-term agenda for action during the coming decades, and aspirational goals for the world community."

Sustainable development is the key concept of the Brundtland Report, being defined as "development that meets the needs of the present without compromising the ability of future generations to meet their own needs".

It should be noted, however, that although the concept of sustainable development was widely made known by this United Nations initiative, it had indeed been introduced earlier and used previously as a basis for dealing with environmental issues.   In fact, other authors have used different terminology to illustrate concepts similar to sustainable development. Just to mention a few, reference can be made to Lester Brown, the current President of the Worldwatch Institute in Washington D.C., who makes use of the concept of "sustainability" (BROWN 1981), to Sachs, who refers to "eco-development" (SACHS 1984), and to the World Bank terminology when referring to "environmental ethics" (GOODLAND and LEDEC 1985).

The author, already fifteen years ago, also made use of a similar concept as the basis of a comprehensive analysis of water resources management problems (DA CUNHA et al. 1974 and DA CUNHA et al. 1977). This analysis considered the trade-off between consumer goods and environmental values, as given by Curve A in Fig. 1. From this curve, which corresponds to the "transformation curve" currently used in economic analysis, it can clearly be seen that if we want to assume an increase C in the level of consumer goods by going from point 1 to point 2 along curve A, we will have to relinquish part of the environmental values E_1. However, if steps are taken to replace curve A by another curve B, which corresponds to resorting to a more rational use of resources and a better integration of science and technology in the development process, we might then be able to go from point 1 to point 3 ensuring an increase in consumer goods without damage to the environment. We are indeed in this case in a situation of sustainable development. An even more favourable situation will be reached if we can succeed in setting a transformation curve such as curve C. Then we still have an increase in consumer goods ΔC, at the same time enabling an increase ΔE_2 of the environmental values.

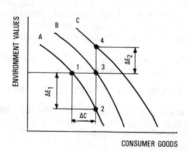

Fig. 1: Trade-off curve between consumer goods and environmental values

The concept of sustainable development is also associated with that of quality of life, introduced in the '60s. Sustainable development would then be a development which preserves the quality of life in its three components (see, e.g., SAINT-MARC 1971): standard of living (measured by the per capita consumption of consumer goods, such as food, clothing, housing, etc.); living conditions (defined by the standards of education, culture, health, security, leisure-time, distance between home and place of employment etc.); and quality of the environment (assessed through indices of green zones, free spaces, clean or polluted air and water etc.).

Sustainable development aims to ensure human survival and improvements in welfare and standards of living. Sustainable development should also imply an idea of equity in water resources use, to be achieved at several levels: a social equity between individuals at a given time (intra-generation equity) and between individuals throughout time (intergeneration equity), as well as an equity between countries and world regions (by managing international conflicts), and between man and his environment (by protecting the other species and nature in general).

Sometimes the term "sustainable" is used in relation to resources rather than to development. The dichotomy between sustainable and non-sustainable resources is considered by some authors to make more sense than the traditional dichotomy between renewable and non-renewable resources, as all resources may be considered as renewable on some time-scale.   TILTON and SKINNER 1987, for example, state that "the difference between resources is one of time-scales of replenishment" and further explain that the growth rates of sustainable resources are similar to or greater than use rates. "With wise management, sustainable resources can be used in perpetuity and need not become more costly as time passes even in the absence of cost-reducing technological change.   Over- or improper use can of course lead to depletion, but the concept of a sustainable resource is that with effective management the rate of resource withdrawal can be balanced by the rate of resource replenishment.   Water, soil, tidal power, and air are examples.   In contrast, non-sustainable resources are those for which withdrawal greatly exceeds the replenishment rates of the deposit types that have traditionally been exploited.   Examples of non-sustainable resources are petroleum, coal, copper, lead, and silver. For most metals and for petroleum, for example, rates of use exceed rates of replenishment by factors of thousands or more."

In general terms, the use of sustainable resources has generated less concern than that of non-sustainable resources, because it is considered they are not depleted.   However, these resources can also suffer from depletion, as the cost of obtaining additional supplies of the resource rises sharply.

Fig. 2 presented by TILTON and SKINNER 1987 represents three supply curves for a given resource showing the variations of the marginal cost of the resource as a function of the output.   If one considers curve S_1, for instance, it is apparent that as the rate of exploitation of the resource increases the supply curve at some point turns upward very quickly.   As Tilton and Skinner point out, this is partially due to the fact that the best sources for the resource tend to be exploited initially and, as the demand increases, lower quality and thus more costly resources must be brought to use.   One example is the case of groundwater where, as the rate of water withdrawal increases, deeper wells and more powerful pumps may be necessary.   The inadequate use of a sustainable resource or its careless management can also reduce or destroy the replenishment systems and determine that curve S_1 is replaced by a curve such as S_2 or S_3 implying even higher costs.   An adequate management of the water resources may make it possible to avoid going from curve S_1 to S_2 and S_3, and to operate on the more favourable parts of the first curve.   Then a sustainable development will be achieved.

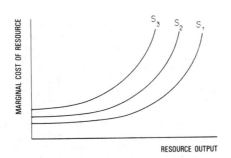

Fig. 2: Supply curves for sustainable resources

Sustainable development of water resources problems are in some way comparable to the problems of other natural resources, but there are important differences to take into consideration, which result from special characteristics of water resources.

One of the special characteristics of water resources is their renewability implying that water use does not correspond to a definite and irreversible consumption, as is the case with other natural resources.

While the stock of most mineral resources is limited, the water resources stock is not limited in this sense as it can, to a degree, be recovered yearly by the hydrological cycle.

Another important specific characteristic of the water resources is mobility. This characteristic implies that the several water uses (involving or not water withdrawal, effective consumption or pollution) taking place at different locations of the same river basin cannot be considered as being independent. The interdependence of the several water uses in a given river basin has very important social and economic consequences, both from an interregional and international standpoint.

There is a third important difference between water resources and other natural resources. For other natural resources, when their prices started to increase, due to growing demand or exhaustion of stocks, in many cases substitutes were found which proved capable of fulfilling an identical role at a lower cost, in some cases with even better results. In the case of water the situation is different, as although it is possible to improve efficiency in water resources use and even, at least in theory, to obtain as much water as is needed by making use of expensive solutions such as water desalination or long distance water transportation, so far no substitute for water has been found.

The specific characteristics of water resources obviously condition their sustainable development. For example, the mobility of water implies that the basic unit for water resources management should be the river basin or a set of river basins. Moreover, the sustainable development of water resources should embrace the several forms of occurrence of water in the hydrological cycle - inland, estuarine and coastal waters -, consider jointly surface and groundwater and their interactions, and always take into consideration the mutual influence of water quantity and water quality.

The strategies recommended by the Brundtland Report, as summarized by MUNN 1988, include:

- reviving growth and changing the quality of growth;
- meeting essential needs for jobs, food, energy, water and sanitation;
- ensuring a sustainable level of population;
- conserving and enhancing the resource base;
- reorienting technology and managing risk;
- merging environment and economics in decision-making.

These objectives apply fully to the environment and natural resources in general, and thus also to water resources. However, it appears that, in the Brundtland Report, water was not given the individual attention it deserved as a fundamental element of a sustainable development strategy. The references made to water appear as occasional or indirect and, in most cases, express only some concern with the water pollution issues. The Report contains several chapters on such topics as population, food, security, species and ecosystems, energy, industry, oceans, space, and the Antarctica but not a special chapter on water resources. At the last World Congress on Water Resources organized by the International Water Resources Association (IWRA) in Ottawa in 1988, this was fully recognized. Following the discussions taking place during a special session organized by the "Committee on Water Strategies for the 21st Century" (of which the author has been a member since its creation in 1985), a statement was prepared in relation to the consideration given to water in the Brundtland Report. This statement is reproduced as an Annex, as it discusses in detail certain issues which are complementary to the ideas expressed here.

This paper is intended to respond to the need of an additional chapter to the Brundtland Report and to be a preliminary contribution to the paper which is due to be prepared by the Committee on Water Strategies for the 21st Century of IWRA. It should be noted that another member of the Committee has previously prepared a paper dealing with this topic (FALKENMARK 1988) in the context of a discussion of the Brundtland Report

held in Sweden (SGSNRM 1988).   FALKENMARK 1988 centres her main concern on
the fact that the report is "heavily biased towards present thinking in the
temperate zone" and does not consider with adequate relevance the case of
Third World countries and "pays no attention to the galloping water
scarcity now developing in Africa". Falkenmark also recognizes the need of
a water chapter in the Brundtland Report when she states that had the
Commission in charge of preparing the report "been more aware of the
implications of water-related problems, particularly in the arid tropics,
it would probably have concluded that water - being as complex as energy -
would have deserved a chapter of its own in the book."

2 - WORLD WATER RESOURCES SITUATION

A broad water resources assessment is presented in this section by
comparing water availability and demand, even if there is full awareness of
the somewhat simplistic character of this type of overall analyses.

Annual precipitation over the continents (including Antarctica and
Greenland) corresponds to around 110,000 km^3 of water, 65% of which is
returned to the atmosphere by evapotranspiration, the remaining 39,000 km^3
corresponding to runoff. This is the theoretical maximum limit of the
annual average availability for meeting human water requirements.

The distribution of the average annual runoff given in Table 1, partly
based on data in LVOVICH 1979, shows that there are considerable
differences between continents.

TABLE 1

Distribution of world runoff by continents

	Natural runoff (km^3/yr)	Equivalent water depth (mm/yr)	Per capita runoff volume (m^3/day)
Europe	3,100	316	13
Asia	13,190	293	18
Africa	4,225	142	34
North America	5,950	292	52
South America	10,380	576	153
Australia and Oceania	1,965	231	298
All Continents[1]	38,830	305	30

(1) Excluding Antarctica and Greenland

Table 2 shows how large are the differences between the continents in
terms of water available per unit of area of the territory (equivalent
depth) and per capita. If the future growth of the population is taken
into consideration, Asia and Africa appear clearly as the more water
stressed continents. For example, Africa's runoff per unit area is
approximately one quarter of South America's and half of the world average.
The non-uniform geographical distribution of the runoff length is also well
expressed by the distribution of discharges among the various rivers of the
world, as shown in the following table.

The only river in the first group is the Amazon, with an average
discharge of 175,000 m^3/s, which accounts for almost one-fifth of the total
discharge of world rivers, while less than 70 rivers account for 60% of the
total direct runoff.

One way of giving an approximate idea of the area with water surplus
and deficit is by comparing the values of precipitation and potential
evapotranspiration. Fig. 3 (FALKENMARK and LINDH 1976) shows the world

TABLE 2
Distribution of world rivers according to river discharge

Discharge (m³/s)	Number of rivers	Percentage of the total discharge of world rivers
Q 100,000	1	18%
10,000 Q 100,000	15	27%
500 Q 10,000	50	15%
Q 500	remainder	40%

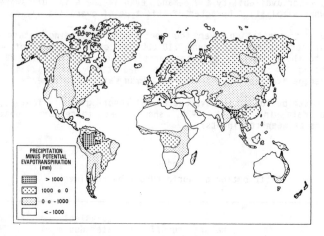

Fig. 3: World distribution of water surplus and deficit

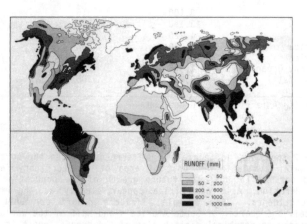

Fig. 4: World distribution of total river runoff

regions with water surplus and deficit defined as indicated. It is easy to recognize that, in general terms, the natural water situation in the developed countries is far more favourable than in the developing countries. Fig. 4 shows the world distribution of the annual average river runoff according to LVOVICH 1979.   The marked differences in the geographical distribution of annual average runoff from more than 1000 mm in certain regions to less than 50 mm in others indicate clearly how water stress can vary in different parts of the world.

Moreover, variability of available water resources is also important. In many regions, although the average available water volume may be quite enough to cover consumption needs, irregular distribution in time leads both to droughts and floods, which may be a cause of difficulties or even major catastrophes and economic damage. Since a world map indicating runoff variability does not, to the best of our knowledge, exist, the geographical distribution of overall precipitation variability is used to illustrate this aspect. Fig. 5, prepared by TREWARTHA 1968, gives the distribution of the precipitation variability expressed in percentage of average values. It becomes apparent that the highest values of variability correspond roughly to the regions with lowest average water runoff, making the situation even more critical in these regions.

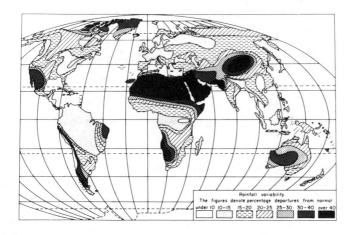

Fig. 5: World distribution of annual precipitation variability

The 39,000 km^3 of runoff previously referred to, correspond to a theoretical maximum limit of the annual average value of water available for meeting human water requirements. Floods correspond to approximately 70% of the total natural runoff and, as a consequence, only 12,000 km^3/yr correspond to stable natural runoff, i.e., to a reliable year-round source of water for the various water uses, though the same water may be used successively several times. The stable natural runoff is increased by the storage capacity of reservoirs. According to LVOVICH 1987, the present 2,600 world large reservoirs (i.e., with a volume greater than 100 million m^3 each) correspond to a useful capacity of slightly more than 3,000 km^3/yr, thereby increasing the stable natural runoff to a value of approximately 15,000 km^3/yr or to an average per capita value of 3,000 m^3/yr. By the beginning of the next century, also according to LVOVICH 1987, one could expect to see this stable runoff increased to approximately 20,000 km^3/yr. If one further considers that approximately 25% of this stable runoff occurs in sparsely populated regions such as the Amazon or similar, this would correspond approximately to an average per capita value of approximately 2,000 m^3/yr.

The evolution of water withdrawal in the various continents is given in Fig. 6, which is based on data provided by SHIKLOMANOV 1986. The figure shows that water withdrawal at the beginning of the century had doubled by 1940 and again in 1980, this value being expected to be 2.5 times higher in the year 2000. Fig. 6 also shows clearly the differences in growth characteristics for the various continents.

Table 3 gives the values of water withdrawal and the effective water consumption for the year 2000 expressed in absolute values and in percentage of the natural runoff. The table shows that in some continents, such as Europe and Asia, water use corresponds to a comparatively large

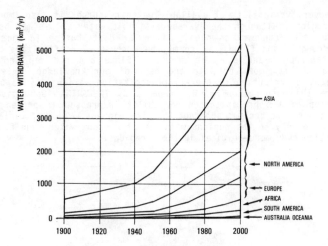

Fig. 6: Evolution of water withdrawal in the 20th Century

TABLE 3

Distribution of water withdrawal and consumption by continents in 2000

	Natural Runoff (km3/yr)	Withdrawal in 2000 (km3/yr)	(%)[1]	Effective Consumption in 2000 (km3/yr)	(%)[1]	Coefficient of Effective Consumption(2) in 2000 (%)
Europe	3,100	673	22	22	7	33
Asia	13,190	3,140	24	2,020	15	64
Africa	4,225	317	8	211	5	67
North America	5,950	796	13	302	5	38
South America	10,380	216	2	116	1	54
Australia and Oceania	1,965	47	2	22	1	49
All Continents	38,830	5,189	13	2,621	7	51

(1) in relation to natural runoff;
(2) ratio of effective consumption to water withdrawal

part of available resources whereas in others, such as South America and Australia, it is still quite low. The table also gives the coefficient of effective consumption, expressed by the ratio between effective water consumption and total water withdrawal. The values vary between 67% in Africa and half of this in Europe.

The major water user is agriculture followed by industry and municipal uses. Table 4, also, according to SHIKLOMANOV 1986 gives the evolution of percentages of water withdrawal and effective consumption for the three types of uses.

Table 4 indicates that even if water withdrawal for agriculture has been regularly decreasing to the advantage of industrial and municipal uses, if one considers effective consumption the importance of agricultural use is still extremely relevant and this situation is not likely to change significantly in the future. Industry is the second water user accounting for about 25% of the total water withdrawal but only 4% of effective consumption. The major industrial water use is cooling of power plants.

TABLE 4

Evolution of distribution of water use by types of use

	Water withdrawal (%)			Effective Consumption (%)		
	1940	1980	2000	1940	1980	2000
Agriculture	84	72	65	96	95	94
Industry	12	22	26	2	3	4
Municipal	4	6	9	2	2	2

Municipal uses for domestic supply are less important both in terms of withdrawal and effective consumption. This apparently easy situation is, however, misleading as not only are we talking here of water with high standards of quality but also urban demands are concentrated in small areas and can easily exert great pressure on the local water resources. The individual consumption varies greatly from urban areas in industrialized countries (where it can reach 1000 liters/day.p) to rural areas in developing countries (where it can be as low as 20 liters/day.p).

In 1980 the water withdrawn in the whole world was 7.5% of the total average river runoff, it being estimated that this value will increase approximately 50%, that is up to 11.5%, by the year 2000. This would seem to be a comparatively small figure. However, the irregularity of occurrence of the water resources in time and space make the situation more difficult. Fig. 7, prepared by SHIKLOMANOV 1986, illustrates this for the 1980 and 2000 situations, by showing the geographical distribution of the water availability. The water availability in these maps is given by dividing the total runoff minus the effective water consumption by the population. These types of maps provides a more realistic perception of the situation than the runoff map shown in Fig. 4. Fig. 7 shows an evolution that leads some regions to critical situations, proving that the current highly irregular space distribution tends to increase steadily with time.

Another important world issue is the possibility of international water conflicts. Fig. 8 shows a map of the 200 international river basins in the world where 40% of the world population lives at present. The distribution of these basins by continent is given in Table 5 (UN 1975).

Considering the very large number of international river basins, actually larger than the number of the existing countries, it is no wonder that multiple conflicting situations involving water quantity or water quality issues may arise. A good review of this evolving situation was made at a recent workshop organized by the IIASA (VLACHOS et al 1986). In a recent study covering the Middle East published by the Center for Strategic and International Studies in Washington D.C. (STARR and STOLL 1987), it is stated that "if present consumption patterns continue, emerging water shortages, combined with a deterioration in water quality will lead to more desperate competition and conflict" and concludes by stating that "by the year 2000 water - not oil - will dominate the resources issue of the Middle East".

The perception of the actual importance of water resources problems on a worldwide scale led to the convening in 1977 of a very important inter-national meeting, the United Nations Water Conference held in Mar del Plata (UN 1977). This Conference approved the so-called Mar del Plata Action Plan containing recommendations to be implemented by the UN Member States and resolutions addressed to organizations of the United Nations system. One of the important decisions of the UN Water Conference was the establishment for the 80s of the International Drinking Water Supply and Sanitation Decade, the aim of which was to stimulate action and investment to provide clean water and appropriate sanitation for all by the end of 1990.

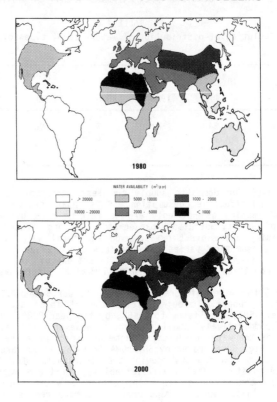

Fig. 7: World water availability in 1980 and 2000

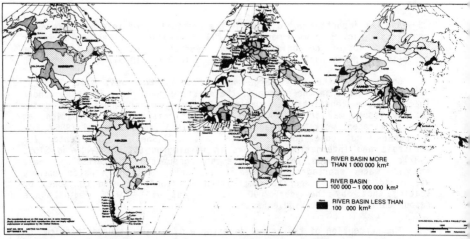

Fig. 8:   International river basins

The major basis for the establishment of this Decade was the recognition that disease, water supply and sanitation are closely associated. The Decade is now approaching its end and it is obvious that the objective will be far from reached.   United Nations data for 94 developing countries indicate that by 1983 the percentage of people with access to safe drinking water was 74% in urban areas but only 39% in

TABLE 5
Distribution of international river basins by continents

| | Number of countries sharing the river basins | | | | | | | | | |
	2	3	4	5	6	7	8	9	10	Total
Africa	33	10	6	-	2	1	-	3	-	55
America	53	5	-	1	-	1	-	-	-	60
Asia	27	8	3	-	2	-	-	-	-	40
Europe	35	7	-	1	-	1	-	-	1	45
Total	148	30	9	2	4	3	-	3	1	200

rural areas. As for access to sanitation facilities, the corresponding values were 52% and 14%. The increase, percentage-wise, since the beginning of the Decade is practically nil in urban areas and modest in rural areas. Of course, absolute figures increased but this was largely absorbed by the increase in population. ROGERS 1987 points out that to meet the UN goal the World Bank estimates that 2,400 million people would have to be serviced, leading to capital investments of $50,000 million for water supply and six to twelve times that much for sewerage." Rogers also indicates that the World Bank has also estimated that the investment costs for households would range between $650 and $4,000 and that the annual cost of service would be from $150 to $600 per household and concludes that "it seems doubtful that a very large fraction of total family income could be spent on such programmes even if it could be justified economically."

3 - PROBLEMS THREATENING SUSTAINABLE DEVELOPMENT OF WATER RESOURCES

Different regions of the world are faced with largely different types of problems associated with water resources' occurrence, use and control, which may endanger a sustainable development of these resources. If one considers the water resources distribution depicted in Figs. 4 and 7, it is evident that some parts of the world are much better favoured with water resources than others. By and large, the more developed countries are located in higher latitudes where the climate is mild and precipitation is sufficient. There are, of course, exceptions, and not only inside large countries like the United States and the Soviet Union can the water conditions change drastically from region to region, but this may also occur in small countries. A good example is Portugal, the country hosting the Symposium where this paper is presented, where, for example, natural average runoff varies from 1000 mm in the north of the country to 50 mm in the south, within a distance of less than 600 km.

The prevalent types of water uses, obviously also condition the types of water problems. To take three examples of large world regions, one can state that in a country such as the United States water quality problems are overall the main concern (even if in some regions water quantity may be an important issue), whereas in countries such as India or China the main concern is related to irrigation and drainage, and in many areas of South and Central America hydropower is a major issue.

But natural conditions of water resources occurrence and forms of water resources use are not the only reasons for different types of problems. The level of social and economic development and the available technology also contribute greatly to the nature of these problems. Good technology and, in particular, good engineering technology is essential for finding the best solution to the problems but is not, by itself, a guarantee that the problems will be solved in the best possible way. In many areas the engineering is good but conditions are not created to ensure an efficient use of water resources. This is often the case with irrigation works, when the social organization does not allow a full use of land and water resources, or the case of water supply and sanitation works, when the educational level of the population does not ensure the full understanding of basic hygiene rules.

Besides the problems which derive directly from the regime of occurrence of water resources or of the use of water by man, some problems can also derive from external influences, as is the case with the cost of energy. In most water supply projects, some kind of pumping is involved and thus the cost price of energy has to be taken into consideration. An illustration of this is the drastic change in the economics of several water uses associated with the oil shocks of the seventies.

For the sake of illustrating the diversity of water resources problems, a brief reference is made below to different types of problems which may threaten a sustainable development of water resources.

a) Limited information on water resources

A basic problem in many regions of the world is an inadequate knowledge both of the natural and potential water resources and of present and forecast water demand. Water resources occurrence is defined by a set of stochastic variables. It is thus essential to know not only their average values but also their space and time distributions. Measurements of these variables should meet the following conditions: (i) suitable geographical distribution and suitable density of measurement points; (ii) suitable frequency of measurements; (iii) sufficiently long periods of measurements; and (iv) accurate measurements. The assessment of water availability is based on information relating to climatic variables (precipitation, evaporation, temperature, etc.) to water quantity (surface and groundwater flow) and to water quality (defined by a number of parameters to be specified in each case). The assessment of water demand is usually made for the several water uses, which may have a variable comparative importance for each specific case: municipal water supply, agriculture, industry, hydropower, irrigation, recreation, environmental uses, etc.

b) Water scarcity

The basic problem preventing a sustainable development of water resources is simply, in many cases, water scarcity which can be the result of natural conditions related basically to low average amounts of available water resources, i.e., aridity or related to extreme events, i.e., droughts, both creating difficulties to normal water supply and/or limiting the length of the growing season. Future climate changes due to global warming can also become a cause for water scarcity in certain regions. Due to the present concern with climate changes this problem is treated separately in Section 4.

Water scarcity can equally be induced by man, as a result of an increase in population and/or per capita consumption associated with social and economic development, or as a consequence of incorrect water use (over-exploitation of water resources, pollution, desertification, etc.). One relevant case is the excessive use of groundwater which tends to take place in some countries, leading to a decrease in water levels and consequently increasing the cost of pumping, and consequently the cost of the water supplied, to levels which are likely to generate non-competitive economic activity, particularly in the case of irrigated crops. Other possible drawbacks of excessive groundwater withdrawal are land subsidence affecting civil engineering works (buildings, road, canals, piplelines, etc.), or causing sea-water intrusion in coastal areas or even fossil groundwater mining.

c) Inefficient water use

Water resources problems are often associated with a lack of efficiency in water use in agriculture, industrial and domestic supply. Agriculture which, as discussed in Section 2, is by far the most important water use activity, is also probably the sector least efficient in water use. Worldwide the efficiency of irrigation systems is estimated at only

40%.   But given sufficient incentives, water withdrawals for irrigation could be cut by 10 to 40% without reducing crop production.   The inefficiency in water use results in many cases from habits created in times when water was plentiful and considered as a free commodity.   Prices paid for water tend to be lower than real cost and, in particular, irrigation water is often heavily subsidized in many countries.   But it should be realized that if we want to have enough water of a sufficiently good quality we have to pay for its true value.

d)   Water Pollution

According to the figures indicated in Section 2, the annual world water withdrawal in 2000 is expected to be approximately 5200 km^3 which corresponds roughly to 13% of the natural runoff (39,000 km^3) and 25% of the stable runoff foreseen for the same year (20,000 km^3).   If one considers effective water consumption instead of water withdrawal these percentages will be reduced to approximately half of these values, i.e., 7% and 13%.

These figures, as presented, give no evidence of reason for special concern, but the situation changes drastically if pollution is taken into consideration.   According to LVOVICH 1979, the annual volume of waste-water rejected in water bodies in 2000 is expected to be approximately 6,000 km^3, these wastewaters mobilizing approximately 38,000 km^3 of river runoff to absorb the corresponding pollution.   Considering that this last figure corresponds approximately to total natural runoff and is twice the stable runoff in 2000, and considering furthermore the marked differences in the world geographical distribution of water, this situation seems to be a reason for serious concern.

These estimates are of course only approximate, but they illustrate a very unfavourable trend and draw attention to what appears to be a potentially alarming situation threatening the sustainable development of water resources, unless the future evolution of water resources management and particularly of wastewater treament technologies is very favourable and its application largely diffused.

Water pollution can be direct or indirect.   Direct pollution is that resulting from rejecting pollutants directly in the water courses or the aquifers, and is usually the result of routine activities or accidental events.   Pollution resulting from routine activities can be of two different types: point source pollution, which is generated mainly by municipal and industrial water use and can be controlled by adequate sewage treatment technologies which are well known but usually expensive (these technologies cover primary and secondary treatment and in some cases tertiary treatment, e.g., when eutrophication problems are important); diffuse pollution (or non-point-source pollution), which results from farming, forestry, urban discharges, transportation, construction-industry and sanitation landfills, its control being closely associated with good agricultural practices, and the implementation of adequate land-use policies.   Accidental pollution results from accidents which, in many cases, are likely to cause great damage.   These accidents have recently increased and will probably tend to increase further in the future. Accidental pollution has to be controlled by planning in advance adequate strategies for earlier detection and subsequent restriction of the region affected and minimization of impacts.

The forms of indirect pollution are those by which waste is originally discharged into the soil or the atmosphere and only later reaches the water bodies.   The indirect pollution of water bodies through the air are a consequence of dry and wet acid deposition as a result of SO_2 and NO_x emissions into the atmosphere, originated by power plants, industrial units and vehicles.   Acid rains in lakes and water courses affect the ecosystems and have been the cause of fish deaths, particularly in Scandinavia and North America.   Indirect water pollution occurring successively through air and soil can also take place, leading to the acidification of groundwater. This acidification increases the mobility of heavy metals, such as cadmium, which may be leached out of the ground and transported to groundwater.   A third form of indirect pollution of groundwater, of special importance in coastal regions, is saline intrusion of sea or brackish waters due to overpumping of groundwater or to careless drilling of wells putting in direct contact fresh water and salted or brackish groundwater levels.

e)   Floods and droughts

Floods and droughts are extreme hydrologic phenomena, which have caused great damage to man's social and economic activities since time immemorial. Impacts of floods and droughts, however, are felt differently as, while flood effects are sudden and immediately felt, droughts come slowly in a creeping way and are only felt after a significant time has elapsed. The accidental character of floods and droughts naturally creates a tendency to forget them once their effects have passed and thus these phenomena tend to be studied less than those relating to other more persistent problems.

Flood problems which deserve more attention are those related to flood prediction, flood control, evaluation of flood impact and potential risks of floods, emergency aid in case of floods, and flood insurance.   As regards droughts, the main research needs relate to their prediction and control.   Drought prediction involves problems related to initiation, duration and intensity of droughts.   Control of drought effects involve measures related to the water availability, the water demand and the minimization of drought impacts.   Both flood and drought effects can be aggravated by man's action.   In the case of floods, this aggravation can be due, for example, to urbanization or to mismanagement or disruption of dams or other hydraulic works.   In the case of droughts, the aggravation can be due to deforestation, inadequate agricultural practices, construction of hydraulic works or poor water management.

Desertification is a process usually considered as being associated with drought even if drought is not the main cause. The basic causes are degradation of the vegetation cover, high water and wind erosion, salinization and waterlogging.   These effects are largely man-induced, particularly in the case of farming practices which favour erosion, excessive tree cutting and indiscriminate use of fire to clear the agricultural soil.   Drought, though not being the major cause of desertification, can intensify it and have a catalytic role increasing the intensity of the effects previously referred to.

f)   Disturbance of aquatic ecosystems

Threats to aquatic ecosystems may result from water pollution, and particularly the chemical loading of water resources, and also from the construction of hydraulic works.

In the industrialized countries, the wastes rejected to water bodies are increasingly overloaded with synthetic chemicals, heavy metals, mining products, etc.   None of these substances is biodegradable and they easily accumulate in aquatic ecosystems creating dangerous imbalances. Preservation of resilient ecosystems with a sufficient diversity of organisms, i.e., able to resist structural and functional displacement and serving as reservoirs of genetic diversity, is an essential aspect of the sustainable development of water resources.

Hydraulic works, and particularly dams, can also affect the aquatic ecosystems.   In fact, dams induce downstream changes in streamflow regimen turbidity, water temperature, salinity or dissolved oxygen level, which may alter the ecosystem's balance with several consequences, e.g., reduction of fish productivity downstream or aquatic plant proliferation upstream, affecting dam operation and reducing available water due to increased transpiration.

g)   Water borne diseases

According to World Health Organization information, approximately 80% of the Third World pathology is related to water. Distinction can be made between three types of water borne diseases: (i) diseases directly trans-mitted by water, such as typhoid, viral and bacterial diseases, coleo hepatitis, poliomyelitis and gastro-enteritis which every year kills 25 million people, the majority being children; (ii) diseases transmitted by an intermediary host living in the water, the most important example being bilharzi; and (iii) diseases caused by a lack of hygiene related to water scarcity, e.g. typhoid, intestinal parasite diseases, leprosy, conjunctivitis and trachoma.

h)   Erosion and Sedimentation

Erosion caused by water is a natural process occurring in the land surface, which involves disaggregation, drag and transport of soil elements to different locations where they are deposited.   Water is the driving agent of this process, particularly in the form of precipitation and surface runoff.   This natural process is often completed by the action of other natural agents such as wind and low temperature causing alternately water freezing and melting. It is also intensified by human activity, in particular by farming, deforestation, urbanization, mining and road building.   Water erosion can have very negative economic and environmental impacts, with considerable effect on soil and land resources and on several uses of these resources.   World annual erosion can be evaluated (DA CUNHA et al. 1980) in 10 to 14 km^3 which corresponds to an average wear of land surface of 0.7 to 1 cm every century.

Soil erosion, especially that stemming from agricultural land, is currently considered as the main cause of diffuse pollution and should thus be considered together with the other causes of water pollution previously referred to in d) above.   Sediments, besides being themselves a pollution agent, can act as carriers for other pollutants such as chemicals and heavy metals which can be adsorbed by the sediments, transported to another location and de-adsorbed later.   Another possible use of negative impact may originate from the use of dredged-up contaminated sludges for building purposes.

Other major negative impacts of erosion can be pointed out, such as: destruction of land, buildings and other structures; instability of river beds and canals; change in flood regimen; limitation of storage capacity of reservoirs; clogging of canals and pipes in water supply and sewage systems.

4 - GLOBAL CLIMATE CHANGE AND SUSTAINABLE WATER RESOURCES

It is presently accepted that significant man-induced climatic changes may occur in the next decades, related to the greenhouse effect and global warming and, to a lesser extent, deforestation.   The main factor responsible for the greenhouse effect is CO_2 production, it being currently assumed that its concentration in the atmosphere would approximately double that of pre-industrial times by the middle of the next century (WMO 1986). CO_2 alone would be responsible for about 50% of the rise in atmospheric temperature, the remainder being due to the increase in concentration of other gases, in particular CH_4, N_2O and the CFCs (chlorofluorocarbons). The results of numerous climatic models foresee an increase in mean atmospheric temperature of 1.5ºC to 4.5ºC.   The rise in temperature induced by the greenhouse effect would cause changes in the world distribution of precipitation and evapotranspiration and thus also in runoff.   Change in precipitation and evapotranspiration may be amplified in runoff changes, especially when runoff coefficients are low.   Climate variability would thus affect semi-arid regions more than other regions.   Another important aspect of the relationship between climate and water resources is that available water being determined by climate is also an intrinsic factor of the climatological processes.

Important consequences for water resources availability may be expected with significant social, economic and environmental impacts.   In fact, through the workings of the hydrologic cycle, climate plays an essential role in the quantity and quality of the available water.   Thus, climatic changes will have direct consequences on water resources availability and characteristics, as well as indirect consequences on the vegetation, the soil and the environment in general.   Water demand will also be affected by climate change since the water requirements for

municipal, industrial and agricultural uses are affected by prevailing
climate conditions.   From a practical standpoint, it is important to try
and evaluate the magnitude of the consequences of climatic changes on the
sustainable   development   of   water   resources,   identify   the   possible
difficulties   that   they   will   raise   for   water   resources   managers   and,
finally, establish if and how the current water resources management and
planning strategies should be modified.   The objective would be to assess
the impact of the foreseen climatic changes on water resources in order to
help water resources managers and planners to take the necessary measures
in due time to alleviate the adverse impacts and to take the best advantage
of favourable impacts.

Fig. 9 gives a view of the relationship between climate, water
resources, water management and economic and social development. Not only
the main relations (continuous line arrows), are shown, but also the
feedbacks (dotted line arrows), the whole giving a simple perspective of
the impact of climate on water. From this diagram it is easy to appreciate
the complexity of the problems under discussion, which involve an
interdisciplinary approach with numerous professional skills, as well as
an intersectoral approach which results from the multiple water users.

It has been suggested that the current design procedures for water
resources works make it possible to ignore climatic change, because
anticipated long-term changes in mean runoff are much smaller than the
changes due to the present variability with which the works are designed to

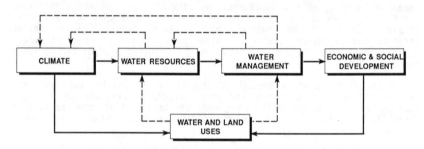

Fig. 9:   Relationship between climate, water resources, water management
and economic and social development

cope.   But change in mean values is simply one aspect of climatic change.
In fact, climatic changes of any importance will determine non-negligible
changes in variability of runoff, determing the magnitude and frequencies
of extreme phenomena (floods and droughts and seasonality of water supply
and demand).

Climatic change can affect both inland water resources (influencing
the amount of water resources available and its geographical, seasonal or
interannal distribution) and coastal waters (mainly in relation to the
problems caused by sea-level rise).   The main possible consequences of
climatic changes related to inland waters are the induced changes in: the
global amount of water resources available and in the spatial and temporal
distribution of these resources; in soil moisture; in floods and droughts;
in water quality; in sedimentation processes; and in water demand.   The
main consequences of climatic change related to sea level rise are:
flooding of urban, industrial and service areas; rendering inoperable civil
engineering works and other facilities;   intrusion of sea water in aquifers
and estuaries; changes in coastal sedimentation processes; and changes in
the coastal environment.   The author has provided elsewhere a detailed
discussion of the several possible consequences indicated (DA CUNHA 1989).

The approaches followed by meteorologists and hydrologists to deal
with the interaction between climatic changes and water resources are quite
different.

Meteorologists tend to prefer large scale tools such as general circulation climatic models or climatic analogue scenarios.   General circulation models have the disadvantage of not ensuring enough resolution to provide detailed climatic information at regional level and/or seasonal level.   Climatic analogue scenarios can be constructed by three different methods, namely by reconstructing particularly warm historic periods, by using statistical models of atmospheric dynamics to estimate climatic changes and by using past instrumented climatic data.

Hydrologists, in contrast to climatologists, like to work with models on the scale of a river basin.   The description of the details of the hydrologic processes is much better done in hydrologic models than in climatic models.   However, in hydrologic models climate is held static. The hydrologic models are quite detailed in reproducing the physical phenomena at the drainage basin levels, but they do not include important climatic-hydrologic feedback effects which are important to describe the highly non-linear climatic processes.

The determination of the impacts of climatic changes on water resources availability can be made by combining general circulation models with regional hydrologic models, either conceptual models or water-balance models.   GLEICK 1986, after comparing different models, has concluded that the water-balance models have important advantages over other methods for assessing the impact of climatic changes on water resources.   It should be noted, however, as indicated by GLANTZ and WIGLEY 1987, that if hydrologic models are to be used to predict the runoff changes that might occur in response to prescribed changes in climate, the results can be very sensitive to model errors.

Five recently used Atmospheric General Circulation Models coupled to mixed-layer ocean models, namely the models GFDL (Geophysical Fluid Dynamics Laboratory - WETHERALD and MANABE 1986), GISS (Goddard Institute for Space Studies - HANSEN et al. 1984), NCAR (National Center for Atmospheric Research - WASHINGTON and MEEHL), UKMO (United Kingdom Meteorological Office - WILSON and MITCHELL 1987) and OSU (Oregon State University - SCHLESINGER and ZONG-CI ZHAO 1989) provide for a CO_2-doubling scenario a range of results which can be expressed by an increase in annual-mean global-mean continent surface air temperature of 2.8^0 to 5.2^0 and an increase in the annual-mean global-mean precipitation of 7.1% to 15.0% of its value before CO_2 increase.   The majority of the models also indicate an almost general trend for desiccation in the Northern Hemisphere continents in summer.   To provide an idea of the geographical distribution of these changes, the more recent results, i.e., those provided by the OSU model (SCHLESINGER and ZONG-CI ZHAO 1989), were selected.

Fig. 10, based on Schlesinger and Zong-Ci Zhao's results, gives the distribution of the changes in precipitation rates for a CO_2-doubling scenario providing average annual values and average for the periods December-January-February (DJF) and June-July-August (JJA).   Fig. 11 gives a similar distribution for the change in soil moisture for the same periods.   Both figures indicate changes in the distribution of the variables represented, expressed either by increases or decreases in different regions.   It should be noted, however, that SCHLESINGER and ZONG-CI ZHAO 1989 indicate that these changes are not statistically significant everywhere.

Fig. 10 shows that the annual precipitation rate has relatively larger increases and decreases between 30^0N and 30^0S, smaller increases pre-dominating poleward in the other latitudes.   However, the figure always shows an increase in the annual average precipitation rate over the continents except for the western part of the United States, the Caspian Sea and a region in Southern Peru and Northern Argentina.   The DJF and JJA maps of Fig. 10 show some differences when compared to the annual map.   For example, the increase in precipitation rate over the Amazon basin is larger in DJF than in JJA, while an increase over Southeast Asia is larger in JJA than in DJF.

Fig. 11 shows that the annual soil moisture tends to decrease in bands centred near 30^0N and 30^0S and to increase in the other regions.   The DJF

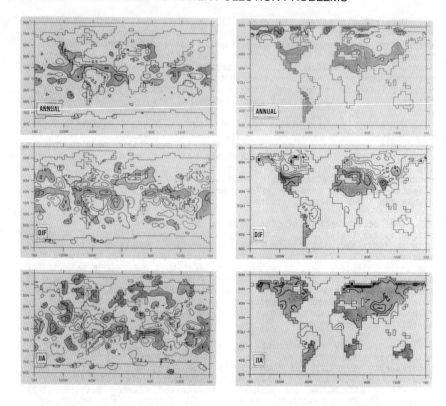

Fig. 10: Changes of average precipitation rates for a CO_2-doubling scenario, annually and in the periods December-February (DJF) and June-August (JJA)(mm/day)

Fig. 11: Changes of average soil moisture for a CO_2 doubling scenario, annually and for the periods December-February (DJF) and June-August (JJA)(cm)

and JJA maps of Fig. 11 show that, in the Northern hemisphere, there is in DJF a decrease in the band 20°-40^oN and an increase upper North, and in JJA a decrease everywhere except in India and Southeast Asia.

5 - HOW TO ACHIEVE A SUSTAINABLE DEVELOPMENT OF WATER RESOURCES

The review previously made in this paper of the world water scene and of the problems threatening a sustainable development of water resources makes it possible now to identify the actions to be taken to try to achieve such a sustainable development. This necessarily involves a concern with future water resources problems.

There are various ways of dealing with the future. One possible way is to scare people with the idea that mankind is in trouble. This line had a remarkable success in the early seventies, when it was adopted by the authors of works like "The Limits of Growth", "Future Shock" and "World Dynamics" (MEADOWS et al. 1972, TOFFLER 1970 and FORRESTER 1971). Another line relies on the application of forecasting techniques without necessarily resorting to Malthusian or apocalyptic views. These future studies can essentially follow two different approaches. The first approach, followed mainly by the Anglo-Saxons, is essentially of a scientific nature, relying on the extrapolation of tendencies detected in the past. The second approach, followed mainly in continental Europe, is basically of a political nature and involves in itself a project for the future of society. This second approach is what one could call a "goal-oriented" approach, i.e., an approach relying on the philosophy that the

future is not predetermined, but is mainly a matter of choice and willingness.

Forecasting efforts related to water resources should attempt to
identify what will be the main future problems, the time at which they will
occur and their location, as well as a definition of the way to cope with
the problems. This means that more than knowing "whether" the problems
have to be solved, we have to try to answer the questions: "what?",
"when"?, "where?" and "how?".

It should be added that forecasting methods help to answer questions,
but do not give the answers. Even when we have adequate techniques to deal
with a set of alternatives, choices have to be made as a consequence of the
assumed political options. Future analysis helps to make better decisions,
but does not make the decision process easier. The consideration of
aspects related to innovation when defining future policies for water
resources sustainable development is also an important issue. Social and
technological inovations continually alter the doors open to water
resources planners and managers. But one should also be aware of the risks
involved. In fact, social and technological innovation is always risky,
since present resources are committed to the future, and therefore are
always uncertain.

Sustainable development of water resources implies the adoption of
three successive steps in water management (DA CUNHA 1985):   (i) the
identification of the system, which involves the specification of the
characteristic features of the water resources system relevant to the
different problems to be faced, these features being of a physical,
economic, social or environmental nature; (ii) the prediction of the
behaviour of the system, which corresponds to finding out the response of
the system to certain actions taken by man that excite the system
(including pollution discharge in the water bodies, urbanization, change in
agricultural practices, or building of works which condition the behaviour
of water resources within the system); and (iii) the management of the
system, which involves the selection of the best alternative to attain
certain objectives, management decisions being based on the previous steps
of identification and prediction.

As demonstrated in previous sections of this paper, the demand for
water of sufficiently good quality and with an adequate distribution, both
in space and time, tends to increase in most countries. As water
availability is frequently limited, several problems may develop as
discussed in section 3, generating conflicts between water users, social
groups, decision-makers and water users, or between regions or even
countries. If it is not possible to solve all these conflicts, there
should at least be an attempt to manage them adequately. This is done by
considering a series of measures of a technical, economic and institutional
nature which may favour water resources sustainable development.

Technical measures can be classified in three groups: (i) measures
leading to better use of available water resources, such as: water
storage, water resources transfer, modification of forest cover; and water
pollution control; (ii) measures leading to augmentation of available water
resources, such as: desalination of seawater; reduction of evaporation and
evapotranspiration; rainfall augmentation; forest and agricultural planning
and use of water from the polar icecaps; and (iii) measures leading to
reduction of water consumption, such as: water reuse; water recycling,
modification of production processes, modification of industrial products;
and limitation of water wastage. As economic measures one can refer to the
implementation of pricing systems for water withdrawal and/or consumption,
or of pollution charges systems as well as subsidies, credits or tax
exemptions to help rational water use and control. Finally, examples of
institutional measures are water resources laws and regulations or the
definition and implementation of new institutional frameworks for water
resources management.   These different measures have been discussed in
detail in DA CUNHA et al. 1980.

The implementation of a set of measures of the same type corresponds
to the formulation of what can be called a water resources management

strategy.   Strategies can be classified as reactive strategies and proactive strategies, the latter tending to assume progressively a greater
importance in relation to the first.   A coherent combination of strategies
corresponds to what can be called a water resources management policy.
Policy design thus consists of the generation of strategies and the
screening of these strategies, based upon their engineering, social and
economic feasibility.   The selected strategies, combined in a coherent
manner, give rise to an alternative water resources policy.

The selection of policies aimed at a sustainable development of water
resources is the task of the decision-makers based on alternative scenarios
of the water resources system, and on the assumptions on the water
resources system itself.   The alternative scenarios correspond to the
specification of certain consistent combinations of the more relevant
factors, corresponding to future conditions of economic and social
development, technological progress, standard of living and public
policies. The system assumptions describe the characteristics of the water
resources systems, and are aimed at including the behaviour of users and
management of each system.

One important point is the definition of the criteria used for
comparing various alternative water resources policies.   Clearly, the
criteria deriving from general social and economic policy in the country or
region considered are determinant.   Thus, the general economic objectives
as regards inflation, employment and balance of payments are to be
considered.   Within this general context the prevailing criteria for
comparing the alternatives should be those based on efficiency and equity.

Due to the numerous existing interactions of a physical, chemical and
biological nature, the substainable development of water resources should
always be considered in the broader framework of the sustainable development of natural resources.   The relationship among the environment, natural
resources and the economy has been the object of attention in
industrialized countries. the work carried out by the OECD for several
years deserving a special reference (e.g., OCDE 1985 and OCDE 1989).   The
interrelationship among the various natural resources - water, land, air,
aquatic biota and terrestrial biota -, man being in the centre because of
his capacity to influence the various interrelationships, are indicated in
Fig. 12a (DA CUNHA 1988 adapted from FALKENMARK 1983).

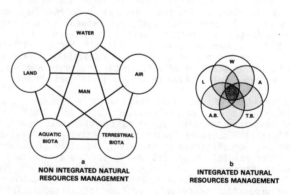

a
NON INTEGRATED NATURAL
RESOURCES MANAGEMENT

b
INTEGRATED NATURAL
RESOURCES MANAGEMENT

Fig. 12: Interrelationship among the various natural resources

The interaction between the different domains indicated in the figure
are usually considered through an assessment of mutual impacts and
identification of possible associated conflicts. Usually the assessment of
impacts is made simply by preparing lists of impacts of the projects under
consideration and by identifying the associated conflicts.   The purpose of
water resources management, as part of natural resources management, should
be not only to assess the impacts but to manage the impacts and not only to

identify the conflicts but also to manage them and to attempt, as much as possible, to solve them. To achieve this, efforts should be encouraged to make the different domains of interest converge and partially overlap, so that zones of common interest can be identified as indicated in Fig. 12b. The current approach for impact assessment should be replaced by an integrated approach where the existing problems are identified in areas by superimposing two or more of the domains under consideration. This applies to many familiar examples such as water use in agriculture, soil erosion by water, wild rivers, reservoir behaviour, river channelling, acid rain, etc.

In particular, water and land use are closely related and this makes it necessary to plan in a co-ordinated way the use of water and land resources in each river basin or region. The policy often adopted in the past, by which land-use planning decisions were made first and the corrective measures to face the impacts of these decisions on the water resources were only taken later, must be abandoned. Instead, water resources planning and land use planning must be integrated processes pursued simultaneously at all levels, right from the beginning.

To conclude, the main actions required to ensure a sustainable development of water resources are briefly referred to below, brief comments being provided in each case.

a)   Water resources planning

Water, as a basic factor in economic and social development, should always be taken into consideration in national and regional planning. The water resources planning process should obviously comply with the general methods for economic planning processes, of which detailed reference is not appropriate in this paper. A brief allusion to some aspects which should specifically condition planning for a sustainable development of water resources is, however, justified. These aspects are the following:

- the interdependence of the different types of water sources and of the different water uses (this makes it necessary to consider the river basin as the basic unit for water resources planning);

- the irregularity of the annual and interannual occurrence of water resources (this makes it necessary to consider not only the mean values but also other characteristic values of the distribution, for instance, the extreme values corresponding to floods and droughts);

- the circumstances that rivers - often the main source of water supply - are also the natural collectors of polluted waters (this makes it necessary to bear in mind, jointly, the problems of water quantity and quality);

- the distinction that has to be made between consumptive and non-consumptive water use, and the uses which condition simultaneously quantity and quality of the water, or only one or the other;

- the possible importance of economic, social and environmental impacts of hydraulic projects which have to be duly assessed and kept to a minimum.

The relationship between water resources and other natural resources management reinforces the need of assuming both an economic and an environmental perspective in water resources planning. The water resources policies should also be duly co-ordinated with other regional and national policies with emphasis on agricultural, forestal, industrial, energetic and public health policies. This co-ordination should normally take place within the framework of a land-use planning policy and, of course, of the general social and economic planning.

b)   Assessment of water resources

The water resources planning process requires the knowledge of existing and required water resources, and on how the various parts of water resources systems interact. This assessment must cover water availability, both natural and modified by man's action, it must include both surface waters and groundwater, consider the aspects of quantity and quality and refer to the present situation and to the future. Moreover, as regards international river basins, an effort must always be made to carry out the assessment by joint action of countries sharing river basins.

As regards the assessment of demand, it must not only include projections of water consumption and of pollution discharge, but also a definition of the evolution of water quality goals. Classification of watercourses according to their water quality may form part of this definition.

A reliable assessment of the water resources is an expensive process as it requires the existence of a good hydrological monitoring network relating to the climate and to water quantity and quality, and proper data processing, storage and retrieval systems.

A different approach for assessing water resources is trying to evaluate them in monetary terms. This raises the issue of the usefulness of establishing national systems of natural resources accounts in general, and particularly for water, as some countries have recently tried, namely France and Norway.

c)   Increased efficiency in water use

As mentioned in Section 3, there is much room for increased efficiency in water resources use. Irrigation is, in this respect, a major issue, as water is often supplied to farmers at a cost well below the cost of supply, this being a particularly important issue in developing countries. If we could significantly improve the equity in water resources use and the reliability of the irrigation projects in the developing countries, the food production of these countries would greatly benefit. Efficiency should also be increased in industrial water uses, particularly by resorting to clean technologies and recycling practices. Finally, the efficiency of domestic water supply, which deals with highly-valued treated water, should also be increased by resorting to conservation measures, water recirculation and, in cases where consumption is excessive, progressive pricing schemes.

d)   Water quality control

Water pollution control is crucial for a sustainable development of water resources. Persistent efforts should be made to decrease or at least avoid the increase of surface water pollution in many parts of the world. Pollution problems are particularly important in relation to groundwater because of the very slow movement of groundwater in the soil, which is likely to determine long-term pollution effects that are difficult and slow to reverse. The extremely low velocities of groundwater movement can imply that water pollution is only detected after a number of years and once the aquifers are contaminated, several decades could pass before an ever-expensive regeneration could be achieved. Also pollution by water carried sediments through accumulation of toxic chemicals and heavy metals should be controlled because of possible consequences for water quality.

e)   Impact of climate changes

There are a number of possible alternatives of coping with the consequences of climatic change on water resources including prevention strategies, changes in water resources management and planning, abandonment of certain areas and compensation for affected water areas. Further research is also needed on the linkages between climate, meteorology and hydrology, especially by getting better models which simulate the physical reality in the river basins and which take into consideration the regional aspects of climatic change.

The current practice of setting some scenarios based on the results of global circulation models but without being able to assign probabilities to these scenarios is somewhat limited. This limitation will only be overcome when it will become possible to establish a direct relationship between the climatic modelling and the hydrologic modelling.   Linking global circulation modelling and regional water-balance modelling is likely to be the best way to obtain new and better information on the impact of global climatic change on regional water resources.

In the future, water resources planning should always include the consideration of plausible climatic change scenarios when doing the sensitivity analysis and deriving social, economic and environmental impacts which are part of the planning process.

(f) Legislation relative to water resources

Sustainable development of water resources requires an adequate legal support that may take the form of basic laws and regulations. Examples of aspects that call for regulations are the following:

- definition of the legal right to ownership and use of the water resources;

- definition of the possible uses of the various reaches of the watercourses and of the quality goals of their water, according to existing uses and those foreseen in the future;

- definition of the maximum admissible pollution levels in receiving waters and of the standards for effluent characteristic parameters;

- definition of the conditions for licensing water use, establishment of pricing systems for water use, in particular for water withdrawal charges and pollution charges, and of fines and other penalties for disrespecting the standards.

g)   Education and Training

Education in relation to water resources is needed at all levels from the water project designer to the water user. Water resources management is the result of teamwork by a large number of people with different qualifications and levels of education, such as economists, jurists, sociologists, environmentalists, administrators, engineers, chemists, biologists and ecologists. These specialists must have, besides the basic knowledge acquired during their studies, specific training to enable them to play a proper part in the various activities of water resources management. Personnel education and training is indeed one of the most difficult, but one of the most important, tasks and must be regarded as a priority, because the success of a water resources management policy largely depends on having skilled professionals.

h)   Research and Development

The main aim of water resources Research & Development is to anticipate future water problems and indicate the best ways of studying them. System-atization, according to research topics of the various water resources problems and co-ordination of research, is extremely important and helps to ensure rational water resources management by providing solutions for problems as they arise. The various problems illustrated in Section 3 may be considered as corresponding to the main research needs in water resources R&D, to be investigated on a regional, national and international scale.

Science and Technology is essential to ensure a sustainable development of water resources and can help to clarify many of the relevant technical, economic and institutional issues.   But R&D can only be effective if the necessary steps to implement its results are taken, this being a major difficulty in some countries.

i)   Information Systems and Public Participation

Information systems are essential for public awareness of water resources problems and management principles and procedures which is a condition for a sustainable development of water resources.

Public participation in water resources management is vital to ensure access by decision-makers to the points of view of the public and consequently to help in reaching the public. Public participation should be as great as possible involving all the possibly interested social groups. It is essential that the decision- makers, i.e., the government representatives as well as the representatives of the various financial interests, come down to a local level to face public opinion where it actually exists rather than the other way round. The future of water resources should be discussed in regional and local fora and not in the capitals where power is centred.

j)   International Co-operation

The international co-operation in relation to water resources can be ensured through bilateral or multilateral agreements between countries sharing surface or groundwater resources or through international action resulting from political or scientific initiatives. These international initiatives may, in some cases, have a global dimension as with the issues related to climate change referred to in section 4. Combining international experience of countries with different levels of development may also be mutually beneficial as the developing countries can learn from the experiences of the more developed countries taking the lessons of their successes and failures, and the more developed countries have the opportunity to use their skills to help sustainable water resources development in the developing countries.

International co-operation in water resources use should aim to ensure an equitable share of water between nations and regions. In fact, a condition for sustainable development of water resources would be to manage international water resources as if they belonged to the same country, i.e., respecting the water mobility within each river basin and ensuring maximization of benefits provided by water. This objective should be primordial, particularly when water is scarce, it being understood that adequate criteria to share the benefits among the different countries in the river basin should be established and applied.

k)   Institutional problems of water resources management

Considering the interdisciplinary and intersectoral character of water resources problems, it is essential for the achievement of a sustainable development of water resources that adequate institutional frameworks for water resources management are established in each region and each country. There are different types of framework adopted for different conditions, such as conditions of water occurrence, political systems, stage of development, but the adoption of an adequate water resources framework in each case is essential to deal with the issues previously referred to. These institutional aspects are discussed in detail by DA CUNHA 1977 and DA CUNHA 1980.

REFERENCES

BERGER, A. (Ed.) (1989) - Climate and Geosciences: A Challenge for Science and Society in the 21st Century. NATO Advanced Science Institutes Series. Kluwer Academic Publishers, Dordrecht.

BROWN, L.R. (1981) - Building a Sustainable Society. Norton, New York.

DA CUNHA, L.V. et al. (1974) - Fundamentos de Uma Nova Política de Gestao das Aguas en Portugal (Fundaments for a New Water Management Policy in Portugal). Direccao Geral dos Servicos Hidraulicos, Lisbon.

DA CUNHA, L.V. et al. (1977) - Management and Law for Water Resources. Water Resources Publications, Fort Collins, Colorado.

DA CUNHA, L.V. et al. (1980) - A Gestao do Agua: Principos Fundamentais e sue Aplicacao en Portugal (Water Management. Basic Principles and its Application in Portugal). Fundacao Calouste Gulbenkian, Lisbon.

DA CUNHA, L.V. (1985) - Methodologies for Water Resources Policy Analysis in River Basins. In LUNDQVIST et al. 1985, pp. 329-335.

DA CUNHA, L.V. (1988) - Water Resources Management in Industrialized Areas. Problems and Challenges. International Workshop on Water Resources Awareness in Societal Planning and Decision-making, Skolkloster.

DA CUNHA, L.V. (1989) - Climatic Changes and Water Resources Development. In BERGER 1989.

FALKENMARK, M. (1989) - Urgent Message from Hydrologists to Planners. Water a Silent Messenger Turning Land Use into River Response. Symposium on "Scientific Procedures applied to the Planning, Design and Management of Water Resources Systems", Hamburg, International Association of Hydrological Sciences, Publ. no. 47.

FALKENMARK, M. (1988) - Sustainable Development as Seen from a Water Perspective. In SGSNRM, 1988, pp. 71-84.

FALKENMARK, M.; LINDH, G. (1976) - Water for a Starving World. Westview Press, Boulder, Colorado.

FORRESTER, J.W. (1971) - World Dynamics. Wright-Allen, Cambridge.

GLANTZ, M.H.; WIGLEY, T.M.L. (1987) - Climatic Variations and their Effects on Water Resources. In McLAREN and SKINNER 1987, pp. 625-641.

GLEICK, P.H. (1986) - "Methods for Evaluating the Regional Hydrologic Impacts of Global Climatic Changes". Journal of Hydrology, 88, Amsterdam, pp. 97-116.

GOODLAND, R.; LEDEC, G. (1985) - Neoclassical Economics and Principles of Sustainable Development. The World Bank, Office of Environmental and Scientific Affairs, Washington, D.C.

HANSEN, J. et al. (1984) - Climate Sensitivity: Analysis of Feedback Mechanisms. In HANSEN and TAKAHASHI 1984, pp. 130-163.

HANSEN, J.; TAKAHASHI, T. (Eds.) (1984) - Climate Process and Climate Sensitivity, Maurice Ewing Series, 5, American Geophysical Union, Washington, D.C.

LUNDQVIST, J. et al. (Eds.) (1985) - Strategies for River Basin Management. Reidel Publishing Company, Dordrecht.

LVOVICH, M.I. (1979) - "World Water Resources, Present and Future", Geo-Journal 3(5), 1979, pp. 423-433.

LVOVICH, M.I. (1987) - Ecological Foundations of Global Water Resources Conservation. In McLAREN and SKINNER 1987, pp. 831-839.

McLAREN, D.J.; SKINNER, B.J. (1987) - Resources and World Development. John Willey & Sons, West Sussex.

MEADOWS, D.H. et al. (1972) - The Limits to Growth. Universe Books, New York.

MUNN, R.E. (1988) - Towards Sustainable Development: An Environmental Perspective. International Conference on Environment and Development, Milan.

OECD (1985) - Environment and Economics. Organization for Economic Co-operation and Development, Paris.

OECD (1989) - Integrated Water Resources Management. Organization for Economic Co-operation and Development, Paris.

ROGERS, P. (1987) - Assessment of Water Resources: Technology for Supply, In McLaren and SKINNER 1987, pp. 611-623.

SACHS, J. (1984) - "The Strategies of Ecodevelopment". Ceres. Review on Agriculture and Development, FAO, 17(4) pp. 17-21.

SAINT-MARC, P. (1971) - Socialisation de la Nature. Editions Stock, Paris.

SCHLESINGER, M.; ZONG-CI ZHAO (1989) - "Seasonal Climatic Changes Induced by Doubled CO_2 as Simulated by the OSU Atmospheric GCM/Mixed-layer Ocean Model". Journal of Climate, 2, May.

SHIKLOMANOV, I.A. (1986) - Dynamics of World Water Consumption and Availability. State Hydrological Institute (translated from "Vodnye Resursy", Nov.-Dec., pp. 119-139), Plenum Publishing Corporation.

STARR, J.R.; STOLL, D.C. (1987) - U.S. Foreign Policy on Water Resources in the Middle East. The Center for Strategic and International Studies, Washington, D.C.

SGSNRM (1988) - Perspectives of Sustainable Development. Some Critical Issues Related to the Brundtland Report. Stockholm Studies in Natural Resources Management, Report No. 1, Stockholm.

TILTON, J.E.; SKINNER, B.J. (1987) - The Meaning of Resources. In McLAREN and SKINNER 1987, pp. 13-27.

TOFFLER, A. (1970) - Future Shock. New York, Random House.

TREWARTHA, G.T. (1968) - An Introduction to Climate. McGraw-Hill, New York.

UN (1977) - Report of the United Nations Water Conference. Mar del Plata, 14-25 March 1977, United Nations, New York.

VLACHOS, E. et al. (1986) - The Management of International River Basin Conflicts. Proceedings of a Workshop held at the International Institute for Applied Systems Analysis, September.

WASHINGTON, V.M.; MEEHL, G.A. (1984) - "Seasonal Cycle Experiment on Sensitivity due to a Doubling in CO_2 with an Atmospheric General Circulation Model Coupled to a Simple Mixed-Layer Oceans Model". Journal of Geophysical Research, 89(D6), pp. 9475-9503.

WETHERALD, R.T.; MANABE, S. (1986) - "An Investigation of Cloud Cover Change in Response to Thermal Forcing". Climatic Change, 8, pp. 5-23.

WCED (1987) - Our Common Future. The World Commission on Environment and Development. Oxford University Press.

WILSON, C.A.; MITCHELL, J.F.B. - "A Double CO_2 Climate Sensitivity Experiment with a Global Climate Model including a Simple Ocean". Journal of Geophysics Research, 92(13), pp. 315-343.

WMO 1986 - International Conference on the Assessment of the Role of Carbon Dioxide and of other Greenhouse Gases in Climate Variations and Associated Impacts. Report on the Villach Conference (October 1985), World Meteorological Organization, Geneva.

ANNEX

STATEMENT ON THE WORLD COMMISSION ON ENVIRONMENT AND DEVELOPMENT REPORT "OUR COMMON FUTURE", BY THE "COMMITTEE ON WATER STRATEGIES FOR THE 21ST CENTURY" OF THE INTERNATIONAL WATER RESOURCES ASSOCIATION ON THE OCCASION OF THE VITH WORLD CONGRESS OF WATER RESOURCES HELD IN OTTAWA, MAY-JUNE 1988

1. The presence of water distinguishes the Planet Earth from other planets. Biomass production is driven by water and most of man's activities are water-dependent. Water is circulated in and provided

from a global system, the hydrological cycle. The type of natural disasters affecting the largest numbers of people are droughts and floods, both water-related. Water flows from fundamental components of ecosystems on all scales. Water, in view of its great versatility as a substance, forms an extremely complex part of both natural systems, societal systems and everyday life.

2. All these facts add up to the basic fact that human life is constrained by the limits posed by the global water cycle, and the natural laws governing that circulation system. This simple fact is illustrated by an increasing scale of environmental pollution, generated by water's mobility and chemical activeness. Water is a unique solvent, always on the move in the visible as well as the invisible landscape. The same fact is illustrated by the water-scarcity driven problems in the arid and semi-arid tropics. In fact, water-scarcity is expected to develop into a first rate issue within a few years' time only. The Ottawa World Water Congress even concluded that in the 1990s water will replace oil as the major crisis-generating issue on a global scale.

3. In view of what has just been stated, the World Commission on Environment and Development in its report "Our Common Future", although aware of the fact that water is fundamental for soil productivity, and that water may be subject to resource limits, tends to severely under-estimate water-related problems involved. The report is indeed strongly misleading in that respect. In spite of the evident ambitions to cover the specific problems of each Third World region, the Commission pays no attention to the galloping and multidimensional water scarcity now developing in Africa.

4. The fact that so many developing countries are situated in an arid and semi-arid climate is thought-provoking in itself. A report discussing a sustainable world development without reference to the specific conditions of these arid and semi-arid regions indeed lacks in credibility. Had the Commission been more aware of the implications of water-related problems, particularly in the arid and semi-arid tropics, it would probably have concluded that water - being as complex as energy - would have deserved a sub-chapter of its own in the book. It is in fact remarkable that there are sub-chapters on the oceans, space, and the Antarctics but not one on fresh water, the blood and the lymph of the geophysiological system that we call the biosphere.

5. Such a chapter would have brought up the many parallel basic functions of water and the many links produced, creating dilemmas for human activity and global development: the hydrological link between the water consumed for plant production and the water remaining for feeding terrestrial water systems in aquifers and rivers, i.e., the water available for human use; the chemical link between moving sub-surface water and the quality that it attains; the mobility link between land use activities on one hand, and river response flow, variability and quality of both groundwater aquifers, rivers and lakes, estuaries and coastal waters on the other.

6. Although there are some scattered references to water-scarcity-related problems, the Report as regards water resources is heavily biased towards present thinking in the temperate zone. The broad geographical composition of the Commission and the substantial input from worldwide hearings were evidently not enough to counter this intellectual inertia.

7. The absence in the World Commission on Environment and Development Report of a credible discussion of sustainable development as seen from a water perspective caused great concern, even dismay, at the World Water Congress. The Congress therefore urged the "Committee on Water Strategies for the 21st Century" to produce an IWRA (Inter-national Water Resources Association) paper on sustainable development as seen from a water perspective. That paper will analyse the complex

involvement of water in relation to sustainable development, the various criteria and components involved, and main processes and activities that are threatening the sustainability. The report will discuss key concepts such as limitations posed by the environment, sustainable yield, systematic effects etc. Given the fact that life is based on myriads of water flows - through every single plant, every animal, every human body - and that it is globally circulated in the water cycle which man continuously interacts by land use and by societal activities, sustainable development must be a question of sustainable interaction between human society and the water cycle, including all the ecosystems fed by that cycle.

8.   The fact that water is subject to the natural laws controlling the hydrological cycle has distinct consequences for the degrees of freedom under arid and semi-arid conditions. The effects of such limitations in the hunger crescent in sub-Saharan Africa can be seen in the breakdown of the environmental fabric of the area, and in population dislocations, driven by the spread of a feeling of insecurity among the zone societies. The area is involved in a population-driven risk spiral, composed of several types of water scarcity, added on top of each other and related during intermittent drought years.

9.   Fundamental strategy changes are needed to address the massive sustainability problems in the realm of water. Institutional systems have to be resource-oriented and integrated, rather than user-oriented and sectorized. Changes in water attitudes are crucial to avoid the present biases and make sustainable water development possible. The legal framework has to take water as a major element in a comprehensive land use law.

For the Committee on Water Strategies for the 21st Century: Malin Falkenmark (Sweden), Luis V. da Cunha (Portugal), Hiroshi Hori (Japan), Yahia Mageed (Sudan).

Chapter 2

THE INTERNATIONAL HYDROLOGICAL PROGRAMME OF UNESCO AND THE WATER POLLUTION PROBLEMS

N Da Franca Ribiero Dos Anjos (UNESCO, Paris, France)

ABSTRACT

The activities of the International Hydrological Programme related with the water pollution problems since 1974 are described in this paper, as well as the future activities planned for 1990-1995.

Key Words : hydrology, international programme, Unesco.

1. THE INTERNATIONAL HYDROLOGICAL PROGRAMME

The International Hydrological Programme (IHP) is an integral part of the efforts made by the United Nations system as a whole to promote a rational policy for the development and management of water resources around the world.

The solution lies in rational management of water resources that should take supply and demand into consideration. Grounded on solid scientific and technical foundations and of interdisciplinary approach to the ecological economic and social problems associated with the development resources, such management should aim to promote the use of water resources in such a way as to ensure the satisfaction of society needs while still preserving them for the future.

The realization of the significance of a scientific basis of hydrology in the development of water resources began in the years following the second World War. In 1950, Unesco launched a programme of research on the world's arid zones in which hydrology played an important role, followed in 1964 by the launching of the International Hydrological Decade (IHD) which made a significant contribution to the understanding of the processes occurring in the water cycle, the assessment of surface and groundwater resources, and the adoption of a more rational attitude towards water use.

As a result, in 1974, the General Conference of Unesco launched the long-term International Hydrological Programme (IHP) with the aim of finding solutions to the specific problems of different geographical conditions and different levels of technological and economic development. The main purpose established for the IHP, which is the major component of Unesco's water resources programme, is to develop a scientific and technological basis for the rational management of water resources, both as regards quality and quantity.

The IHP is based on the active participation of the 138 National Committees and focal points of the IHP. They provide the link with the water resources specialists in the countries.

Organizations of the UN system such as the World Meteorological Organization (WMO) and the United Nations Environmental Programme (UNEP) co-operate in the execution of the IHP. More than six world-wide international scientific non-governmental organizations and over 100 water specialists participate as working group members or rapporteurs in the execution of the international activities of the IHP. Moreover, several thousand specialists participate, on a national scale, in the execution of activities linked to the different projects of the IHP.

The IHP is planned, co-ordinated and evaluated by the Intergovernmental Council of the IHP which meets at least every two years. Its 30 members are elected by the General Conference of Unesco. The secretariat of the IHP is provided by the Division of Water Sciences at Unesco Headquarters, in Paris, with the support of regional hydrologists in the Regional Offices for Science and Technology.

During the phases of IHP, several scientific publications have been issued in the various water-related series of Unesco. A number of international symposia and workshops in the field of hydrology and water resources have either been convened by Unesco or organized with its support. Greater emphasis has been given to education and training activities.

2. WATER RESOURCES PROBLEMS IN THE WORLD

One of the most important resource in need of careful management is water.

Consumption of fresh water for domestic, agricultural, energy and industrial purposes is expected to increase greatly by the end of the century for a number of reasons, among which are: population growth, the extention of irrigation for the purpose of increasing agricultural output, the foreseeable development of water distribution systems, and the expansion of water-consuming industrial activities.

Although in global terms water resources are well in excess of foreseeable demand, they are unevently distributed. Some regions are already experiencing severe water shortages which are in danger of worsening in the future. Instances of an insufficient supply of water in relation to demand are periodically and, in some cases, endemically complicated by major fluctuations in supply. The exceptional flood or drought thus often results in a disastrous loss of human life.

It can have some serious repercussions on the economy and the environment. Add to this the fact that although some countries and regions realize they have serious problems - many don't know what to do, or where to begin. There are tremendous international implications.

In many cases, the situation is further aggravated by the poor quality of water, which makes it unsuitable for various purposes.

The solution to the problems of water resources lies in rational management, which should be concerned both with supply and demand. Grounded on solid scientific and technical foundations and necessitating an interdisciplinary approach to the ecological, economical and social problems arising in this respect, such management should aim to promote use of water resources in such a way as to ensure the satisfaction of society's needs while preserving those resources for the future.

3. THE IHP AND THE WATER POLLUTION PROBLEMS

During the first phase of IHP (1975-1980), several studies, having not only scientific value, but also clear practical impact, were accomplished. Among them, may be cited those on changes in the hydrological regime due to various human activities, hydrological problems related to energy development, the dispersion of pollutants in aquatic media, socio-economic aspects of urban hydrology, aquifer contamination and protection and on land subsidence due to ground-water abstraction.

During the second phase of IHP (1980-1983), the studies concerning man's influence on the hydrological cycle which were started during the first phase of IHP was further developed to arrive at an applicable methodology for the assessment and prediction of changes in the hydrological regime. A workshop on the application of mathematical models for the assessment of changes in the quality of water was organized in La Coruña, in 1982.

The programme of IHP-III (1984-1989) represented significant departure from the earlier IHD/IHP efforts. Whereas a strong emphasis on the traditional hydrologic sciences, the increasing importance of rational water management has required that a much more broadened view of the programme be accepted. This concept of IHP-III, then, led to a much greater degree of consideration for applications of results. It has also required that the scope of the programme consider areas and audiences which, up till this time, would not have normally been thought of as part of a programme of hydrology.

The general title of the third phase of the IHP is **Hydrology and the Scientific Bases for the Rational Management of Water Resources for Economic and Social Development.**

One of the sections of IHP deals with *the influence of man on the hydrological cycle.* This section covers scientific studies on the influence of man on the hydrological cycle, including water quantity and quality. Activities of man are considered to include direct actions such as land-use changes, consumptive use of water, physical operations on river systems, addition of contaminants of various kinds, as well as those of a more indirect nature, such as, for example, man-induced climatic changes. These studies have included the effect of changes in the hydrological cycle on social, environmental and ecological aspects relative to water resources.

One of the themes of section two deals with the *methods for assessing the changes in the hydrological regime due to man's influence.* Man's development of water resources has resulted in great economic and social benefits by providing favourable environments for agriculture, industry and living conditions. However, if not carefully planned adverse effects may also occur. The assessment of the influence of man's management activities on the hydrological cycle therefore is central to the problem of modern hydrology, especially considering the increasing scales of river runoff regulation, ground-water utilization and general changes in the environment. Developments projected over the next few decades are likely to be much greater than those already realized. Simple prudence requires that changing parameters of water quantity and quality be understood both locally and globally, that systematic observations and studies be carried out, and that resulting accurate predictions of the consequences be made and disseminated.

The IHD and IHP Councils have continuously given priority to this subject. During the IHD and IHP-I, II and III, a number of reports were published, including:

- *Man's influence on the hydrological cycle*, published by FAO, (1974);

- *Hydrological effects of urbanization* (1974);

- *Effects of urbanization and industrialization on the hydrological regime and water quality*, Proceedings of the Amsterdam Symposium (1977);

- *Impact of urbanization and industrialization on water resources planning and management*, Proceedings of Zandworst Workshop (1979);

- *Case-book on methods of computation of quantitative changes in the hydrological regime or river basins due to human activities* (1980);

- *Aquifer contamination and protection* (1980);

- *Dispersion and self-purification of polluants in surface water regimes* (1982);

- *Study of the relationship between water quality and sediment transport* (1983);

- *Relationship between natural water quality and health* (1983);

- *The application of mathematical models of water quality and polluant transport: an international survey* (1984);

- *Hydrologic aspects of land disposal of radioactive waste* (1987);

- *Metals and metalloids in the hydrosphere: impact through mining and industry, and prevention technology*, Proceedings of the Bochum Workshop. (1987);

To complete the activities of the Third Phase of IHP, a workshop on the application of mathematical modelling to the assessment of changes in water quality will be held in Tunisia, in 1990, in the framework of IHP-III Project 6.2.

4. FUTURE ACTIVITIES

As the world's human population continues to grow, it is evident that we must all think very carefully in terms of development that meets the needs of the present without compromising the ability of future generations to meet their own needs and aspirations, i.e. truly sustainable development. It is clear, however, that some quite unsustainable development policies and practices, particularly concerning water management have been followed. Economic and social change necessitate development of water resources based on sound environmental principles. Rapid change increases the urgency. A sound scientific understanding should form the foundation upon which rational decisions regarding water resources management should be made.

However, the essential role of science in the continued socio-economic development--an area in which the water resources are essential-- is not a simple one. For in managing our resources it is evident that nature not only affects man, but man's activities can sometimes have devastating results on nature. It has become evident now, for example, that some of man's activities appear to be leading to possible major climate changes. The probable consequences are not yet known with certainty, but it is clear that a climate change would result in a redistribution, in time and space, of our water resources. In order to understand this change and be able to cope with it, we must have a much stronger scientific understanding of the processes involved. The approaching problems, coupled with the many existing environmental stresses (including, for example, land and water pollution, erosion and sedimentation, natural and man-caused hazards) emphasize the need for continued development of human potentials --education, training and public understanding-- as an essential element in a major international effort.

We must come to appreciate more fully that our rapidly changing and increasingly vulnerable environment will only be protected if it is given outstanding legal, organizational and scientific efforts. If we look only to the short-term future there will inevitably be increasing conflicts between development and the social/environmental aspects. If, on the other hand, greater consideration is given to an understanding of the need for scientifically based integrated development of our resources, with consideration of socio-cultural aspects and careful thought to the problem of protecting the natural heritage for future generations, then rather than being points of conflict, those aspects become extremely important considerations. Without doubt, the need for science as a base for integrated management will increase in importance.

The responsibility of water scientists and engineers, then, must be --within a full consideration of the changing environment-- (1) to develop and maintain information on the availability of water resources; (2) to assess, monitor and predict the resulting quality of water bodies and the water-related environment; (3) to develop a bettter scientific understanding of those effects of man's activities that influence hydrologic regimes (especially those resulting from climate change); and (4) to provide decision-makers with the necessary information in succint, properly constructed formats such that they will understand the problems, and the importance of the hydrological sciences as a basis for proper environmental management --especially the water resources-- and be thus enabled to react appropriately.

In so doing, water experts must also lead in the development of a better understanding within society of the importance of society's own role in protecting water resources. This calls for a coordinated programme to provide each country with the knowledge needed so that it can manage its water resources in an environmentally sound manner consistent with its own social and cultural identity and aspirations.

The plan (1990-1995) of the International Hydrological Programme **Hydrology and Water Resources for Sustainable development in a Changing Environment**, opens a new era in the development of water sciences and management, and has been designed to provide an international focal point for a broad co-ordinated effort in hydrology and the scientific bases for water management. It represents the combined efforts of national, regional and international governmental and non-governmental organizations with the one overriding goal of helping nations to help themselves.

In sub-programme **Hydrological research in a changing environment**, Theme H-3 deals with *the changes in water quality through the hydrological cycle* and two projects wil study pollution aspects: *Project H-3-1:* Prediction of hydrological, chemical and micro-biological processes of contaminant transport and transformation through the hydrological cycle and *Project H-3-2:* Hydrological, chemical and micro-biological processes of contaminant transformation and transport in river and lake systems.

In sub-programme **Management of water resources for sustainable development**, Theme M-3 deals with *the evaluation of social and environmental aspects of fresh water systems and prediction of impacts of man's activities.* Two projects have special interest in water pollution problems: *Project M-3-1:* Hydro-ecological models and bio-monitoring for environmental evaluation and prediction of impacts of natural and man-made changes and hydro-ecological classification of fresh water bodies, and *Project M-3-4:* Flood plain pollution control management of the rivers Vistula (PR Poland) and Main (Federal Republic of Germany).

Chapter 3

ENVIRONMENTAL MEDIATION AND THE NEED FOR NEW ENVIRONMENTAL STRATEGIES

J B Wood (Wood Communications Group, Wisconson, USA),
G Chesters (Water Resources Center, University of Wisconsin-
Madison, USA)

ABSTRACT

In a world of limited financial and human resources, the prioritization of environmental objectives and the resulting funding and implementation of environmental programs is a process which is inevitably defined by both scientific and political judgements. Failure to recognize the multiplicity of interests and judgements required to generate sound environmental management has led to a plethora of incomplete, poorly funded, and inadequately managed environmental projects; an equally devastating body of good ideas that were never explored; and a clutter of contradictory, if not bad, public policy.

Over the past decade and a half, policy makers, philanthropists, scientists and environmentalists have experimented with what is generically referred to as "environmental mediation." The mediation and consensus development techniques utilized under this rubric have been tested in a variety of circumstances that have helped to illuminate both strengths and weaknesses. Equally important, past experience has helped to delineate the criteria required for a successful application of these techniques.

This paper reviews some of the lessons learned and makes the argument that the economic, social and political circumstances which now condition the effort to develop and implement sound water management policies dictate the need to evaluate, adapy and adopt these techniques as a major weapon in the environmental management arsenal.

Key Words: environment, mediation, enviropol, policy

1 - INTRODUCTION

In the years dividing the two World Wars, the generals of France committed their nation to a breathtaking strategic defense As the forces of fascism loomed ominously on the horizon, thousands of France's men, women and children toiled from sun-up to sundown, attacking the hard soil of their native land with picks and shovels. Slowly, but surely massive bulwarks and towers began to spike the border. Even more slowly, but just as surely an elaborate supply line began to seam the countryside linking the towers. And, when it was complete, France waited...serene and confident in the generals' belief that this Maginot Line would render their nation invulnerable. When the battle finally came, millions of young French soldiers discovered that what was supposed to have made them invincible in fact made them vulnerable. Death and defeat, not safety was to be the reward for their labors.

A hundred years earlier, perhaps even twenty years earlier, the Maginot Line might have worked. But, critical technological advances in the art of war turned their bulwarks and towers into little more than stationary targets for the new German airforce. Disease turned the supply lines from the open channels along which men and supplies were to travel quickly and easily into clogged sewers of human death and misery. The concentration of French military power in the long thin line limited the generals of tactical flexibility and allowed the Germans to seize the initiative when and where they wished. And so, despite the victories won along the Line every now and then, one must conclude that the Maginot Line was a strategic error.

More than a half century later, the political environmental community is dangerously close to committing a similar strategic mistake. In a noble effort to protect our precious natural resources, these well-intentioned men and women have built their own Maginot Line. Nearly thirty years ago when the world finally heard Rachel Carson's Silent Spring (1962), the environmental strategists concluded that the battle to save the world's resources had to be fought in the political arenas as well as in the laboratories. On the positive side of the ledger, this decision has resulted in powerful political organizations like the Greens, the Sierra Club, the Audubon Society and others who in turn have led the successful fight for environmental protection laws around the world. But, like the bulwarks and trenches of the Maginot Line that at first glance appeared so impressive, the political environmental bulwarks and trenches have crippled our strategic flexibility and created major flaws in our defenses.

The political environmental strategy was premised upon the idea that the scientific community would generate a vision of what needed to be done and that the political wing of the community would package it and sell it to the public, thereby creating political pressure which would encourage elected officials to take the necessary legislative action. What was not foreseen was the emergence of a symbiotic relationship between the political environmental community and the political community itself which gradually metamorphosed the political environmentalists into politicians, what one writer has termed "enviropols," without changing politicians into scientists. The upshot of this transformation is that public environmental policy is now determined less on the basis of scientific reality than it is on the basis of political feasibility.

Twenty years ago, for example, when new frontiers of environmental protection could be established by over-riding the concerns of corporations that had no real constituency, the distinction between scientific mandates and political feasibility did not seem important. It was relatively easy to enact clean air and water standards which affected chemical producers and manufacturers. And, it did not take the enviropols long to discover that it was far easier to organize and build the large public grassroots organizations they needed when the issues were easy to understand and the villains were clearly identified. Digging in, like the generals who designed the Maginot Line, the enviropols constructed organizational bulwarks and towers against the narrow constituencies they could most easily defeat. The "intellectual supply lines" that were to provide the vital link with the scientific community slowly became clogged with political considerations and with a developing institutional recalcitrance towards scientific findings that complicated or threatened their political positions.

Today, we must seek, develop and implement strategies that will re-empower the scientific community. Strategies that will allow the current body of scientific knowledge to inform and assist the interest groups affected by that data so that future environmental policy is the product not of political convenience, but of scientific realities. Over the past decade and a half, policy makers, philanthropists, scientists, and environmentalists have experimented with one strategy that may offer some hope. Generically referred to as "environmental mediation," the mediation and consensus development techniques utilized under this rubric have been tested in a variety of circumstances that have illuminated both strengths and weaknesses. Past experience has also helped to delineate the criteria required for a successful application of these techniques. A review of the lessons learned in environmental mediation suggest that, given the economic, social and political circumstances which now condition the effort to develop and implement sound water management policies, serious consideration should be given to evaluating, adapting and adopting these techniques as major components of a new environmental strategic offensive.

2 - ENVIRONMENTAL MEDIATION: WHAT IS IT?

In its most basic form, environmental mediation is nothing more, nor less than the application of a set of agreed upon presentation and negotiation procedures by an inclusive, defined set of interested parties, operating with the assistance of a neutral third party, to a specific, defined environmental challenge for the express purpose of achieving a mutually acceptable response to the challenge.

From an evolutionary point of view, environmental mediation tends to emerge as a part of a multi-phased development, which may include: a) emergence of the problem; b) identification and quantification of the problem; c) expansion of interested party awareness of the problem; d) public awareness of the problem; e) pressure to respond to the problem; f) increased sensitivity to the pressure by interested parties; g) public/political/legal identification of the interested parties; h) decision to utilize environmental mediation; i) selection of a mediator, j) procedural identification of interested parties; k) inclusion of all appropriate and/or necessary interested parties; l) adoption of operating rules; m) definition and presentation of positions by the interested parties; n) negotiation and mediation; o) resolution or dissolution.

Initially environmental mediation was modeled after traditional labor relations in the United States. Most such labor disputes tend to be contract disputes involving two or three parties. On the other hand, many environmental disputes begin with no specific language or document and involve much larger numbers of parties. Consequently, environmental mediation techniques and procedures have continuously evolved to accomodate these important differences. In labor disputes, for example, identification of the interested parties is a relatively simple matter, usually dictated by who the legal parties to the contract are. Identification of the interested parties in an environmental dispute or issue is a more complicated challenge and requires identifying those who have a health, job, social, aesthtic, economic and/or political interest as well as those who have a scientific interest in the issue.

Similarly, while existing contract language and/or statutes usually dictate the appropriate procedures and relationships in a labor dispute, parties to an environmental dispute normally must develop and agree upon procedures that are mutually acceptable to those at the table. Because each situation tends to embody its own peculiarities in terms of issues and players, procedures may have some generic similarities, but are often functionally unique. Participants, for example, will typically agree that procedures must guarantee equality, but how that equality is insured may vary from situation to situation.

3 - ENVIRONMENTAL MEDIATION: DOES IT WORK?

Environmental mediation has been used successfully around the world to help resolve disputes that ranged in scope from the siting of local solid waste dumps to the siting of national nuclear waste depositories, and that dealt with issues affecting air quality, water quality, and land use. Environmental mediation has also failed in an equally impressive array of situations, and the differences between the successes and the failures have begun to help us understand how, when and where it is of most value and is most likely to produce a successful resolution.

At the risk of oversimplifying the experiences of those who practice this complex art, there appear to be seven key criteria that enhance the chances for success:

a) *The issue needs to be defined and that definition needs to be accepted by the interested parties.*

This criteria does not mean that different parties cannot have different perceptions of the ramifications of the issue, but it does mean that they must agree on what the focal point of the problem is. Take, for example, an issue involving the run-off of a chemical used by both farmers and manufacturers in a particular area. Manufacturers may, for example, see the issue as one of cost. Farmers might see the same issue as a matter of inconvenience or increased requirements for machinery they cannot afford. Environmentalists may see it as one of health and water quality. They may all bring those attitudes to the table provided they can agree that the focal point of the problem is the introduction of that particular chemical to that particular body or source of water.

b) *The interested parties must feel a roughly equivalent need to resolve the problem.*

Pressure on the interested parties may come in many shapes. The general public may have become so exorcised about an issue that the parties feel the need to act. The political/governmental community may be on the verge of implementing a solution that is not acceptable to any of the parties. The problem may have reached the point of impending economic or scientific disaster. Whatever the source of the pressure, however, it must exist and all the parties must perceive, therefore some potential benefit in resolving the problem.

c) *The identification and involvement of the interested parties must be inclusive.*

One of the major public policy benefits of environmental mediation is that it has the capacity to generate "permanent" agreed upon solutions rather than "temporary" contested responses. When that major benefit is clearly perceived as being available because all the parties are involved, pressure on the parties increases and the political/governmental sector is more likely to agree to implement the response. When the process appears to be exclusionary because some parties are ignored, the process is more likely to be viewed as coalition building and the product is more likely to be viewed as a "temporary" contested response.

d) *The operating procedures must insure that all parties feel that they are equally empowered in the process.*

There are a great many ways to organize and operate in an environmental mediation. But, any procedures agreed upon must respond to the legitmate concern that financial abilities, access to expertise, numbers, etc. not be used to outweigh merit.

e) *Those who come to the table must have the authority to speak for and commit their organizations .*

The integrity of the process requires that actions taken by the *ad hoc* group engaged in environmental mediation are an accurate reflection of the parties represented and of the organizations they represent. To that end, it is essential that the "negotiators" have the power to negotiate.

f) *There should be agreement from the appropriate governmental, corporate, or whatever other body is legally empowered to act on the issue that an agreed upon resolution will be implemented.*

While understanding that such agreements may not always be available, the odds of success do seem to increase when such guarantees can be obtained in advance.

g) *The mediator must be a skilled practitioner of his/her art.*

There are a limited number of internationally recognized mediators, but those who seek success in environmental mediations must seek to find them. These individuals are not "touchy-feely" facilitators, but knowledgeable, hard-nosed, pragmatic practitioners of a complex art.

4 - ENVIRONMENTAL MEDIATION: DO WE NEED IT?

Improved technologies and advanced scientific exploration have revealed new chasms of risk that suggest that the environmental frontiers that must be conquered today must touch the lives and interests of large, established constituencies in a multiplicity of ways. But, if science has exposed new risks, it has also shed light on new opportunities. Groundwater research, for example, has not only heightened awareness of the presence of devastating toxins, but also offers hope in new biotechnological remediation techniques. Nonpoint source research indicates dangerous run-off problems in both rural and urban areas, but work with "best management practices" suggest that low cost responses may be available. In short, what the science of the past decade begins to suggest is that we have some flexibility in how, when and where we tackle the problems confronting us. We have the technology to identify and quantify the risks. Now we need the wisdom and the courage to manage those risks in a responsible manner.

The enviropols have neither understood, nor welcomed the intricate nature of this modern challenge. We need massive public support if we are to move forward. To win that support, we need to give the public understanding and hope. Dug in along their Maginot Line, the enviropols have instead given the public fear and dispair. The residents of Los Angeles, for example, are told that they must significantly alter their work and life styles in order to halt the destruction of the ozone layer. Workers in New Jersey are informed that they must lose their jobs so that clean air standards can be met. Farmers in Wisconsin and Minnesota are told that they must change their farming practices in order to curb the degradation of drinking water sources.

This is not to say that there are no "victories" along the Maginot Line. The enviropols can still produce phosphate detergent bans, even in areas with tertiary treatment plants. They can still whip up support for clean-ups of dump sites, for more limited, but stringent air emission standards, and for recycling. But while precious legislative capital is expended achieving phosphate detergent bans,   critical nutrient management plans languish in committee. While some air emission standards are improved, acid rain legislation falls on deaf ears. While recycling moves slowly forward, toxins continue to build up in our groundwater.

The reason for these seeming anomalies is that enviropols do not create advocates or problem solvers. They tend instead to identify victims and villains, and in so doing build blocs of political opposition that   make it impossible for us to achieve the gains we need.   The worker, for example, who can be persuaded that a phosphate detergent ban or a recycling bill is a good idea becomes far more skeptical about the need for legislation that affects his/her job. The farmer who favors erosion control will oppose linking farm subsidies to what he/she perceives as arbitrary regulation of crop production. And, because the enviropols have failed to understand the need to bring these powerful constituencies into the process...because they chose instead to operate along the same old Maginot Line...because they chose to identify the farmers, or workers, or urban dwellers as "villains"...because they failed to offer any hope or involvement...because in short they have systematically rejected the reality of the need for, and the scientific community's ability to provide, responsible risk management, they have succeeded in creating the public misperception that people must choose between their livelihood and the environment.

Put simply, we live in a world of limited financial and human resources. The prioritization of environmental objectives and the resulting funding and implementation of environmental programs is a process which is inevitably defined by both scientific and political judgements. But because the mentality of the Maginot Line has failed to recognize the multiplicity of interests and judgements required to generate sound environmental management, it has led to a plethora of incomplete, poorly funded, and inadequately managed environmental projects; an equally devastating body of good ideas that were never explored; and a clutter of contradictory, if not bad, public policy.

The science that should guide our search for solutions suggests with increasing intensity that if Man is to survive we must begin to engage in a serious critique of the outmoded environmental Maginot Line. That critique must result in a dynamic new partnership between Man and science...a partnership that acknowledges both the immutability of the laws of nature and the capacity of science to preserve and enhance society's ability to provide food, jobs, and basic human comforts within the boundaries of those immutable laws. Those who would secure the environmental future must devise new strategies, strategies that recognize the need for man and nature to coexist, but do not subjugate the mandates of science to the temporary conveniences of politics. Achieving that goal will require that we have the courage and the vision to abandon the environmental Maginot Line in favor of strategies that seek not to exclude, but to involve; not to dictate, but to educate. Strategies, in short, that lead not to confrontation, but to resolution.

Environmental mediation has the capacity to be such a strategy. It is based on the premise that those affected must be involved in the search for a solution. It's operational guidelines require equality and democracy. Correctly established, it has the ability to influence public opinion and public policy directly. It does, in short, involve, educate and resolve. And, it is available today. Available to help initiate negotiated agreements between farmers, manufacturers, and urban dwellers on how to respond to nonpoint source pollution. Available today to bring together the engineers, politicians, financiers, artlovers, and environmentalists that hold the fate of Venice in their hands. Available today to begin the critical process of redirecting environmental policy away from the simplistic, divisive objectives of the enviropols and toward the massive international environmental problems that demand creative, responsible responses.

The critical nature of the problems we face requires adaptation of the environmental mediation process. Risk management cannot be acomplished without an accurate, detailed understanding of the the nature and magnitude of the risks involved. We are past the point where we can continue to ignore, or compromise the scientific data that must serve as our road map to the future. For this reason, the environmental mediation process must make provisions for institutionalizing scientific data as a non-negotiable element in the negotiating process. Obviously there will be times when the science is unclear or where contradictory data emerges. In those cases, some agreement must be reached amongst the scientists prior to the negotiations and that agreed upon body of "facts" must then serve as the statement of scientific "truth" for the remainder of the process. With this adaptation, the scientific data will become much like the contract and statutes that govern mediation in the labor model. That is, participants in the process may reach any agreements they wish provided they do not violate the scientific "contract."

We find ourselves at a critical point in time, a time when Man's interaction with the environment must be re-evaluated and changed. This paper has been critical of those overly politicized members of the environmental community, the enviropols, but even they have played an important role in advancing our cause. There have been, and are, actors in the environmental drama who willfully ignore society's needs and their own responsibilities. These individuals have earned the inflexible wrath of the environmental community and the enviropols have been effective and appropriate messengers. But, today we must also grapple with the inadvertent, non-malicious assault on our ground and surface waters by farmers, workers, manufacturers and urban dwellers who sought not to violate the environment, but to wrest some basic human comforts from their habitat. These people do not deserve to be labeled, attacked and beaten into submission. They deserve to be informed and helped. They have a right to be a part of the solution and to continue to have the ability to provide for their families. Science can help achieve that goal. And, an energized commitment to utilize an adapted environmental mediation process will enable science to realize its crucial potential in the struggle to protect our environment.

ACKNOWLEDGEMENTS

While he shares no blame for any of the thoughts or conclusions in this paper, we are deeply indebted to Mr. Howard Bellman for his seminal work in environmental mediation and his willingness to periodically share some of his hard-won wisdom.

BIBLIOGRAPHY

CARSON, R. - Silent Spring. New York, Fawcett Crest, 1962.

Chapter 4

EIB'S CONCERN WITH THE ENVIRONMENT

G Toregas (European Investment Bank)

ABSTRACT

This report gives an idea of the EIB's approach to conservation of natural resources and environmental protection. On the one hand, it describes the evolution of the Bank's policy in this field and the role played by the EIB's governing bodies (the Management Committee, the Board of Directors and the Board of Governors). On the other hand, the report shows how this policy is put into practice. Environmental protection is an integral part of the project appraisals carried out by the EIB's staff. The environmental impact assessment is assigned to the Bank's technical advisers who have practical knowledge in environmental matters as well as experience in financial and economic aspects.

Key words : EIB, environmental protection, policy, practice, appraisal

I - THE ROLE AND ACTIVITIES OF THE EIB

1 - General

The European Investment Bank (EIB) was established 30 years ago by the Treaty of Rome, with the aim of providing on a non-profit making basis medium and long-term finance to investments contributing to the balanced and steady growth of the Community economy. Its members are the twelve Member States of the Community.

The EIB is not a political body and has no legislative powers. Its contribution is as a project financing bank.

The tasks assigned to the EIB are spelled out in Article 130 of the Treaty of Rome and are presented in Table 1 below :

TABLE 1
EIB assignments

- contributing towards the economic development of the Community's less privileged regions ;
- harnessing indigenous energy resources and fostering energy saving ;
- assisting the introduction of high technology ;
- helping to endow the Community as a whole with a wider interconnecting mesh of infrastructure (transport and telecommuncations links) ; and
- <u>supporting greater protection and improvement of the environment</u>.

Projects and schemes eligible under the above headings are financed by the Bank through loans which have to be repaid with interest. The EIB is not in a position to grant interest subsidies on such loans. As a general rule, up to 50 % of the investment cost of a project can qualify for a loan from the Bank, except, in the case of certain projects designed specifically to protect the environment.

EIB finances both large-scale investments and small initiatives, the latter-defined as those involving investment costs not exceeding 15 million ECU. In this case, global loans are channelled through intermediaries which then on-lend to suitable borrowers in accordance with the EIB's criteria.

EIB finances projects both in the Community and outside the Community, in Mediterranean countries and countries of the Third Lomé Convention (ie. the states ACP : Africa Caribbean Pacific).

Investments in the various sectors during the period 1984-88 are shown in the Table 2 below :

TABLE 2
Individual loans and global loan allocations signatures/EIB 1984-88

	Individual loans (M)	Global loan allocations (M)	TOTAL (M Ecu)	%
Energy	9662	558	10220	29.4
Transportation	5900	353	6253	18
Telecommunications	3464	0	3464	10
Water and Sanitation	2443	266	2709	8
Urban infrastructures	188	29	217	0.6
Other infrastructures	1089	83	1172	3.4
Industries	3774	5547	9321	26.7
Services	226	1038	1264	3.4
Agriculture	8	187	195	0.5
TOTAL	26754	8061	34815	100.0

2 - Environmental Field

It was in 1973 after the United Nations' Conference on the Human Environment in Stockholm that the EIB began for the first time to lend for environmental protection projects as such (projects specificaly designed to have a significant and positive environmental impact without a direct financial profitability or justification).

In 1983, the EIB's continued interest in environmental protection led to the signing of the Declaration of Environmental Policies and Procedures Relating to Economic Development and the Bank becoming a Member of the Committee of International Development Institutions on the Environment (CIDIE).

TABLE 3
EIB Loans for environmental protection and improvement works
1981 - 1987

YEAR	M ECUs	as % of total loans
1981	20	-
1982	63	1
1983	130	2
1984	154	3
1985	430	7
1986	1556	23
1987	2067	26

Since 1984, when the Bank's Governors made their recommendations on the environment, EIB lending for projects geared to protecting the environment has risen steadily as can be appreciated from the figures presented in Table 3.

II - EIB'S APPROACH TO ENVIRONMENTAL ISSUES

1 - Overall

EIB's support for the conservation of resources and environmental protection is demonstrated both by the growing number of projects in this field receiving EIB finance and also in the ways in which a decision to finance is reached. Detailed appraisals, carried out by the Bank's staff form an essential part of selecting the investment projects which receive EIB support. The environmental impact of the proposed investment plays an important, at times decisive, role in the decision making process.

The EIB's governing bodies wanted to go further. In 1984, a working group of the EIB's Board of Directors examined how the Bank should contribute to environmental protection. A series of recommendations, widening the scope for EIB support of environmental protection, was formulated by the working group and unanimously endorsed by the Board of Governors in 1984. These recommendations are presented in Table 4 below :

TABLE 4
Recommendations of Board of Directors
- extension of the eligibility criteria to projects outside assisted areas helping substantially to protect the environment;
- additional (1) finance of up to 10 % of total costs for projects incorporating antipollution equipment offering greater protection than that required under existing standards;
- strict application of national, international and Community regulations;
- in the absence of binding regulations :
  . encouraging investors to adopt the least polluting design and, in all events, to provide for subsequent incorporation of adequate waste treatment facilities;
  . consideration of the overall impact on the environment when assessing the economic viability of a project, particularly in the case of cross-border pollution;
- outside the Community, the Bank should refrain from financing projects which seriously transgress international standards, allowance being made for the specific ecological problems of developing countries;
- the Bank should continue to join forces with other financing institutions, particularly within the framework of the Committee of International Development Institutions on the Environment (CIDIE).

It is worth noting that improvements of the environment, measures to control different kinds of air and water pollution, as well as energy saving schemes in industry are considered as being of general European interest. Hence, all projects of pollution control and environment improvement - wherever located in the EEC - are eligible for financing.

2 - Specific

The division of responsibility for appraisal follows the EIB's internal organisation into Directorates, each of which has its own specific role. Financial analysts from the Directorates for Lending Operations asses the borrower's financial standing. Economists from the Research Directorate evaluate the economic justification of financing the project under review,

---

(1) Normally EIB loans do not exceed 50 % of total investment costs

whereas the technical viability as well as the environmental aspects of the proposed investment are assessed by the Banks' Technical Advisory Service. The Management Committee is regularly informed of the way the appraisal is progressing and it decides if and when the project will be recommended for the financing to the Board of Directors which authorises lending operations.

The EIB has no separate "environmental unit" to assess environmental effects as distinct from the general technical and economic aspects of a project, although appraisal of these problems by its specialised Technical Advisers forms a very important part of all projects. "Environment" is seen not as a "sector" but as a particular feature of every project and environmental considerations are fully integrated into project appraisal. The Management Committee and the Board of Directors pay particular attention to environmental problems during the decision-making process.

The Technical Advisory Service was reorganized in 1986. The new structure is based on technical considerations only and consists of 4 groups :

1. Chemical and electronic industries
2. Manufacturing industries
3. Mining and energy
4. Infrastructure, civil engineering

Each of these groups is managed by a Group Leader whose special task is to monitor the standard of project appraisal and evaluation, including the assessment of the impact of investment projects on the environment. Within each group, there are specialised teams.

In addition to the 4 groups, there is also a team dealing with agriculture and agro-industry under the direct supervision of the Chief Technical Adviser. The latter is also directly responsible for all general environmental matters such as cost/benefit analyses of anti-pollution measures, environmental assessment procedures and guidelines, training of staff, etc. He therefore acts as a "focal point" for all environmental questions both inside and, above all, outside the Bank.

The span of the fields of the Bank's operations is too large to let the environmental aspects be handled by generalists. The thirty-five technical advisers are all specialists in their respective areas. Knowing their sectors, they are able to identify potential problems and, based on experience gathered from similar projects elsewhere, to offer alternative, least polluting solutions.

Our experience is that many projects have environmental implications which need attention. Only in a few sectors -for example in telecommunications- investments appear to be environmentally neutral.

The environmental impact assessment by the Bank's staff consists of an analysis on the basis of a standardised checklist which takes into account the effects on water, air and soil, noise levels and visual pollution. This checklist is complemented by more specialised and detailed lists and sectoral guidelines, also used by other concerned international institutions such as the Commission of the European Communities, the World Bank, and the United Nations Environmental Programme. During the appraisal process the Bank does not undertake its own extensive independent studies but may require further studies by others.

Once all the relevant data have been collected, the appraisal team will meet again with the project promoter to discuss the outcome. At all times, the Bank guarantees that the existing environmental legislation - whether national, Community, or international - has been met before it decides to finance. In case improvements are necessary or desirable, the Bank will offer advice on which of the various options for reducing environmental pollution are most suitable for the project concerned and which show the optimum cost/benefit ratio. The ultimate aim is to convince the borrower to adopt the least polluting solution for his investment plans, interalia by drawing attention to ongoing Community deliberations which may. affect the project in the years to come.

The EIB's Board of Governors has recommended that schemes should not be financed, in cases where the general economic cost of a project (including damage caused by pollution) would exceed its general economic benefits.

III − EIB'S ENVIRONMENTAL CONSIDERATIONS AND EXPERIENCE DURING APPRAISAL OF PROJECTS

1 − Basic considerations and parameters

TABLE 5

Questions to be addressed

- What aspects of the environment will be altered by the project ? Of these changes, which will improve and which degrade the environment ?
- What irreversible changes may be caused ? (disappearance of animal or vegetable species; impairment of the particular characteristics in the area, etc.)
- What measures can be adopted to lessen the deleterious environmental effects or to enhance any beneficial ones ?

Environmental parameters to be checked
    (before and after the project)

- Quality of water
- Quantity of water
- Quality of soil
- Quality of the air
- Aquatic ecosystems
- Forestial ecosystems
- Undesirable and/or irreversible change
- Vulnerability to natural hazards
- Aesthetic aspects
- Microclimate
- Auditory nuisances

2 − Main areas of concern

The engineers' initial action is in establishing the appropriate technology the project promoter can apply in his scheme. As far as the environmental aspects are concerned, the baseline is that the technical solution at least meets the legal requirements with regard to its external effects (e.g. emission control) as well as internally (e.g. work safety).

The approach to the environmental assessment of a project is wide-ranging with all possible effects outside the project also being examined (e.g. protection of biotopes, bird sanctuaries and final fate of residual products)and if it is necessary efforts are made to ensure that appropriate corrective measures are carried out.

In our experience even large anti-pollution programmes, where a lot of money have been spent for a detailed study of the problem are sometimes not fully coherent from an environmental point of view and EIB makes sure that proper coordination together with a monitoring are also included in the programme.

IV − CONCLUSIONS AND OUTLOOK

Right from the outset, even in those early days before the crystallisation of a fully-fledged policy on the environment, the Bank has endeavoured to finance only those projects which, to its mind, steered clear of occasioning any avoidable damage to the environment. After the UN Conference in Stockholm in 1972 on the Human Environment, the EIB's technical advisers were entrusted with the task of systematically appraising all environmental matters in line with the considerations and parameters referred to briefly in this paper.

The recommendations handed down by the Bank's Board of Governors in 1984 followed on from the signature of the declaration by the international development-aid institutions in support of protecting the environment. Ever since, the EIB has been stepping up its lending throughout Community Member Countries and their regions, both assisted and non-assisted, in support of projects designed first and foremost, if not exclusively, to serve the cause of environmental protection.

The Bank's lending to environmental projects has risen to over 25 % of its total lending reflecting the growing importance of these operations. However, in view of the aggregate capital investment needed to protect our environment, the Bank's contribution is still small. Conscious of this disparity, the EIB is fully prepared to commit itself even more heavily in future, in both absolute and relative terms, towards improving the environment.

Nevertheless, the degree of its activity in this sphere is not so much a question of Bank policy or goodwill but of the readiness on the part of borrowers to accept debts for the sake of protecting the environment. The EIB is not able to provide outright grants or interest subsidies on its loans, which consist of funds raised on the international capital markets and are passed on without any profit margin. At the same time, the Bank is not bound by annual budgets, country or sectoral quotas. It has never been a shortage of available resources that has restricted the EIB's activity but, rather a shortage of, well-conceived projects combined with a reticence to fund through loans.

Chapter 5

INTEGRATED APPROACH TO WATER QUALITY MANAGEMENT

V Novotny (Marquette University, Wisconsin, USA)

ABSTRACT

Present and future water quality management efforts must be integrated and address the pollution abatement of dry and wet weather urban, industrial and agricultural wastewater discharges (point sources), reduction and mitigation of nonpoint sources, land use and its pollution effects. They should include effects of deforestation and conversion to agricultural and urban land use, effect of drainage, and the waste assimilative capacity of receiving water bodies and its enhancement.

The solutions are both structural and nonstructural. Milwaukee River (Wisconsin, USA) and Lagoon of Venice (Italy) water quality management plans were presented as examples of an integrated approach to a particular areawide water quality problem.

Key words: water resources, pollutant loads, point pollution, nonpoint pollution, pollution control.

INTRODUCTION

Water quality, pollution, management, abatement, waste assimilative capacity are technical and legal terms that must be firstly defined before one can summarize and/or suggest integrated solutions to water quality problems.

Water quality generally means a chemical, biological and microbiological composition of water that can be of natural or cultural origin. **Pollution** on the other hand, implies addition of something to water that impairs its downstream use and, generally, is associated with man's activities. However, impairment of water quality or of a beneficial use of the water body has different meanings to different groups of users. For example, agricultural irrigation users can tolerate levels of pollutants that would not be acceptable to those using the water for drinking or commercial fishing. Second, there are groups of people who would not accept any impairment of water quality and no discharge of waste materials would be acceptable to them. In this case the notion of reasonable use or unreasonable interference should be considered. Hence, we may define pollution as a presence of foreign unnatural materials in water which would interfere unreasonably with one or more beneficial uses of the water body.

Waste assimilative capacity is then a capability of a water body to receive certain limited amounts of potential pollutants without causing harm to the biota of the receiving water body or without causing an unreasonable impairment of downstream water uses. Adverse water quality requiring management may therefore be caused by pollution (human activities) or by natural causes.

Water pollution abatement usually refers to a reduction of pollution inputs while **water quality management** means a comprehensive solution to an adverse water quality situation which includes both a reduction of the pollution loads reaching the receiving water body and enhancement of the waste assimilative capacity.

Pollution sources are generally classified into those that are concentrated, usually conveyed by a man made conduit to a single place of disposal, and those that are of diffuse character, difficult to identify and hydrologically highly variable. The former category is commonly referred to as **point sources** of pollution while the latter encompasses **nonpoint sources**. Urban sewage and industrial wastewater sources and concentrated animal feedlot operations are examples of point source pollution while runoff from fields is considered nonpoint pollution. The legal system of the United States provides for regulation and enforcement of abatement of point sources of pollution, however, pollution abatement of nonpoint sources is difficult to enforce and relies on voluntary approaches (Novotny 1988). A recent rule by the U.S. Environmental Protection Agency has reclassified urban stormwater pollution sources as point pollution, hence, subject to enforcement by the National Pollutant Discharge Elimination Permit System (U.S. EPA 1988). Presently, such regulation and enforcement is considered only for urban areas with more than 100 000 population.

SOURCES OF POLLUTION

Watershed in transition

The most severe, profound and often irreversible water quality modifications and adverse effects occur in a watershed that is in transition. The transition of a watershed from a pristine unaffected watershed to a fully urbanized basin (the final stage) progress in the following phases:

1) **Deforestation** increases the hydrological activity of the watershed and erosion potential. Deforestation is a worldwide problem with many adverse effects, including worsening of water quality and increased flooding, increase of atmospheric carbon dioxide levels with associated "greenhouse effect," increased soil loss and others. The pollution is primarily of nonpoint character, highly variable and related to meteorological parameters such as precipitation.

2) **Conversion to intensive agriculture** deprives the watershed of protective vegetation cover. This increases the erosion potential by one to two orders of magnitude. In addition to the higher sediment loads, the use of fertilizing chemicals and pesticides has increased dramatically within the last 20 years. In developing countries the increase of agricultural production is primarily by deforestation and reclamation of pristine lands. In contrast, in developed countries it is a result of the use of chemicals while the total agricultural land area remains the same or is reduced due to urban pressures. For example, according to the U.S. Census of Agriculture, fertilizer use in Iowa has tripled since 1955 and the pesticide use has increased by 600 percent (Figure 1). In the Venice (Italy) Terra Ferma (mainland) watershed of the Venice lagoon nitrate-nitrogen concentrations in the canals draining agricultural areas have occasionally exceeded 50 mg/l during spring seasons when fertilizing chemicals are applied. Such high nitrogen concentrations are then a cause of prolific algal growths in the canals and in the lagoon and make water then unsuitable for municipal uses.

Concentrated animal operations are the most polluting agricultural operations. Loehr (1972) reported that barnyard and feedlot runoff had extremely high BOD_5 concentrations (1 000 to 12 000 mg/l), COD (2 400 to 38 000 mg/l), 6 to 800 mg/l of organic nitrogen, and 4 to 15 mg/l of phosphorus. Feedlot runoff is concentrated point source pollution while most of the pollution from agricultural operations is mostly of diffuse - nonpoint character.

Erosion and soil losses by surface runoff are commonly considered as the predominant source of pollution from croplands. As reported by

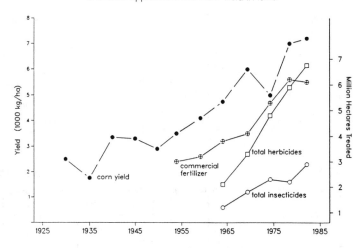

Figure 1. Corn Yields and Agricultural Chemical Application in Iowa
Source: U.S. Census of Agriculture

Alberts, Schuman and Burwel (1978) for U.S. conditions, over 90 percent of the nutrient losses are associated with soil loss. Although the nutrient losses by surface runoff usually represent only a small portion of the applied fertilizer, their pollution impact almost always exceeds the standards accepted for preventing eutrophication of surface water bodies.

However, erosion as the primary mechanism of nutrient transport from agricultural lands is significant only for sloped fields. By simulation with a hydrological model, the author has found that agricultural cropland with slopes less than 3 percent and with a medium texture or coarse soils, produce minimal pollution loading's carried by surface runoff. In these locations, the shallow subsurface flow is the primary transport mechanisms by which the nutrients (primarily nitrate-nitrogen) and other pollutants are transferred from the soils into drainage flows and, subsequently, to the receiving water bodies. This is typical for flat watersheds. For example, Bendoricchio (1988) has reported that the total nutrient loads to the Lagoon of Venice in Italy ranged between 10 000 to 15 000 tons/year for nitrogen, and 1 000 to 2 000 tons/year of phosphorus, respectively, from which agriculture was responsible for more than 50 percent. Almost all of it was carried by subsurface and drainage flows.

Agricultural runoff, as specified previously, is also a source of organic chemicals (pesticides and herbicides).

3) **Watershed during development.** The soil loss from construction sites can reach magnitudes of over 100 tons/ha. In urbanizing watersheds, few percent of the watershed under construction can contribute a major portion of the sediment carried by the streams. Often the streams themselves are affected and irreversibly changed by urbanization, regulation, straightening and lining of the streams which destroys their natural aquatic habitats and the streams no longer can sustain fish and other biotic population. The hydrological activity of the watershed is increased by increased imperviousness which makes the surface runoff more flushy and higher in volume. On the other hand, groundwater recharge is reduced. Other profound and adverse changes in hydrology and water quality are caused by draining wetlands. Pollution is mostly of nonpoint character unless a drainage conveyance system is in place and the drainage basin is located in a large urban area and/or industrial zone, which would legally reclassify the runoff pollution as a point source.

4) **Nonsewered suburban and urban development.** Disposal of sewage into soils (septic tanks and soil adsorption) eliminates or reduces only pollutants that can be filtered out or adsorbed onto soil particles.

Figure 2. Comparison Snowmelt and Stormwater Quality in Milwaukee

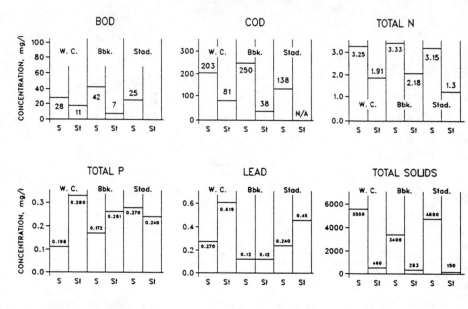

S -- Snowmelt
St -- Storm runoff
WC -- Wood Center watershed – commercial land use
Bbk -- Burbank watershed – medium density residential
Stad -- Stadium freeway interchange

Mobile pollutants such as nitrates can cause severe contamination of groundwater and, subsequently, of base flow of the streams and other water bodies. When the adsorption capacity of the soils disposal system sis exhausted, contamination of surface waters by organics and fecal microorganisms will occur. However, in comparison to other land uses, the pollution potential of established low density residential and commercial areas without storm or combined sewer systems is generally low and not much above the natural loadings from a pristine prairie watershed.

Pollution from developed urban lands

Pollution from urban lands is diversified but generally can be categorized as concentrated - point source pollution by sewage and industrial wastes which is less related to rainfall, and diffuse - wet weather pollution caused by erosion and wash-off of pollutants from urban surfaces that is then conveyed to storm drainage.

In analyzing the pollution impact of wet weather overflows from urban drainage systems the type of drainage is important. In the U.S. most sewer systems are separated, having two conveyance systems, one that carries sewage, infiltration and limited amounts of surface runoff from cross-connections, the other carriers mostly surface runoff with some infiltration. In the majority of European urban areas and in some older U.S. cities the sewer system is combined. In spite of the diffuse origin of wet weather urban pollution loads, both sewer systems in the U.S. are legally considered as point sources.

Measurements of the pollution potential of urban wet weather flows both in the U.S. by the Nationwide Urban Runoff Program (U.S. EPA - 1983) and in Europe (Malmqvist-1984, Ellis-1986, Krejci-1988) showed that the total pollution potential of overflows from combined sewer systems is of the same order of magnitude as that of raw sewage while the pollution

Table 1
Ranges of average concentration of pollutants in
urban runoff sewer overflows, and sewage.
(Mean of the data in parenthesis)

Type of Wastewater	BOD_5 mg/ℓ	Suspended Solids mg/ℓ	Total Nitrogen mg/ℓ	Total Phosphorus mg/ℓ	Lead mg/ℓ	Total Coliforms MPH/100 mℓ
Urban stormwater[a]	10-280 (27)	3-11000 (630)	2-5 (2.5)	0.2-1.7 (0.8)	0.03-3.1 (0.4)	10^3-10^8
Combined sewer over-flows[a]	60-200 (120)	100-1100 (410)	3-24 (11)	1-11 (4.3)	(0.4)	10^5-10^7
Light industrial area[b]	8-12	45-375	0.2-1.1	NA	0.2-1.1	10
Roof runoff[b]	3-8	12-216	0.5-4.0	NA	0.005-0.03	10^2
Typical untreated[c] sewage	(300)	(350)	(55)	(15)	NA	10^7-10^9
Typical biologically[c] treated effluent	(20)	(20)	(35)	(10)	NA	10^5-10^7

( ) = mean, NA = Not available
Adapted from: (a) Novotny and Chesters, (1981) and Lager et.al 1974
(b) Ellis 1986
(c) Imhoff and Imhoff 1986

potential of urban stormwater only is about the same as that of treated sewage from the same area. However, it has to be realized that long term averages can be quite misleading. During a wet weather period the wet weather pollution component can be much larger than a dry weather raw sewage load as it is demonstrated in Table 1. This is especially true for suspended solids for which the wet weather loads exceed the dry weather sewage loads by several orders of magnitude. Hence, the wet weather pollution should be considered as a shock to the receiving water body.

Wet weather pollution of urban runoff is a resultant of several diversified input sources and processes. These sources can be categorized as follows (Novotny and Chesters, 1981):

* Wet and dry atmospheric deposition
* Street refuse deposition, including litter, street dust and dirt, and organic residues from vegetation and urban animal population
* Traffic emissions and impact
* Urban erosion
* Road deicing
* Solids deposition and slime growths in combined sewers during antecedent dry periods.

Even though the research on average strength of wet weather urban pollution loads has been extensive the actual loads can be only speculated and/or determined by a calibrated and verified hydrological urban runoff quantity-quality model. Reviews of available models and modeling technology have been most recently presented by Ellis (1986), Huber (1986) and Novotny (1988).

Pollutant loads by sewage. The magnitude of the so-called dry weather (sewage) pollution is also variable but not in such great degree as it is for wet weather pollution sources. The pollution loads are commonly expressed by so-called **population equivalent unit loads** that are similar for both Europe and the U.S. (Table 2). There is however, a substantial difference in quantities of sewage between the U.S. and other countries. The municipal water use and, hence, sewage volumes of U.S. cities is around 400 l/cap-day, which is about twice of that for typical European urban centers.

Table 2
Population equivalents (unit loads) for
municipal wastewater[*]

Constituent	g/cap-day
BOD$_5$	60
Suspended solids	80
Total nitrogen	15.5
Phosphorus	3.0

[*]After Novotny et.al 1989

Pollutant loads for combined sewer systems. The capacity of combined sewers has commonly been designed to accommodate flows that are
four to eight times the peak dry weather flow. In Europe, a critical
rainfall of 15 l/sec-ha (5.4 mm/hr) plus the dry weather flow is now
being implemented in the design of combined sewer systems. If the rainfall exceeds the critical design rate an overflow of sewage without any
treatment is anticipated. Using this criterion for a typical city located in Switzerland, out of about 1000 hours of rainfall in a typical average year, overflows would occur for about 80 hours (Krejci-1988). In
Milwaukee, Wisconsin, there are over 40 overflows per year from the
combined sewer systems of the older sections of the city. A permit
issued by the Wisconsin Department of Natural Resources which regulates
pollution discharges in the state, specified that the overflow frequency
had to be reduced to two per year. As a result, an expensive underground
storage tunnel is being built 100 meters below the city surface to store
the excess overflows and discharge them for subsequent treatment.
    The pollution strength of the overflows is about the same or only
marginally less than the strength of raw sewage. Since solids can accumulate in the sewers during the antecedent dry period and slime can develop
on the sewer walls the first portion of the overflow carries dispropor-
tionate amount of pollutants - so called first flush. Abatement of wet
weather pollution from combined sewers should first focus on the abatement of the first flush of the large storms. Roesner (1988) stated that
for most U.S. areas east of Mississippi River focusing abatement on the
first 2.5 cm of the rainfall would capture and treat 91 percent of the
total rainfall falling on the area. Similarly for conditions in Switzerland (Krejci 1987), designing combined sewers overflow abatement for rainfalls of 15 l/s-ha (0.5 cm/hr) would result in 90 percent capture.

Effect of overland drainage on pollution loads

    The source strength, that is the amount of pollution generated at
the source, may not be the same as the pollution load reaching the receiving water body. This is especially true for diffuse pollution loads.
Consequently, pollution control planners must locate the sources that
will have the most profound water quality impact. The water quality impact then depends on the source strength, pollution attenuation on t'
route from the source area to the receiving water body and on the wa
assimilative capacity of the receiving water body.
    The source strength for point sources is determined by measurem nts
for existing sources and, for example, by population equivalents or by
comparison with a similar existing source for planned sources. For rural
and urban diffuse sources the source strength, i.e. sediment or pollutant
load, is usually determined by a model or by an equation such as the Uni-
versal Soil Loss Equation (Wischmeier and Smith, 1965).
    Several studies beginning with Roehl (1962) have revealed that up-
land erosion and pollutant generation estimated either from a hydro-
logical-erosion model or from extrapolations of measurements made on
small plots, do not equal the sediment or pollutant yield measured in the
receiving water body. To account for these differences, the delivery
ratio factor (DR) was introduced:

$$DR = \frac{Y}{A} \qquad (1)$$

where Y = basin sediment or pollutant yield and A = upland erosion or
pollution generation potential.

It should be pointed out that most of the models for estimating
pollution potential (source strength) of diffuse sources are capable of
considering only the surface runoff components. Modeling subsurface
transport of pollutants is still in infancy at best and unreliable.

It has been documented (Dickinson and Pall, 1982) that sources of
stream sediment do not necessarily coincide with major soil erosion
areas, due to the capacity of different parts of the watershed to transport sediments. A source with a high soil loss located far from established channels may not contribute as much pollution to the stream as a
less erosive source located close to the stream. Using lumped DRs will
a priori put a higher weight on sources of high erosion over sources
with high pollution potential.

Modeling watershed sediment transport. As sediment particulate
loadings from a watershed are very difficult if not impossible to measure
hydrological modeling is the only viable solution.

Pollution loadings from diffuse sources can be modeled by distributed or lumped parameter hydrological models.

Walling (1983) criticized the spatial lumping of the DR and presented cases of a distributed modeling concept in which the watershed is divided into small cells and the sediment is routed from each cell separately.
An example of such an approach is a hydrologic-sediment movement model
ANSWERS developed by Purdue University researchers (Beasley and Huggins,
1980 and Beasley et al., 1982). In the ANSWERS model, the watershed is
divided into small square elements (Figure 3) ranging from 1 to 4 ha.
Sediment yield from an area is modeled by equations which predict soil
detachment (erosion) due to rainfall impact and overland flow. The detached soil, when combined with the amounts contributed from the adjacent
elements, becomes available for movement from the element. The distributed parameter concept does not require the formulation of the delivery
ratio because delivery from one cell to the next downstream cell is determined by the sediment carrying capacity of the overland flow within the
cell. The problem with the distributed parameter concept lies in the

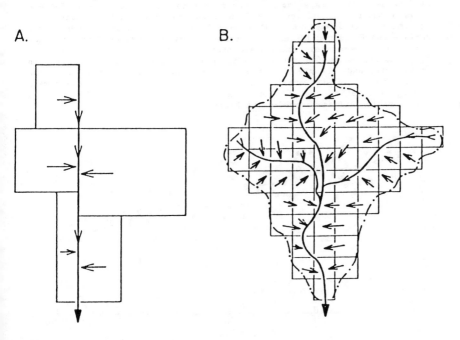

A.   B.

Figure 3.   Basic representation of hydrological - nonpoint polllution
            models. A. represents a lumped parameter model and B. a
            distributed parameter model.

uncertainty and inaccuracies of the available equations for overland sediment transport which contain several parameters, such as roughness, vegetation, slope and others, that must be calibrated in a distributed fashion.   Furthermore, even the smallest size of the computational elements,
the 1 ha cell, may not be uniform.   Baun (1985) analyzed data from experimental watersheds in south central Wisconsin and concluded that out of
1024 1-ha cells located within the watersheds only 153 contained a single
soil type.   In addition, slopes, vegetation cover, and surface roughness
added to the variability and nonuniformity of parameters even on the 1-ha
cell size scale.

The lumped parameter concept is more used and accepted.   In this
concept the parameters are averaged over a large portion of the watershed
which results in simpler and smaller models.   Regarding the sediment delivery there is only one unknown factor, the lumped DR.   However, this is
not an advantage.   As stated by Walling (1983), the problem of spatial
lumping is further highlighted by the fact that in humid regions only a
fraction of the watershed contributes surface runoff and, hence, the sediment and pollution associated with it.   These areas are called "hydrologically active areas" and their extent depends on the magnitude of rainfall
and rainfall losses.   A lumped parameter model is not usually suited for
determining the extent of hydrologically active areas.

Measured ranges of the sediment delivery ratio. Figure 4 shows
the geomorphological lumped delivery ratios for eastern U.S.   From 1975
to 1980, a number of small experimental sub-basins were monitored in the
Menomonee River basin located in metropolitan Milwaukee (Novotny et al.
1979).   A hydrologic model, calibrated and verified by field measurements
from very small, uniform sub-basins within the watershed was used to estimate upland loadings of sediment and phosphorus.

The simulation showed that DR for sediment ranged from 0.01 for pervious portions of the basin and for developed residential nonsewered
lands to 1.0 for urban land with a good storm or combined sewer systems.
Thus DR was dependent largely on the degree of storm sewering in the sub-
basins (Table 3).

The profound effect of drainage on nonpoint pollution loadings can
also be seen when comparing loadings from rural and urban areas.   Table 4
shows a compilation of loadings from urban and rural small Wisconsin experimental watersheds.   Note that only long term averages of storm water
surface runoff is considered.   The rural values were measured from Wisconsin priority watersheds which are mostly agricultural basins with a serious nonpoint pollution problem.   Essentially, average soil loss within
these watersheds is greater than 10 tonnes/ha-year.   Judging from the
average soil loss estimates and measured pollutant yields, one could make
a rough estimation that DR for rural watersheds is 0.1 to 0.2, not an
unreasonable estimate according to Figure 4.

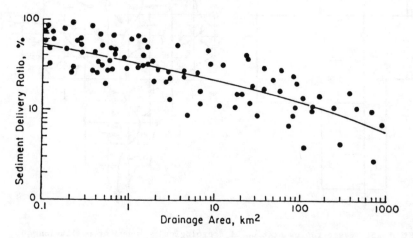

Figure 4.   An example of the relationship between delivery ratio and
watershed size (after Roehl, 1962).

Table 3
Estimated sediment DR from pervious areas* for
various land uses in sub-basins of the
Menomonee River (Wisconsin) experimental basin
(Novotny et al., 1979)

Sub-basin type	Impervious area (%)	Degree of storm sewering (%)	Sediment DR
Agricultural	<5	0	0.01 - 0.3
Developing construction	<5	20-50	0.2 - 0.5
Low density residential, unsewered	<20	0	<0.1
Parks	<10	0	<0.03
Medium density residential, partially sewered	30 - 50	50	0.3 - 0.7
Medium density residential, sewered	30 - 50	50	0.7 - 1.0
Commercial high density residential, sewered	<50	80-100	1.0

*Delivery ratio of sediment from impervious connected areas - 1.0.

Urban sewered sites on the other hand have a much lower strength, yet overall because of much higher DR, the loadings are approximately the same as those for rural watersheds. If urban sites were not storm sewered, loadings would have been much lower, approximately equaling the background loadings from well-protected, undisturbed lands. In unsewered areas, swales and grassed waterways can attenuate pollutants (Oakland, 1983).

WATER QUALITY PROBLEMS AND WASTE ASSIMILATIVE CAPACITY

Pollution Discharges and Dissolved Oxygen. According to most interpretations in the U.S. and throughout Europe, water quality impairment is perceived by planners as oxygen depletion during hydrologic conditions of severe drought that _a priori_ excludes any significant surface storm related input. In such cases, the point pollution, i.e. treated (or sometimes untreated) sewage and industrial wastewater effluent discharges, are the primary and often only pollution inputs that are considered in waste assimilative capacity studies. In the U.S. there are no enforceable standards for sediment concentrations and the toxic standards are difficult to implement for diffuse sources. The BOD_5 standard is enforced in Europe but not in the U.S.

If the diffuse pollution load, as it is commonly done, is expressed as sediment or nutrient losses originating from urban and rural lands during periods of rare storm events there is no apparent link between the sediment and sediment adsorbed pollutants and the DO depletion. Such sediments have a relatively low organic content and the organics are not easily biodegradable. Hence, several authors have claimed and documented that diffuse pollution loads occurring during high flows do not directly impair the DO regime of the receiving water bodies nor do they result in violation of DO standards (Field and Turkeltaub 1981, Keefer et al. - 1979).

However, the recent study of the Metropolitan Water Reclamation District of Chicago (Lue-Hing et al. - 1989) has documented a very significant impact of the Chicago deep tunnel and reservoir project (TARP) on water quality. The TARP components collect combined sewer overflows from a large portion of Greater Chicago, convey them to a large underground tunnel and surface storage reservoirs, where the polluted water is temporarily stored and subsequently pumped for treatment. 25 percent improvement in average monthly DO concentrations (higher during summer low DO conditions), 50 percent or greater improvement in Total Suspended Solids, up to 60 percent improvement in Sediment Oxygen Demand, and 60 to 90 per-

Table 4

Comparison of water and pollution yields from
Wisconsin experimental watersheds, 1980-81
(Source - Wisconsin Department of Natural Resources)

Watershed type	Watershed area ha	% Imperviousness	Runoff[1] coefficient	Pollution yield, kg/ha-year[2]		
				Suspended solids	Total P	Total lead
Urban - Storm sewers						
Commercial I	11.7	77	0.58	718	1.48	1.53
Commercial II	18.2	81	0.69	1197	1.50	3.90
Residential I	14.6	57	0.40	487	1.12	0.90
Residential II	25.3	51	0.33	272	0.62	0.28
Residential III	13.3	50	0.33	161	0.54	0.21
Residential: 10% under construction	522	47	0.31	767	0.75	0.21
Suburban - Low density, residential, partly sewered	4974	7	0.10	217	0.30	0.12
Agricultural I	3900	<5%	0.06	752	1.11	id
II	1528	<5%	0.14	743	0.74	id
III	2615	<5%	0.09	470	id	id
IV	2144	<5%	0.06	386	0.63	id

[1] Annual runoff volume/rainfall volume

[2] Excluding winter

id Insufficient data

cent improvement in sediment TKN concentrations were observed in the streams previously receiving the overflows and in the Illinois River. Since the TARP project is not yet fully completed better result and greater water quality benefits are expected during full operation.

While discharges of raw or partially treated sewage will affect the Dissolved Oxygen regime directly, apparently, sediment discharges may affect the Sediment Oxygen Demand (SOD) indirectly and in a delayed fashion. The first scientific measurements of SOD by Fair and his co-workers (1941) related the oxygen demand of river sludges (primarily sewage solids) to the rate of deposition of organic sludge solids from raw sewage onto the bottom. After primary and secondary treatment has been installed in most U.S. point discharges, dominance of diffuse urban and rural sources has become evident. Recent investigations have revealed that the SOD includes oxygen demand from several separate biochemical and chemical oxidation processes (Bowman and Delfino - 1980, Rudd and Taylor - 1980).

As stated previously, typical sediment composition from diffuse sources (with exception of combined sewer overflows) such as solids in urban runoff and from soil loss of agricultural lands, has a relatively low organic content and a very low BOD demand and may not result in appreciable SOD. It should also be pointed out that sediment, especially its organic and clay fractions, can adsorb and make biologically unavailable many pollutants, including ammonia, phosphorus and pesticides. Most of the SOD is due to organic deposited solids from raw sewage discharges in overflows and effluents and form organic solids that develop in the riverine systems.

Nutrient problems and Eutrophication. Damaging effects of diffuse and residual point sources on water quality have been best documented for impact of nutrients (nitrogen and phosphorus) on stratified lakes, reservoirs and estuaries. A large number of papers have been published that deal with the lake and reservoir trophic status or productivity.

Eutrophication is a process by which organic mass of the lake increases, reaching in an advanced state of eutrophication prolific and objectionable amounts. Traditionally, eutrophication has been associated with the income of inorganic nutrients (nitrogen, phosphorus, carbon), their excessive concentrations in water and prolific algal growth that have resulted from them. Cooke et al. (1986), Likens (1972) and Wetzel (1983) have also emphasized that eutrophication also means a loss of volume by siltation and release of nutrients from sediments and by mineral-

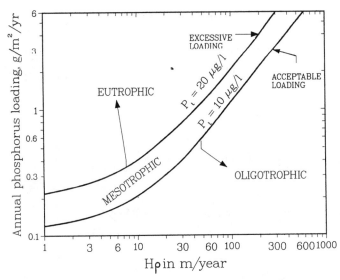

Figure 5.   Relationship between the nutrient (phosphorus) load of a
water body and its trophic status. H is the average depth
of the water body and ρ equals 1/average residence time
(in years). After Vollenweider (1975).

ization of decaying organics present in the water body.   An eutrophic
water body has several other problems in addition to algal growths such
as loss of dissolved oxygen, siltation, increase in turbidity, loss of
aestetic values, loss of recreation potential, change in biological population with a trend from a diversified biota at lower trophic levels to
monospecific algal blooms at higher trophic levels.   Figure 5 shows a
classic Vollenweiders' relation of the nutrient levels to the trophic
status of lakes.

The eutrophication problem is not limited only to lakes.   Some newly
constructed reservoirs may become almost instantly eutrophic from the
decomposition of materials left in the flooded basin or from washoff of
nutrient rich soils into the reservoirs (Cooke et al. 1986).   Similarly,
streams can show severe symptoms of eutrophication (excessive algal growths, dissolved oxygen problems, loss of aesthetic values and impairment
of beneficial uses) similar to lakes and reservoirs.   Novotny and Bendoricchio (1989) documented the impact of nutrients on eutrophication in the
Milwaukee River and canals of the Lagoon of Venice in Italy.   The most
severe impact occurs when a shallow productive stream reach is followed
by a deep impoundment or estuary section.

Theoretically, based on a classic Stumm and Morgan's (1962) representation of the productivity and decomposition processes, 1 atom of Phosphorus or 16 atoms of Nitrogen will generate about 80 grams of organic
matter and 160 grams of oxygen by photosynthesis.   Again theoretically,
the same amount of oxygen will be used during decomposition.   Figure 6
shows the nutrient-organic matter- oxygen cycle.

At first glance, the nutrient- organic matter- oxygen cycle appears
oxygen neutral, the same amount of oxygen is used as it is produced.   In
reality, the cycle is oxygen deficient.

A well known fact is that algae respire and when put in dark BOD
bottles they use oxygen.   Figure 7 shows a relationship between the chlor-
ophyll-a concentration and BOD measurement by standard laboratory techniques for Milwaukee River upstream from the Metropolitan Milwaukee area.
Therein the nutrient load is primarily from diffuse sources however, it
also receives treated sewage effluents from three small towns.

The particulate organic matter (dead algae and organic detritus from
macrophytes) that has developed in upstream reaches may be deposited in
slower sections of the stream or in headwaters of lakes, reservoirs and

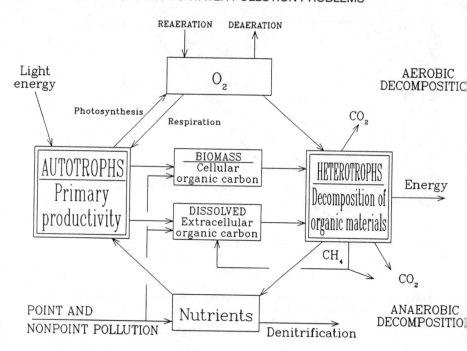

Figure 6. Nutrients - organics - oxygen cycle in receiving
water bodies.

estuaries. This organic matter will exert a dissolved oxygen demand in
spite of the fact that most of the decomposition in the benthic layers
may be anaerobic.

The decomposition processes in the benthic layer may be both aerobic
(in a relatively thin sediment-water interface provided that the overlay-
ing water has enough oxygen to support an aerobic layer) and anerobic.
The aerobic decomposition exerts a direct SOD. The thickness of the aerobic interfacial layer is in millimeters.

During anaerobic decomposition in the benthic (sediment) layer the
organic materials are broken down. The major products of anerobic decomposition are methane, carbon dioxide and ammonia. However, anaerobic
decomposition (fermentation) is a complex process that progresses in
several stages. The intermediate products of anaerobic decomposition are
fatty and organic acids. Methanogenesis, i.e. formation of methane in
addition to carbon dioxide and other by-products is the final process of
the anaerobic decomposition of organics in the sediment. In fresh water
sediments, these products are formed from the methyl carbon of acetate.
Methane is released from sediments into the overlying water by diffusion
or ebullition where it can be further biochemically oxidized and become a
part of the BOD. A significant fraction of the deposited sediment is
either inert or refractory (decays in a very slow to negligible rate).
Consequently, nutrients are tied up by this nondegradable organic matter
and, in this way the bottom sediments are the sink of the nutrients. The
inert and refractory organic fraction of sediment should also be included
in the overall oxygen balance since they will not cause appreciable SOD.

Most of common water quality models presently used are designed to
simulate transport, biochemical reactions and physical processes dealing
with biodegradable organics (BOD) and, as such, can be used for modeling
effects of pollution from point sources. Modeling in-stream effects of
pollutants from nonpoint sources is difficult and requires expertise and
several models. The results are speculative. Most models are not capable to describe and handle processes taking place in sediments and in
the sediment-water interface.

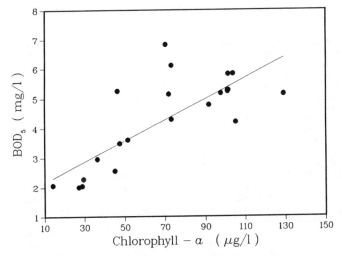

Figure 7.   BOD$_5$ - Chlorophyll a relationship for Milwaukee
River upstream from the Milwaukee metropolitan
area. Data from the Milwaukee Metropolitan
Sewerage District.

Nitrogen transformation and removal in sediments. In addition to
the methane production and its oxidation, ammonification (decomposition)
of sediments releases ammonia ion or gas (if pH is above 8). Conditions
at the sediment-water interface can have a major impact on the N balance
of the system. Unless the bottom waters are completely anoxic, nitrification should proceed just above the sediments. If the NO_3^- thus formed
diffuses into the anaerobic sediment, denitrification is almost certain
(Keeney-1973, Keeney, Schmidt and Wilkinson - 1975). This is commonly
referred to as a simultaneous nitrification-denitrification process by
which nitrogen may be lost from the system. This process uses oxygen at
quantities of 1.7 grams of O_2 per one gram of ammonia. Up to 30 percent
of the nitrogen input can be lost from a lake or a lagoon by this process
(Keeney et al. 1975).

INTEGRATED WATER QUALITY MANAGEMENT

Traditional emphasis by environmental engineers has been directed
towards treatment of point source discharges. Billions of dollars were
spent in the U.S. on the point source clean-up mandated by the Clean
Water Act passed by Congress in 1972 with several subsequent amendments.
Similar policies are in place in several advanced industrialized
countries of Europe. As a result of these policies, marked improvements
of water quality of some water bodies have recently been noticed. River
Thames in London which for decades exhibited anoxic foully conditions is
alive again and can support a viable recreational fishery. The dissolved
oxygen levels of the Potomac River near Washington, DC, have risen from
near zero to half saturation during summer. The successes and failures
of the environmental efforts implemented in the U.S. under the Clean
Water Act have been documented by Wolman (1988) who noted that focusing
on the point source abatement only has maintained more-or-less a status
quo in the majority of water quality monitoring stations throughout the
U.S. The secondary biological treatment mandated by the Act for point
source discharges has resulted in improvements of Dissolved Oxygen in
about 10 percent of stations while D.O. declined in another 10 percent.
Similar trends have been noticed for phosphorus while nitrogen
concentrations have increased at 20 percent of water quality monitoring
stations.

The recent experience of point source abatement indicated that
single focus on one type of pollution (municipal and industrial waste-

water) may not be efficient and that an integrated approach that would address both point and diffuse sources is needed.

Such approach should also include the waste assimilative capacity and its enhancement as a part of the solution. For example, in Milwaukee (Wisconsin) after spending over $2 billion on point source clean up and sewer rehabilitation, the receiving waters of the lower Milwaukee River would still remain unacceptably polluted due to diffuse rural pollution sources from upstream and from the Milwaukee metropolitan urban area. The water quality study of the harbor (lower Milwaukee River) suggested in-stream measures (dilution, in-stream aeration) that would cost only about 0.2 percent of the cost of the point source clean-up and would bring the water quality within the standards prescribed for this receiving water body. The dilution water is brought into the Milwaukee harbor from Lake Michigan by so called flushing tunnels that were built in 1888 and have become an engineering historical landmark (SEWRPC - 1987).

Karl Imhoff in the 1920's was probably the first prominent environmental engineer who recognized that a water pollution abatement plan should not be limited only to point source clean-up. In the Ruhr area of Germany several very small rivers carry the entire burden of water supply and wastewater disposal from one of the most industrialized regions of the world with a resident population of about eight million. Karl Imhoff suggested and subsequently implemented several waste assimilative capacity enhancement measures in addition to building municipal and industrial treatment plants. Such in-stream measures included (Imhoff and Imhoff - 1986):

* dilution (low flow augmentation) by increasing flows from upstream reservoirs and by back pumping the Rhine River into the Ruhr River;
* lining the streams to increase velocity and in-stream aeration;
* building river reservoirs that act as polishing post treatment;
* in-stream aeration by turbines and floating aerators;
* reassignment of stream use where some streams are used primarily for wastewater disposal (River Emscher) and some other for water supply and recreation (Ruhr and Lippe rivers);
* dredging of accumulated sludges to reduce SOD and improve water quality; and
* cutting weeds in reservoirs affected by higher nutrient input.

As stated in the preceding paragraph, dilution by clean Lake Michigan water was also used more than 100 years ago (and still used today) in Milwaukee.

Pollution control measures for urban areas. The primary decision factor is the type of urban drainage. The sewer separation policies and practices advocated twenty to forty years ago will not resolve the pollution problem nor will they result in reduced cost. For example, due to the fact that melting of urban snow is a relatively slow process, heavily polluted snowmelt from urban centers with combined sewers will be directed to the treatment plant while it will be discharged untreated by storm sewer systems. On the other hand, combined sewer systems without adequate in-line and off-line storage will result in frequent overflows of untreated mixture of sewage and polluted urban runoff during off-winter storm and rainfall events. However, a significant portion of the rain water - sewage mixture in the combined sewers will receive treatment even when no special storage-treatment facilities are incorporated. In these cases the pollutional impact of both systems is about the same and combined sewers may be actually less polluting if storage-treatment facilities are included.

The fundamental premise of any urban pollution control policy is to minimize the loads of untreated sewage and industrial wastewater. Hence, a treatment of wastewater cannot be excluded or replaced by any other measure. The degree of required treatment, however, especially when advanced treatment would be required (such as for nutrient control or removal of toxics) should not be limited only to point source pollution abatement. Furthermore, typical secondary treatment is not effective for removal of nutrients. Table 5 shows typical effluent characteristics from municipal wastewater treatment plants employing different unit processes.

Table 5
Typical effluent characteristics
from various treatment processes[*]
(After Imhoff and Imhoff - 1985)

Type of effluent of wastewater	SS mg/ℓ	BOD$_5$ mg/ℓ	Total N mg/ℓ	NH$_4$-N mg/ℓ	Total P mg/ℓ
Raw municipal sewage[*]	350	300	55	40	15
After secondary biological treatment	20	20	35	30	20
Biological treatment with nitrification	15	12	25	5	10
Activated sludge with denitrification (200% recycle)	20	15	20	10	9
Activated sludge with simultaneous precipitation	20	12	28	12	1

[*]European data, U.S. sewage is more diluted.

In an integrated approach, determination of waste assimilative capacity is a part of the process and the design and plan of abatement should be made compatible with the goals of the use of the water body.

Urban Runoff Control (wet weather pollution). Urban stormwater management affects primarily pollution during higher flows. The methods, modeling and available technologies have been a subject of several specialized conferences organized by the joint urban stormwater management group of IAWPRCF and IAHR (Yen-1982, Balmer, Malmqvist and Sjoberg-1984, and Gujer and Krejci-1987). These conferences and proceedings dealt with pollution of overflows from both the combined and separate sewer systems. The pollution of urban runoff from separate systems (storm sewers) and its abatement was a subject of specialized NATO conference (Torno, Marsalek and Desbordes-1986).

Pollution mitigation measures for urban runoff can be divided into several categories (Novotny - 1984). They can be either structural, nonstructural, or a combination thereof. Runoff quantity-quality control measures can be also categorized as to where the measure is implemented, namely, in on-site source control measures, hydrologic modification of urban watersheds, control and reduction of delivery of pollutants from the source to the receiving water body, and end-of-pipe final control.

Source control measures include control of atmospheric deposition (a source of toxics) and/or limitations on gasoline (source of lead), litter control programs, street sweeping, and erosion control of pervious erosive lands.

As mentioned previously, animal (pet) fecal deposits and fallen leaves can be a great source of pollution that can be controlled at the source. Urban erosion is a problem only during construction and if soils are left unprotected by vegetation and grasses.

Hydrologic modification or urban watersheds implies a reduction of surface runoff. These measures can be divided into:

a) practices that increase permeability and enhance infiltration such as the use of pervious pavements or vegetation infiltration strips;

b) practices that will increase hydrological storage such as temporary ponding of parts of streets and parking lots, restriction of stormwater inlets. As a water quality control measure in areas served by combined sewers these practices will reduce the frequency of overflows. In areas served by storm sewers, their water quality benefit is marginal; and

c) practices that will reduce surface areas directly connected to sewers, such as disconnecting roof drains from sewers and letting surface runoff overflow on adjacent pervious areas.

Reduction of delivery involves measures implemented between the source and the receiving water body. Urban areas with good storm sewer drainage have a delivery ratio close to one. Reduction of the magnitude of the delivery ratio can be accomplished by incorporation of grass fil-

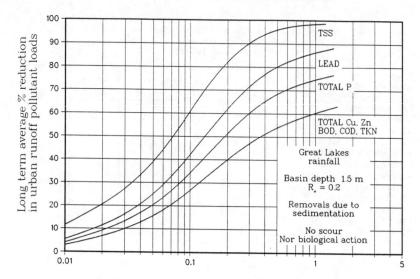

Basin area as % of contributing catchment area

Figure 8. Efficiency of wet ponds for removal of pollutants from urban
runoff. After Driscol (1983).

ters and grassed waterways into the drainage system and by the use of
catchbasins.

Detention-retention facilities include ponds and retention basins, wetlands and in-line and off-line storage facilities on combined
sewer systems.

Dry and wet ponds are now being proposed and used throughout the U.S.
Efficiency of a wet pond (a retention facility always containing water)
to remove some pollutants from urban surface runoff is shown on Figure 8.
A dry pond can be modified by installation of a seepage (infiltration)
bed to improve its water quality control efficiency. The ponds, if properly designed provide sufficient removal and are not followed by treatment.

Wetlands are now being more and more considered for water quality
control, especially for post treatment and polishing of treated effluents
and for control of overflows. A wetland act as a filtration - biological
adsorption and removal unit. Mitsch and Gosseling (1986) summarized the
ecological considerations in the use of wetlands for water quality abatement while Kadlec (1988) focused on removal of nitrogen in wetland systems receiving both treated and untreated effluents and overflows. A
special manual for wetland and aquatic macrophyte pond design for wastewater and overflow treatment and control has been prepared by the U.S.
EPA (1988).

In established densely populated, older urban zones, the use of low
cost systems may be impossible. Extensive storage-treatment systems consisting of oversized underground tunnels followed by treatment are being
now implemented, for example in Chicago, Illinois and Milwaukee, Wisconsin. The Chicago tunnel is now (1989) partially completed and operational and, as stated previously, the water quality benefits are substantial
(Lue-Hing et al.-1989). Several alternatives are possible in lieu of the
expensive underground storage schemes. For example, reducing the hydrological activity of the watershed as shown in the preceding section will
reduce the volume of surface runoff and slow down the flow, hence less
storage volume will be needed. The same effect will be accomplished by
disconnecting roof top drains. An experimental project in Japan described by Fujita (1984) employing infiltration and storage of surface runoff
by specially designed stormwater inlets, infiltration trenches, pervious
pavements and surface storage has accomplished the same flow and
pollution reducing effect.

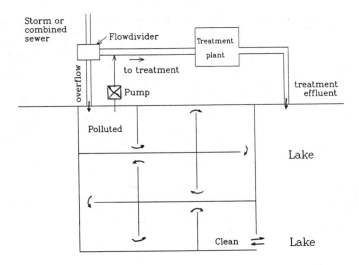

Figure 9.   Dunkers flow balancing storage - treatment system for urban
runoff and combined sewer overflows. Replotted from Soderlund (1984).

An inexpensive flow balancing method for storage and treatment of
urban stormwater and combined sewer overflows has been developed by Karl
Dunkers in Sweden. This system consists of floating relatively inexpensive surface storage tanks shown on Figure 9 was described by Soderlund
(1984).

Pollution control measures for suburban areas. The low density
of suburban population does not allow for expensive structural practices.
If the area is developing, incorporation of drainage and pollution mitigation practices into the development plans is the most effective control.
The delivery of pollutants from the source area is related to the type of
drainage and the natural (swale) drainage yields the lowest loadings.
The road and drainage system with sediment traps and settling ponds for
sediment should be in place before actual construction begins.

Pollution of subsurface flows by nitrates from septic tanks leaching
fields is a problem. Nitrates are not adsorbed by soils and are mobile
in the soil environment. If nonsewered suburban or lakeside communities
are located near nutrient sensitive water bodies the pollution of subsurface flow must be abated. Building sanitary sewer bypasses and limiting
minimum lot sizes has been the most common solution in the Wisconsin Lake
Renewal and Rehabilitation program. In Florida, abatement of runoff and
subsurface pollution in lake watersheds that are rapidly developing has
been a high priority.

Pollution control approaches for rural (agricultural) areas. The
pollution generated by agricultural operations can be carried by surface
runoff (pollutants from soil loss by erosion) or by subsurface flow.
Slopes and soil types must be evaluated before a decision is made on the
best management practices. Improper plowing and planting is the source
of excessive soil loss while overuse of fertilizers is the reason why
nitrate levels in subsurface flows have reached alarming levels in many
agricultural watersheds.

Logically, erosion and soil loss is a problem on high slope fields
(slope greater than 3 percent which in the U.S. has been classified as
soil slope category B and higher). On flat lands (soil slope category
A), subsurface flow from drainage is a problem.

While feasible and efficient management practices have been developed for control of soil loss, control of polluted subsurface flow from
farms is very difficult. Recognizing that subsurface water pollution is
a result of overuse of land and oversupply of chemicals used on the land,
control of the input appears to be the only feasible solution. Norris

and Shabman (1988) have proposed an input substitution on fields coupled with buffer strips along the watercourses as a strategy to control both surface and subsurface pollution. They and other authors (Knisel et al. 1982) have recognized that soil conservation practices aimed at soil and water retention on fields without a curtailment of nutrient and pesticide input may result in greater percolation and subsequent leaching of nitrates and other chemicals to subsurface flow.

The input substitution strategy seeks to reduce the total amount of commercially manufactured (and leachable) fertilizer applied to cropland. This can be accomplished by changing the timing of fertilizer applications, substitution of winter crops and rotation changes, and use of legumes and animal manure to provide the source of nitrogen. One of the primary benefits of input substitution is to build up soil fertility.

Incentives are needed for input substitution and conversion of highly erosive lands to woodland or for installation of buffer strips (Dosi and Stellin - 1988). Such incentives are emerging both in the U.S. and Common Market countries of Europe. Herein, pressures and agricultural policies are geared towards taking some agricultural land out of production. Targeting such conversions to water quality improvement goals would be the next logical step.

Integrated Approaches - Examples

The Milwaukee River Priority Watershed Program is an example of how problems of excessive nonpoint pollution can be addressed on a watershed wide basis (Gayan and D'Antuono - 1988). It has to be stated that the program focuses primarily on control of pollution loads from soil loss and by runoff from concentrated animal operations. The integrated resources management plan has nine management components and objectives which include:

* A reduction of nutrient and sediment contributions to surface and groundwater throughout the watershed by 50 percent. This will be accomplished by soil conservation practices such as strip cropping and no-till planting in agricultural zones, construction erosion control and by sedimentation ponds in urban zones.
* A reduction of toxic contaminants in surface and groundwater. Most of the toxics originate from existing and abandoned landfills.
* An enhancement and protection of wetlands. Existing wetlands will be protected and drained former wetlands adjacent to the watercourses will be restored.
* Five percent of existing agricultural lands will be converted to perennial grassland to provide protective buffer strips along the watercourses.
* Some highly erodible lands will be converted to woodland.

The Food Security Act of 1985 provides incentives to farmers to install soil conservation practices and also provides compensations for taking highly erodible lands and lands adjacent to watercourses out of production. Additional funding for the program is provided by the State of Wisconsin.

The Milwaukee Harbor Study (SEWRPC - 1987) specified that the influx of organic (primarily algae) from the upstream reaches of the Milwaukee River is the primary cause (more than 50 percent) of the water quality problem of the lower reaches of the river. It is, however, not clear whether the measures undertaken in the upper Milwaukee River Priority Watershed will suffice to bring about a sufficient reduction of nutrient input in the river system which would reduce primary productivity. A substantial portion of the nutrient load is due to subsurface flows which have not yet been addressed. The Harbor Study has proposed in-stream water quality measures, including in-stream aeration and reactivation of old flushing tunnels which will bring clean dilution and rich oxygen water from Lake Michigan into the impounded harbor reaches.

The combined Milwaukee River water quality abatement plan consisting of:

- **The Priority Watershed Program** aimed at nonpoint sources of pollution in the upper, mostly rural and suburban parts of the basin,

- **Milwaukee Metropolitan Sewerage District Water Pollution Abatement Program**, an ambitious $2.5 billion action plan focusing on point source pollution abatement from the urban metropolitan area, and
- The **Milwaukee Harbor Plan** proposing in-stream measures, could

be considered as an example of an integrated approach to the water quality problem.   In an ideal case such three components of a watershed wide planning effort would be concurrent and integrated.

Lagoon of Venice Water Quality Abatement Plant. In a similar effort, state, regional and local agencies in Venice (Italy) are preparing their Pollution Abatement Program (Bendoricchio - 1988) for the Lagoon of Venice.   In contrast to the Milwaukee River watershed, most of the agricultural nonpoint pollution loads are carried by subsurface drainage and erosion input is minimal.   Also the major water quality problem is excessive eutrophication in some portions of the lagoon while the water quality problems of the Milwaukee River are low dissolved oxygen levels and toxics in the lower harbor-estuary reaches.

For the agricultural sector which is responsible for about 50 percent of the total nutrient load to the lagoon, the action plan is envisioning the following actions:

(1)   Optimization of fertilizer application rates to minimize losses and matching the applications to the nutrient uptake by the crops.

(2)   Optimization of temporal distribution of fertilizers.

(3)   Use of soil drainage to increase denitrification in soils.

(4)   Limitation and optimization of irrigation to reduce nutrient losses in the irrigation (surface or subsurface) return flow.

(5)   Substitution of organic and slow release fertilizers.

(6)   Increasing organic content of soils.

(7)   Soil conservation practices on erodible lands to reduce loads of phosphorus and soil adsorbed nitrogen.

(8)   Suitable choice of the set-aside lands and conversions to less polluting crops according to the guidelines of the European Common Market.

The Pollution Abatement Program is also considering use of wetlands situated along the boundary of the lagoon for abatement of diffuse nonpoint pollution loads.

The point source abatement is focused on both dry weather and wet weather pollution loads.

CONCLUSIONS

In general, pollution abatement planning today must be an integrated effort that will address the pollution abatement of dry weather and wet weather point sources, reduction and mitigation of nonpoint pollution, land use and its pollution impact, impact of drainage on pollution attenuation, and the waste assimilative capacity of the receiving water bodies and its enhancement.   Although in some parts of the world raw sewage discharges are still unabated emphasis may now be shifting from removal of organic, oxygen demanding pollution from urban and industrial wastewater sources to controling nutrients and toxics, both from point and nonpoint sources.

The solutions are both structural and nonstructural.   Traditionally, structural solutions are more common for urban and industrial point sources while nonstructural management is applied to rural, primarily agricultural nonpoint sources.   In many cases, enhancement of waste assimilative capacity and reduction of delivery of pollutants are feasible and the efficiency of the entire integrated approach will be increased if such measures are considered and included.   Such approaches should be created by planners and implemented, considering also economical externalities the pollution abatement programs are faced with.   Without addressing the externality problem of pollution abatement programs by the political-legislative bodies equitable and efficient solutions are not possible.

It is apparent that the 1990's and the beginning of the next millennium will see a new and increased emphasis on the solutions of the severe and large scale pollution problems, such the pollution of the northern

Adriatic Sea, North Sea pollution, pollution of many large European and U.S. rivers, contamination of the U.S.-Canada Great Lakes, effects of deforestation, severe degradation of surface and subsurface water resources by agricultural, industrial and solid waste deposition practices and urban stormwater runoff pollution. These large scale problems will require coordinated integrated approaches and a change in the way the land is used. Losses of chemicals into water bodies in agricultural areas can be reduced by switching to sustainable agriculture. At the same time, attention at the local levels is and will be on protecting water resources and providing safe water for various beneficial uses. New integrated approaches will have to emerge under a favorable political and economical climate.

REFERENCES

Alberts E.E., G.E. Schuman and R.E. Burnwell (1978), Seasonal runoff losses of nitrogen and phosphorus from Missouri Valley loss watersheds, J. Env. Quality, 7:203-208.

Balmer P., P.A. Malmqvist, and A. Sjoberg ed. (1984), Proc. of the Third Interntl. Conf. on Urban Drainage, Chalmers University of Technology, Goteborg, Sweden.

Baun K. (1985), Modeling sediment delivery from fields to channels: A case for field scale data and hierachial data structures, Proc. Nonpoint Poll. Abatement Symp., Marquette University, Milwaukee.

Beasley D.B. and L.F. Huggins (1980), ANSWERS (Areal Nonpoint Source Watershed Environmental Response Simulation) - User's Manual, Agr. Eng. Dept., Purdue University, West Lafayette, IN.

Beasley D.B., L.F. Huggins and E.J. Monke (1982), Modeling sediment yields from agricultural watersheds, J. Soil and Water Conservation, pp. 113-116.

Bendoricchio G. (1988). The NPSP abatement program for the Lagoon of Venice. In "Nonpoint pollution: 1988 - Policy, Economy, Management and Appropriate Technology," pp. 241-260. Amer. Water Res. Assoc., Bethesda, MD.

Bowman G.T. and J.J. Delfino (1980), Sediment oxygen demand techniques: A review and comparison of laboratory and in-situ systems, Wat. Res., 14(5):491-500.

Cole J.J., G.E. Likens and D.L. Strayer (1982), Photosynthetically produced dissolved organic carbon: An important carbon source for planktonic bacteria. Limnol. Oceanogr., 27(6), 1080-1090.

Cooke G.D., E.B. Welch, S.A. Peterson and P.R. Newroth (1986), Lake and reservoir restoration. Butterworth Publishers, Stoneham, MA.

Dosi C.M. and G. Stellin (1988), Influencing land use patterns to reduce nitrate pollution from fertilizers and animal manure: A case study. Paper presented at XVIII European Seminar on Agr. Economics: Aspects of Environmental Regulation in Agriculture, Copenhagen, Nov. 1-4.

Driscol E.D. (1983), Analysis of detention basins in EPA NURP program, In Stormwater Detention Facilities, (W. DeGrott ed.), Am. Soc. Civ. Eng., New York, NY.

Ellis J.B. (1986), Pollutional aspects of urban runoff. In Urban Runoff Pollution (Torno, H.C., J. Marshalek and M. Desbordes, ed.) pp. 1-38. Springer-Verlag, Berlin.

Fair G.M., E.W. Moore and H.A. Thomas (1941), The natural purification of river muds and pollutional sediments, Sew. Works Journal, 13: 270-307, 756-778.

Field R. and R. Turkeltaub (1981), Urban runoff receiving water impact: Program review, Journal Env. Eng. Div., ASCE, 107(EE1):83-100.

Fujita S. (1984). Experimental sewer system for reducing of urban storm runoff. In Proc. of the Third Interntl. Conf. on Urban Drainage, (Balmer P., P.A. Malmqvist, and A. Sjoberg, ed.) pp. 1211-1220. Chalmers University of Technology, Goteborg, Sweden.

Gayan S.L. and J. D'Antuono (1988). The Milwaukee River east-west branch watershed resource management plan. In "Nonpoint pollution: 1988 - Policy, Economy, Management and Appropriate Technology," pp. 165-172. Amer. Water Res. Assoc., Bethesda, MD.

Gujer W. and V. Krejci, ed. (1987) Topics in urban storm water quality, planning and management. Proc. IVth Int. Conf. in Urban Storm Drainage, Ecole Polytechnique Federale, Lausanne, Switzerland.

Huber W. (1986), Deterministic modeling of urban runoff quality, In Urban Runoff Pollution (Torno, H.C., J. Marsalek and M. Descordes, ed.) pp. 167-242. Springer-Verlag, Berlin.

Imhoff K. and K.R. Imhoff (1986), Taschenbuch der Stadtenwasserung, Oldenbourg Verlag, Munich.

Kadlec R.H. (1988), Denitrifiction in wetland treatment systems, Paper presented at the 61st Annual WPCF Convention, Session 26, October 2-5, Dallas, TX.

Kaplan L.A. and T.L. Butt (1982), Diel fluctuations of DOC generad by algae in a Piedmont stream, Limnol. Oceanogr. 27(6): 1091-1100.

Keefer T.M., R.K. Simons and R.S. Quivey (1979), Dissolved oxygen impact from urban storm runoff, EPA 600/2-79-156, U.S. Environ. Prot. Agency, Cincinnati, OH.

Keeney D.R. (1973), The nitrogen cycle in sediment-water systems, Journal Environ. Quality, 2(1), 15-19.

Kenney D.R., S. Schmidt and C. Wilkinson (1975), Concurrent nitrifictindenitrification at the sediment-water interface as a mechanism for nitrogen losses from lakes, Tech. Rep. No. 75-07, Water Resources Center, University of Wisconsin, Madison, WI.

Krejci V. (1988), New stategies in urban drainage and stormwater pollution control in Switzerland, In Nonpoint Pollution: 1988 -Policy, Economy, Management and Appropriate Technology, pp. 203-211. Amer. Water Res. Assoc., Bethesda, MD.

Lager J.A. and W.G. Smith (1974) Urban stormwater management and technology - An Assessment. EPA-670/2-74-040, U.S. Environmental Protection Agency, Cincinnati, OH.

Likens G.E. (1972), Eutrophication and aquatic exosystems. In Nutrients and Eutrophication, Proc. Symp. on Nutrients and Eutrophication: The Limiting Nutrient Controversy, Amer. Soc. Limnol. Oceanogr., Lawrence, KS.

Loehr R.C. (1972), Agricultural runoff - characterization and control, J. Sanitary Eng. Div. ASCE, 98:909-923.

Lue-Hing C. et al. (1989), Improvements in surface water quality associated with the operation of Chicago's tunnel and reservoir system, Rep. No 89, Metropolitan Water Recl. District of Greater Chicago, IL.

Malmqvist P.A. (1984), Urban stormwater pollutant sources. Chalmers University of Technology, Goteborg, Sweden.

Milsch W.J. and J.G. Gossling (1986), Wetlands. Van Nostrand Reinhold, New York, NY.

Norris P. and L.A. Shabman (1988), Reducing nitrogen pollution from crop production systems: A watershed perspective, In Nonpoint pollution: 1988 -Policy, Economy, Management and Appropriate Technology, pp. 29-38. Amer. Water Res. Assoc., Bethesda, MD.

Novotny V. (1980), Delivery of suspended sediment and pollutants from nonpoint sources during overland flow, Water Res. Bull. 16:1057-1065.

Novotny V. (1984), Efficiency of low cost practices for controlling pollution by urban runoff. In Proc. of the Third Interntl. Conf. on Urban Drainage, (Balmer P., P.A. Malmqvist, and A. Sjoberg, ed.) pp. 1241-1250. Chalmers University of Technology, Goteborg, Sweden.

Novotny V. (1985), Urbanization and nonpoint pollution, In Proc. Vth World Congress on Water Res., Paper 17, Aspect 6, IWRA, Brussels, Belgium.

Novotny V. (1988), Modeling sewer flow quality, Proc. Bilateral Conf. US - Italy, Cagliari, Italy, July 1988.

Novotny V. (1988), Stormwater permits - a point source umbrella for nonpoint pollution problems. In Nonpoint Pollution: 1988 -Policy, Economy and Appropriate Technology, Symp. Proceedings, pp. 173-182. Amer. Water Res. Assoc., Bethesda, MD.

Novotny V. et al. (1979), Simulation of pollutant loadings and runoff quality, EPA 905/4-79-029E, Great Lakes National Program Office, Chicago, IL.

Novotny V. and G. Chesters (1981). Handbook of Nonpoint Pollution: Sources and Management. Van Nostrand Reinhold, New York, NY.

Novotny V., G. Simsiman and G. Chesters (1986), Delivery of pollutants from nonpoint sources, Proc. Interntl. Symp. Drainage Basin Sediment Delivery, Publ. No. 159, Interntl. Assoc. on Scientific Hydrology.

Novotny V. and G. Chesters (1987), Delivery of sediments and pollutants from nonpoint sources - Awater quality perspective, Rep. No. 1, Milwaukee River Nonpoint Pollution Research Project, University of Wisconsin, Madison and Marquette University, Milwaukee, WI. Also submitted to Journal of Soil and Water Conservation.

Novotny V. and G. Bendoricchio (1988), Linking diffuse (nonpoint) and residual point source pollution to water quality deterioration, Paper presented at the 61st Annual WPCF convention, Dallas, TX, Session 55, October 2-5, also submitted to Journal WPCF.

Novotny V., K.R. Imhoff, M. Olthof and P.A. Krenkel (1989), Karl Imhoff's Handbook of urban drainage and wastewater disposal. J. Wiley and Sons, New York, NY.

Patrick R. (1975), Some thoughts concerning correct management of water quality. In Urbanization and Water Quality Control, (W. Whipple, ed.), Amer. Water Res. Association, Minneapolis, MN.

Roehl J.E. (1962), Sediment source areas, delivery ratios and influencing morphological factors. Intern. Assoc. Hydrol. Sci., 59:202-213.

Rudd J.W.M. and C.D. Taylor (1980), Methane cycling in aquatic environments, In Advances in Aquatic Microbiology (M.R. Dropp and H.W. Jannach, ed.). Academic Press, New York, NY.

SEWRPC (1987), A water resources management plan for the Milwaukee harbor estuary, Southeastern Wisc. Regional Planning Comm. Rep. No. 37, Waukesha, WI.

Soderlund H. (1984), Flow balancing method for stormwater and combined sewer overflows, In Proc. of the Third Interntl. Conf. on Urban Drainage, (Balmer P., P.A. Malmqvist, and A. Sjoberg, ed.) pp. 57-66, Chalmers University of Technology, Goteborg, Sweden.

Stumm W. and J.J. Morgan (1962), Stream pollution by algal nutrients, Transact. 12th Annual Conf. on San. Eng., University of Kansas.

Torno, H.C., J. Marsalek and M. Desbordes, ed. (1986). Urban Runoff Pollution. Springer-Verlag, Berlin.

U.S. Environmental Protection Agency (1983), Final report of the Nationwide Urban Runoff Program, Vol. I. U.S. EPA, Washington, DC.

U.S. Environmental Protection Agency (1988), Constructed wetlands and aquatic plant systems for municipal wastewater treatment, EPA/625/1-88/022, Washington, DC.

U.S. Environmental Protection Agency (1988), National Pollutant Discharge Elimination System - Permit Application regulation for stormwater discharges: Proposed rule, Federal Reg., Dec. 7, Washington, DC.

Viessman W., Jr. and M.J. Hammer (1985), Water supply and pollution control, 4th ed. Harper and Row, Cambridge, Philadelphia, PA.

Vollenweider R.A. (1975), Input-output models with special reference to the phosphorus loading concept in limnology, Schweiz. Z. Hydrology, 37:53-84.

Walling D.E. (1983), The sediment delivery problem, J. Hydrol. 65:209-237.

Wetzel R.B. (1983), Limnmology, 2nd ed. Sauders, Philadelphia, PA.

Wischmeier W.H. and D.D. Smith (1965), Predicting rainfall erosion losses from cropland east of Rocky Mountains, Agr. Handbook No 262, U.S. Dept. of Agriculture, Washington, DC.

Wolman M.G. (1988), Changing national water quality policies, Journal WPCF, 10:1774-1781.

Yen B.C. (1982), Urban Stormwater Quality, Management and Planning. Water Res. Publications, Littleton, CO.

Chapter 6

INTEGRATED APPROACH FOR WATER MANAGEMENT IN THE NETHERLANDS: THE POLICY ANALYSIS FOR THE NATIONAL WATER MANAGEMENT PLAN 1990

J Stans, S Groot (Delft Hydraulics, The Netherlands)

ABSTRACT

In 1990 the Third National Water Management Plan for the Netherlands will be presented to the Dutch Parliament. This plan will describe the policy for the Dutch water management on a national scale from 1990.
The chosen policy results from a number of alternatives. The impacts of the various alternatives on user categories have been studied in the policy analysis, as quantitatively as possible, with the help of a number of mathematical models. This paper deals with the policy analysis, an integrated approach of the water management problems of the Netherlands with emphasis on the conceptual framework, the reduction of emissions and the water quality modelling for the network of national waters.

Key words: water management, emissions, planning, policy analysis

1 - INTRODUCTION

In 1990 the Third National Water Management Plan for the Netherlands will be presented to the Dutch Parliament by the Minister of Transport and Public Works. This ministry has prepared the Plan in close cooperation with the Ministry of Housing, Physical Planning and Environment and the Ministry of Agriculture and Fisheries.

The Plan will describe the national policy for water management, very concretely for the years between 1990 and 1995, and more generally for the years between 1995 and 2020. The Plan contains:

- the outlines for management on a national level, including an indication of the functions of the surface waters belonging to the national system and for regional systems where national interests are involved;
- an indication of the desired future development, functioning and protection of the water resources systems under consideration, including water quality standards, and the corresponding time horizons;
- a description of the scope and volume of measures required to arrive at the desired situation;
- an indication of the financial, economic and physical consequences of those measures.

The preferred policy has been derived from several alternative options, which have been investigated in the policy analysis that supports the choices made in the Plan. The policy analysis has been managed as a project and has been executed under the responsibility of the Public Works Department of the Ministry of Transport and Public Works.

An important contribution for the necessary computations and also for the analyses previous to these computations was made by Delft Hydraulics. Also other institutes and consultants were involved in parts of the studies.

This paper will concentrate on the general framework used for the study and on the description of the modelling of the water quality in the national water system. The modelling of the eutrophication processes in the coastal waters of the Netherlands is discussed in another paper (GLAS, 1989).

2 - FRAMEWORK FOR A POLICY ANALYSIS

The purpose of a policy analysis is to analyze systematically the measures for the solution of problems in present and future situations, and to combine these measures into policy alternatives.

The complexity of the problem is enormous, since the analysis has to take into account the large amount of often conflicting interests of water use categories (agriculture, electricity generation, navigation, drinking water and industrial water supply, recreation and fisheries, aquatic and terrestrial ecology) which compete for the scarce resources (water quantity, water quality, money and time for the execution of beneficial measures). It is even more complex in comparison to previous planning studies, as more interests are taken into account. Not only fresh water but also estuaries and coastal areas are taken into account and measures do not only deal with supply, but also with influencing the demand for water.

Therefore application of systems analysis was necessary to obtain a **framework for analysis,** in which the problem has been broken down into feasible analysis steps tuned to each other, and to obtain a **computational framework** containing a schematization of reality which simulates the most relevant phenomena, and which is capable of quantifying the consequences of the several policy alternatives. The framework for analysis is shown in figure 1.

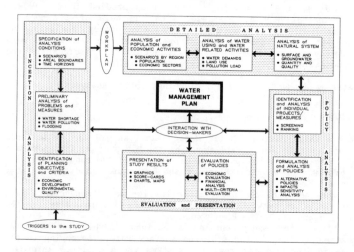

Fig. 1 - Framework for Analysis for Water Resources Management studies

3 - DESCRIPTION OF THE SEVERAL ANALYSIS STEPS IN THIS APPLICATION

3.1 - Inception analysis

The trigger for this study is the mandated planning as described before.

The planning objectives and criteria of the Third National Water Management Plan are:

-   to manage and to develop the water systems in such a way that they reach the assigned ecological objectives and fulfill the assigned functions, based on the potentials of the systems.

- this main objective has been specified as follows:
  - to guarantee a safety level against flooding;
  - to guarantee a safety level against environmental calamities;
  - to protect and promote the availability and the usability of water;
  - to protect the quality of water systems;
  - to develop the water systems.

A preliminary analysis has been executed in which the present and possible future problems of water availability are identified under different hydrological regimes, as well as water quality problems. This was done with the help of working groups, in which national and provincial levels as well as water-boards were represented.

The output of these working groups also assisted in developing possible measures for solving or mitigating the problems that were identified.

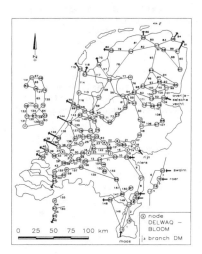

Fig. 2 - The network for the   Fig. 3 - The schematization of the
         national water system   country into districts

During this step analysis conditions were established as well. For the several water user categories likely and possible developments were set, which were called scenarios. The years for which the analyses have to be executed have been established: 1995 and 2020. The policy has to be established for the years 1990-1995 with a global perspective through 2020.

The system to be investigated was determined: the national water bodies (fresh water as well as estuaries and coastal waters) are subject of the study, using the water quantity and quality of the inflowing rivers (Rhine, Meuse, Scheldt and some smaller rivers) as input. For the impact analysis, as well as for the analysis of water demand from agriculture, the total area of the Netherlands had to be taken into account, simultaneously and interactively with the national network. This was done by schematizing the country by districts. A district is an area with a specified water management (water inlet a well as water discharge system).

The national network as well as the districts are shown in the figures 2 and 3 respectively. The national network and the districts are coupled in the mathematical models that compute the water distribution.

3.2 - Detailed analysis

An analysis of population and economic activities was executed. For economic scenario's the data from the Central Planning Bureau were used. Data on cattle, land use as well as discharges of pollutants were collected as well, partly from the Central Bureau for Statistics, which frequently uses questionnaires to collect data. Apart from these data also data on meteorology and hydrology were collected and analyzed.

These data were used as input for the models of the natural system for the year 1985. The computational framework was shown in figure 4 and comprises:
- programmes to compute the emissions of the chosen substances for the reference situation as well as for the policy alternatives;
- studies on the discharge of nutrients from agriculture inside districts;
- a mathematical water quality model to compute the behaviour of substances in the surface water in the districts and the concentrations that enter the national network (DIWAMO);
- the water quantity models DEMGEN (Demand Generator) and DM (Distribution model), which compute the water balances of the districts, the water demand of the districts and the water distribution in the national network;
- the water quality model for the national network (DELWAQ-BLOOM).

As an illustration, results of the calibration of the water quality model for the national network were described in chapter 5, together with results of the computations for different policy alternatives.

3.3 - Policy analysis

The measures identified in step A (3.1) were screened partly qualitatively, partly quantitatively using the computational framework, on their effectiveness to solve or mitigate problems. Also potential side-effects were taken into account. This reduced the amount of measures by about 30% to still more than 100 measures.

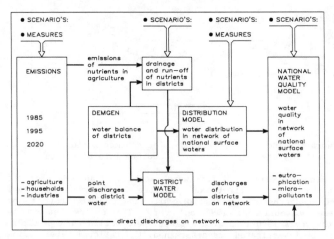

Fig. 4 - The computational framework

Then potential policy alternatives were formulated.
This was done by iterations. During these iterations the number of alternatives was reduced and the remaining alternatives were specified in greater detail.
A policy alternative consists of a policy for the water distribution in the national network as well as a policy for emission reduction.
The main alternatives comprised:
- Water distribution:
  - the distribution as it existed in 1985 (reference situation);
  - the distribution in 1995, if solely the already planned policy measures are and will be implemented (autonomous development);
  - a distribution that aims at a maximum provision of water for the high grounds in the Netherlands (the eastern part of the country), in other words, giving the water coming in from abroad a maximum residence time in the Netherlands;
  - a distribution that aims at minimizing the residence time for water coming in from abroad (rivers Rhine and Meuse).

- Emission reduction:
  - emissions as occurred in 1985;
  - emission reduction for 1995, according to the Rhine Action Plan and the North Sea Action Plan. These plans involve a reduction of 50% compared to 1985 of the loads and concentrations of several substances;
  - a more drastic reduction, ranging from 50% to 99%, depending on the substances, of the antropogenic load on the national network.

Apart from water distribution and emission reduction measures a third category was taken into account: structural measures, which could be necessary to meet chosen target levels, in combination with improved water quality. Examples comprise: fish corridors in sluices, river banks without protection works of stone, shallow areas near river banks to allow water plants to grow and to provide shelter to fish species, cleaning up of the contaminated bottoms of surface waters.

The mentioned target levels are planning objectives for the separate water systems, derived from the main objective (see 3.1). For these target levels the corresponding water quality levels were defined. With this information emission reduction levels would then have to be found meeting the desired water quality levels.

3.4 - Evaluation and presentation

After the computation of the water quantity and water quality changes resulting from the application of a certain policy alternative, a working group for every user category analyzed the impacts for its sector.

For each policy alternative scorecards were made, showing the scores of the alternatives for several criteria, determined beforehand and consistent with the objectives of the study.

An evaluation of a round of analyses was executed by a large group of representatives of the different parties involved during workshops. During the workshops especially the scorecards were subject of discussion, but also the executed analyses with the computational framework, the assumptions made during the computations and the formulation of the alternatives, as well as the applied scenarios.

4 - EMISSION REDUCTION MEASURES

A lot of effort was put in setting up the data files for the emissions in 1985 and for the years 1995 and 2020. This analysis was executed separately for several sources of emissions, and for two levels: first the loads on the districts was computed; then the loads on the national network was determined.

Based on the reference data file for 1985 the consequences of emission reduction measures were computed.

For the year 1995 implementation of the Rhine and North Sea Action Plans is taken into account, for which purpose measures were formulated. For the year 2020 target levels for the reduction of the antropogenic loads are as follows:
- 50% reduction: nitrogen compounds;
- 75% reduction: phosphorus and heavy metals;
- 99% reduction: PCB's;
- 90% reduction: organic micro-pollutants, chloride, calcium.

These reductions are too large to enable a formulation of a complete and sufficient set of measures with present technology. Therefore the following approach was followed: when the formulated set of measures did not produce the desired emission reduction, emissions were reduced further until the desired levels of emissions were obtained. For this approach it was necessary to make a distinction between sources which could and sources which could not be reduced. The reducible sources then were reduced by 60% to sometimes even 100%, resulting in emission reductions meeting the target levels, except for chloride (70%), calcium (57%) and PCB (88%).

Although the average level of emission reduction meets the targets, this will not be true if the separate districts are considered. Figure 5 shows the contribution in the total load on the districts after the Rhine and North Sea Action plans have been implemented, for ammonia, nitrate and

Compound Location	Ammonia W	 E	Nitrate W	 E	Ortho-P W	 E
Source						
Storm water disch.	0-20	0-20	0-20	0-10	0-10	0-10
Atm. dep.	0-20	10-30	0-20	0-10	0-10	0-10
Ground water	30-80	0-40	50-80	70-100	70-100	10-80
Point sources	0-10	0-30	0-10	0-10	0-10	0-60
Treatment plants	10-40	10-70	0-10	0-10	0-20	10-60

Fig. 5 - Contribution (%) from each source to the total load in districts
after implementation of the Rhine and North Sea Action Plans (1995)

ortho-phosphorus. A general distinction was made between the west and the
east of the Netherlands.

From this figure the conclusion can be drawn that in the western part
of the Netherlands the remaining load is predominantly from ground water and
not directly reducible, although also other sources are still relevant. In
the eastern part these other sources contribute more to the total load
(ammonia and ortho-P). Since point sources and discharges from treatment
plants are reducible, in the eastern part a greater reduction of the
emissions may be expected than in the western part of the Netherlands.

5 - WATER QUALITY MODELLING

5.1 - Computational framework for water quality

The model instrument developed for the water quality modelling of the
national system of surface water is based on two different models: DELWAQ
for the water quality and BLOOM-II for the phytoplankton growth. The
phytoplankton model BLOOM-II is used as a submodel from the main water
quality model DELWAQ.

DELWAQ computes the behaviour of water quality constituents based on
water distribution, initial conditions, boundary conditions, waste loads and
kinetics (POSTMA, 1984). DELWAQ has a flexible compartment approach. The
model area is divided into computational elements which may have any shape.
They may be linked together in all directions. Dynamic simulations can be
performed (time-dependent flows, kinetics, waste loads and concentrations)
as well as steady-state simulations. The input of the programme consists of:
- hydraulics and network configuration
- user-defined kinetics
- process parameters
- waste loads
- boundary conditions
- initial conditions

BLOOM-II computes the phytoplankton biomass based on phytoplankton-specific
inputs as production, mortality, stochiometry etc, and system-specific
inputs as nutrient concentrations, solar energy, temperature, depth and
extinction (DHL, 1984). The model considers up to ten phytoplankton species
differing in growth versus light response, temperature dependence, and
nutrient requirements. In theory BLOOM-II is a steady-state model, but in
combination with DELWAQ it can be used for a dynamic description of the
phytoplankton population. The input of the programme consists of:
- concentration of total available nitrogen, phosphorus and silicon
- (background) extinction;
- system-specific characteristics as solar intensity, temperature,
  depth;
- phytoplankton species-specific characteristics as efficiency,
  temperature dependence, mortality and stochiometry.

The phytoplankton model BLOOM-II has been used extensively in many
eutrophication studies (LOS, 1980).

The water quality model (DHL, 1986), composed of the framework DELWAQ coupled to the phytoplankton model BLOOM-II, uses the same network as the water quantity model. The water system of The Netherlands has been divided into 101 elements (Fig. 2). Based on DM water distribution calculation results and user-defined kinetics, DELWAQ, in combination with the phytoplankton model BLOOM-II, calculates the behaviour of 60 constituents with a calculation timestep of 1 hour.

The vast amount of information from model exercises requires a suitable data processing method. To a large extent, the knowledge gained concerning the dynamic behaviour of the system is provided by computer graphics. The results can be presented in several ways, ranging from two-dimensional graphs to a coloured motion picture on a PC. In this way the dynamic behaviour of the network, influence of alien water, nutrient loads etc can be explored to full extent, providing the analysts with a powerful instrument for a profound and quick analysis of water quantity and quality measures.

5.2 - Application of the models at the national water system

Apart from a direct check through water quantity data, the chloride concentration was used to check indirectly the performance of the water distribution model DM, as the quality of the calibration of the water distribution influences the accuracy of the water quality calculations. The monitoring data of 1983 and 1985 were used for calibration and verification. According to the measurements and simulation results for flow and chloride, the description of the water distribution by the DM model was satisfactory.

Based on the calculated water distribution, the numerous water quality variables for the national system were simulated by the DELWAQ/BLOOM water quality model. The chloride description by DELWAQ for 1985 is illustrated in figure 6 for river circumstances (Gorinchem, node 4) and for lake circumstances (IJsselmeer, node 45).

The model calculates the fraction of water originating from the river Rhine for every node of the national network. The results indicate which part of the national system is influenced to a large extent by the river Rhine and which part is merely influenced by the districts (figure 4) through drainage or groundwater inflow. Typical results for the national water system in 1985 are presented in figure 7 for a node that is dominated by the river Rhine (lake IJsselmeer, node 45) and for nodes that are almost independent of the river Rhine (Schermer, node 45 and Friesland, node 54).

The water quality situation in the system depends on both the water distribution (the residence times) and the water quality processes. In case of short residence times (rivers or nodes with small volumes), the water quality is to a large extent dependent on the water quality of the flows to these nodes. In case of large residence times (large lakes), the situation is dependent on the water quality processes and the waste loads for these nodes.

The nutrient concentrations, and especially the nitrogen fractions, are simulated rather correctly by the water quality model, except for some areas where groundwater influence is considerable and loads have to be estimated from scarce measurements of the groundwater quality. Simulations of phytoplankton give satisfactory results as far as biomass is concerned. The chlorophyll description by DELWAQ/BLOOM is illustrated in figure 8 for river circumstances (Moerdijk, node 26) and for lake circumstances (IJsselmeer, node 45). The growth of phytoplankton is limited by the amount of available light (energy) in almost every period of the year, although some lakes do experience nitrogen and phosphorus limitations in summer.

5.3 - Scenario computations for the national water system

One of the most important aspects of models for water management and planning is the simulation of the effects of hypothetical measures on the present water quantity and water quality situation. In this study a number of these water management (or policy) alternatives has been studied, ranging from alternative routing of water to the reduction of the nutrient loading by agriculture, industry, households and by the inflowing river Rhine from Germany and river Meuse from Belgium and France.

The scenario simulations for 1995 and 2020 with reduced emissions indicate that generally the reduced nutrient concentrations will not limit

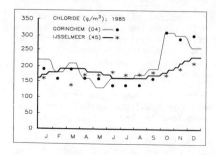

Fig. 6 - Calculated and measured chloride
concentrations for Gorinchem
(node 4) and lake IJsselmeer
(node 45)

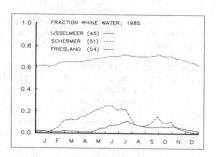

Fig. 7 - Calculated fraction of water
originating from the river Rhine
for lake IJsselmeer (node 45),
Schermer (node 51) and Friesland
(node 54)

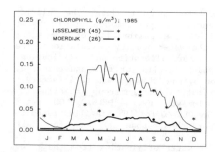

Fig. 8 - Calculated and measured chlorophyll concentrations for Moerdijk
(node 26) and lake IJsselmeer
(node 45).

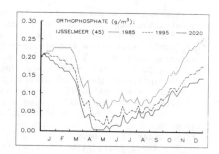

Fig. 9 - Calculated orthophosphate
concentrations in 1985, 1995 and
2020 for lake IJsselmeer (node 45).

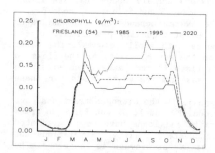

Fig. 10 - Calculated chlorophyll concentrations in 1985, 1995 and 2020
for Friesland (node 54).

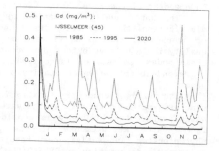

Fig. 11 - Calculated cadmium concentrations
in 1985, 1995 and 2020 for lake
IJsselmeer (node 45).

phytoplankton growth for the 1995 scenario but will limit growth in a number
of nodes for the 2020 scenario. In figure 9 the orthophosphate concentration
in node 45 (lake IJsselmeer) for 1985, 1995 and 2020 is presented,
indicating an orthophosphate limitation for a very short period in the 2020
scenario. Therefore, chlorophyll concentrations in lake IJsselmeer are
almost the same for all scenarios. Nutrients do limit growth of
phytoplankton in node 54 (Friesland) for both the 1995 and 2020 scenarios
due to the reduction of internal loads, as is illustrated in figure 10.

Considerable reduction of nutrient loads is necessary to prevent
present phytoplankton blooms. Analysis of calculation results indicate that

sanitation of the river Rhine and the river Meuse is not enough to solve the eutrophication problems in the water system, especially for a great number of lakes   not dominated by river Rhine water. Due to considerable nutrient loads produced within the water system itself, nutrient emissions from districts will have to be reduced drastically as well, in order to meet water quality standards and to prevent phytoplankton blooms.

The policy alternatives not only cover nutrients, but also describe measures for heavy metals and micro-pollutants. In figure 11 characteristic results for cadmium are presented for lake IJsselmeer. Due to the sanitation of inflows, the improvement of the quality of suspended sediment, and the quality of bottom sediments (not calculated by DELWAQ), cadmium concentrations in lake IJsselmeer will reach acceptable levels in the planning period.

6 - CONCLUSIONS

Some of the conclusions drawn in this study are mentioned here, especially those which had not been anticipated before the study started.

-   The influence of the river Rhine on the water quality of the districts in the Netherlands is less than generally assumed. Although the percentage of Rhine water present in the districts due to water intake during the year (especially in dry periods) is sometimes considerable (30-50% or more), the influence of the river Rhine on the water quality is much smaller, since water quality processes have reduced the concentrations caused by the Rhine, and, more important, the loads within the districts are predominant. This means that emission reduction in the large rivers will have to be supplemented by emission reduction in the districts.

-   A reduction of emissions only according to the Rhine and North Sea Action Plan will not be enough to meet the water quality standards for several components, among which phosphorus and algae, although concentrations have been reduced considerably.

-   Water distribution measures only yield local or small benefits for the user categories. Major improvements result from a combination of emission reduction measures (although only if a considerable amount of reduction is obtained) and structural measures. User categories that benefit most are aquatic and terrestrial ecology.

-   Target levels for the water systems for concentrations of several substances will only be reached after emission reduction levels have been reached as formulated in the policy for 2020.

ACKNOWLEDGEMENTS

The authors are grateful to the project leader of the policy analysis from the Public Works Department, Mr. J.P.A. Luiten, for his permission to publish the results of the study before the official report on the policy analysis is made public, and for his comments on the manuscript.

BIBLIOGRAPHY

DELFT HYDRAULICS LABORATORY - "BLOOM-II a mathematical model to compute phytoplankton blooms", User's manual, WABASIM-project, Report R1310, 1984.

DELFT HYDRAULICS LABORATORY - "Description of the model instruments for water quality in PAWN", Report T0067.09 (in Dutch), 1986.

GLAS, P.C.G.; NAUTA, T.A. - "A North Sea Computational Framework for Environmental and Management Studies: an Application for Eutrophication and Nutrient Cycles", in Proc. of the International Symposium on Integrated Approaches to Water Pollution Problems (SISIPPA 89), Lisbon (Portugal), June 19-23, 1989.

LOS, F.J. - "Application of an algal bloom model (BLOOM-II) to combat eutrophication", in Hydrobiological Bulletin, Vol. 14(1/2), 1980, pp. 116-224.

POSTMA, L. - "A two-dimensional water quality model application for Hong Kong coastal waters", in Wat. Sci. Tech., Vol.16., 1984.

Chapter 7

COMBINING EFFLUENT AND ENVIRONMENTAL QUALITY STANDARDS: THE UK APPROACH TO THE CONTROL OF DANGEROUS SUBSTANCES

R A Agg, T F Zabel (Water Research Centre, Marlow, UK)

ABSTRACT

Dangerous substances discharged to the environment are subject to different control measures depending on the receiving compartment - air, water or land. Following recommendations by the Royal Commission on Environmental Pollution for controls to be based on the "best practicable environmental option" having regard to inter-compartmental effects, the UK is adopting a policy of integrated pollution control (IPC) for selected chemicals (red list substances).

The selection of priority dangerous substances (such as those on the red list) for control is described with particular reference to the protection of surface water quality. Attention is drawn to the need for different approaches to controlling point-sources and diffuse inputs, since many of the substances of concern are pesticides and herbicides which reach water by diffuse routes. The significance of the IPC proposals for future environmental protection is discussed in relation to existing international (PARCOM) conventions, European Community legislation and priority issues such as reducing inputs to the marine environment, discharge to the atmosphere and contamination from land disposal of wastes.

Keywords: pollution control, dangerous substances, aquatic environment, environmental quality standards, limit values

INTRODUCTION

Most of the environmental legislation arising from European Community (EC) directives is aimed at controlling the input of dangerous substances to individual environmental compartments (air, water, land) to eliminate or reduce pollution. One important exception is the EC Impact Assessment Directive (CEC 1985) which for major new projects requires that the inter-relationship between the different environmental compartments must be taken into account.

The control of dangerous substances to the different environmental compartments is based either on limit values or environmental quality standards. Only the EC Dangerous Substances Directive (CEC 1976) controlling the discharge of particularly dangerous substances to the aquatic environment permits alternative approaches of control by either limit values based on "best available technology" (BAT) or environmental quality standards (EQSs). For the control of water quality the UK has favoured the EQS approach as the application of limit values does not guarantee satisfactory receiving water quality since no account is taken of the assimilating capacity of the receiving water, the cumulative effect of multiple inputs or the base load already present in the water from other sources such as diffuse run-off or atmospheric deposition. However, having set water quality standards it is necessary to apply restrictions in the form of discharge consents (corresponding to locally determined limit values) to all point-source inputs, taking account of the intended use of the water, the dilution available in the receiving water and the inputs from diffuse sources. Thus the two approaches are complementary and not, as frequently suggested, in opposition. This is illustrated by the UK Royal Commission standard (ie limit value) for sewage effluent derived in 1912 from the quality objective of avoiding de-oxygenation in rivers affording at least eightfold dilution of the effluent.

Although the UK has argued strongly for the EQO approach as the basis for controlling water quality, emission limit values have traditionally been applied to discharges to the air. There is little spatial variability in the capacity of the atmosphere to dilute and disperse waste inputs or in the assimilative processes such as photodegradation or oxidation. This is in contrast to surface waters where widely different dilution capacities are available and where chemical and biological degradation processes and the toxic effects of substances vary significantly with water type, faunal and floral assemblages and substrate composition. However, whichever control strategy is selected, the control of discharges to one environmental compartment can lead to problems in the others. In 1976 the UK Royal Commission on Environmental Pollution, promoted the concept of "best practicable environmental option" (BPEO) as a way of overcoming the compartmentalisation of pollution control (RCEP 1976). This has led to the establishment of Her Majesty's Inspectorate of Pollution (HMIP) by combining in one organisation the government's inspectors responsible for discharges to all three compartments - water, air and land - and the development of the "Integrated Pollution Control" (IPC) concept. In addition the UK has adopted a more precautionary approach to the control of particularly dangerous substances, the "Red List" substances, and these will be the first subjected to the IPC approach. The paper briefly discusses the BPEO concept and outlines the IPC approach and the selection procedure adopted for the Red List substances and provides proposals for the control of dangerous substances.

BEST PRACTICABLE ENVIRONMENTAL OPTION (BPEO)

Although BPEO has been generally welcomed as a principle of waste management, it has so far proved a difficult concept to implement in practice. Defining the scope and scale on which BPEO should be applied is difficult, particularly when the "practicable" element is interpreted as including costs.

A good example is sewage sludge. The scope for minimising sludge quantities is limited so that the emphasis is on methods of utilisation and disposal. Four disposal routes are used in the UK. Based on 1987 data these were: application to agriculture land (49.1%), landfill (19.1%), sea (26.4%) and incineration (5.4%) (DoE 1988a). Because sludge is produced all year round, the disposal routes need to be guaranteed with sufficient operational flexibility to avoid the need for extensive (and expensive) storage capacity at treatment works. Preferred outlets are site dependent, with for example, sea disposal favoured at large coastal sites with harbour facilities for tankers, and agriculture utilisation in rural areas with easy access to tracts of suitable land. Incineration has been regarded as the most expensive route, but can become the preferred option where "dirty sludges" require disposal and where suitable landfill sites are not available or where sludge has to be transported over long distances (>20 km). Incineration might also cause other environmental problems related to the waste gases and ash produced. The UK is also under intense political pressure to abandon the sea disposal route despite evidence from long-term monitoring that the effects are localised and that contaminant loads attributable to sludge are small compared to the total inputs from all sources, as illustrated in Figure 1 based on data from the Oslo Commission.

Figure 1. Sources of metals entering the North Sea

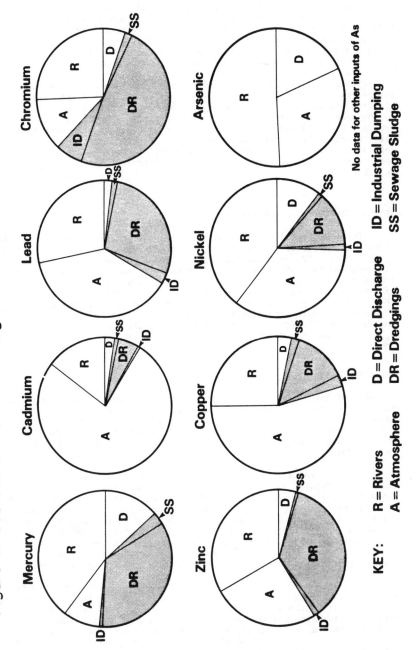

How then should the BPEO for sewage sludge be determined? It could be worked out on a national basis by considering total sludge production, average costs of treatment, transport and incineration, and some global assessment of outlet capacity and environmental impact (provided the detriment could be costed equitably). Such an approach would provide some guidelines for national policy; a more detailed appraisal would be essential on a case-by-case basis for particular areas or even individual works. For example it would almost certainly be unrealistic to base the BPEO for London's sludge (representing approximately 10% of the total for England and Wales) generated in a heavily urbanised area adjacent to the Thames Estuary on a model derived for the UK as a whole.

The prospect for developing BPEO depends very much on establishing parameters of environmental costs which are consistent between widely different impacts. The basis must also be clearly explained if confidence in the system is to be built for both technical and lay audiences. Some progress is being made on the subject of environmental risk assessment, particularly where commercially exploitable resources are affected or the cost of alternative measures can be quantified. However, it is the long-term chronic effects which are both difficult to predict and hard to quantify.

INTEGRATED POLLUTION CONTROL (IPC)

IPC is derived from the BPEO concept and is intended to minimise the production of wastes and to ensure at the same time that the EQSs required are achieved. IPC is a new approach and the required legislation has not yet been adopted; the detailed procedures to implement IPC have not yet been developed. However, the main responsibility for the implementation of IPC will fall on HMIP with a new body, the National Rivers Authority (NRA), being responsible for ensuring that the discharge controls are adequate to protect the uses of the receiving waters and for the monitoring of the aquatic environment. Initially the IPC approach will be applied to the control of the Red List substances.

RED LIST SUBSTANCES

The UK has adopted a more precautionary approach to the control of particularly dangerous substances by regulating the discharge of these compounds by limit values derived by applying "best available technology not entailing excessive cost" (batneec) and by ensuring that the EQS in the receiving water is not exceeded whichever is the more stringent control option.

Several attempts have been made to identify from the long list of chemicals currently in use those which require priority for action in terms of legislation and control. However, a distinction has to be made between "priority for control" which can be assessed for those substances for which sufficient published data on the toxicity and long-term fate are available and "priority for further consideration" which relates to substances for which insufficient published data are available to make an assessment. (For new chemicals separate EC legislation exists of course requiring extensive data to be submitted before a licence is issued.)

The scheme used in preparing the UK Red List of substances has been published (DoE 1988) so that the basis of selection can be scrutinised. Three scenarios (the short-term, long-term and the food chain) were considered in which the key parameters were respectively (a) acute toxicity, (b) chronic toxicity and (c) potential for bioaccumulation. Persistence in the environment also featured, being particularly important for the longer-term scenarios (b and c). In addition an estimate was made of the likely concentration in the environment based on production volume and physical-chemical properties used to assess whether the compound will remain in the aquatic environment or be transferred to other environmental compartments. The 23 substances selected for the Red List are shown in Table 1 (DoE 1989).

Table 1.   The UK 'Red List'

                                                              EC Directive adopted
                                                              (List I status)

Mercury   +
Cadmium   +
gamma - hexachlorocyclohexane (lindane)   +
DDT   +
Pentachlorophenol (PCP)   +
Hexachlorobenzene (HCB)   +
Hexachlorobutadiene (HCBD)   +
Aldrin   +
Dieldrin   +
Endrin   +
PCB (Polychlorinated biphenyls)
Tributyltin compounds
Triphenyltin compounds
Dichlorvos
Trifluralin
1,2 Dichloroethane
Trichlorobenzene
Azinphos-methyl
Fenitrothion
Malathion
Endosulfan
Atrazine
Simazine

All the existing EC List I substances (12) are included except
chloroform and carbon tetrachloride which the selection scheme did not
identify as justifying Red List status. A further 23 substances have been
identified as possible Red List substances and will be considered
individually for designation as red list substances when additional
information is available.

Apart from mercury and cadmium the red list comprises organic
chemicals, more than half of which are likely to reach the aquatic
environment predominantly by diffuse routes as they are pesticides and
herbicides. These substances are designed for widespread use in
environmental control and tend to occur in surface run-off and leachate.

Few of the red list substances are routinely monitored in surface
waters so that information about environmental concentrations is extremely
limited. Action plans are currently being developed by the water industry
in consultation with the UK Department of the Environment (DoE) which will
include three screening surveys this year as a preliminary to more
comprehensive monitoring in future years to monitor the success of the
reduction programme. Information is also being sought from industry about
sources and loads discharged of the substances in river catchment areas to
assist in preparing the action plans for reduction and in targeting
monitoring efforts to best advantage.

Some difficulties are anticipated in meeting the analytical limits of
detection required for load estimation for some substances. Work is in
hand to establish the necessary methodology which will be subject to
analytical quality control to ensure confidence in the assessments of load
reduction.

It is anticipated that for existing EC List I compounds on the red
list the EQS values and limit values laid down in the respective EC
directives for the protection of the aquatic environment will be used as
EQS values and batneec. However, for approximately half of the compounds
EQS values and batneec have not yet been derived. The task of deriving
batneec is designated to HMIP whereas the Department of the Environment is
responsible for setting EQS values.

DERIVATION OF EQS VALUES

There is a well-established procedure in the UK as outlined in Figure 2 for deriving EQS values to ensure that the specified uses (EQOs) are protected (Gardiner and Mance 1984).

This method involves establishing (usually from published data) 'safe' concentrations of a substance related to individual uses. The support of fish life, including food organisms, is usually the most sensitive use. However, for some substances the quality of water used for the abstraction to potable supply is the most critical use but this tends to be related to standards which in some cases are set arbitrarily without considering the toxicity. This is particularly the case for some pesticides and related products were the EC Drinking Water Directive (CEC 1980) lays down a uniform standard for individual substances of 0.1 µg/l independent of their relative toxicities.

In the data used for the derivation of the EQS values for the protection of aquatic life care is taken in selecting those results for which test conditions have been properly reported and reliance is placed wherever possible on measured rather than nominal concentrations. Thus the original papers describing ecotoxicity or bioaccumulation tests are consulted, particularly those reporting critical data or involving sensitive life-stages of test organisms. Besides toxicity and bioaccumulation data the fate in the environment is also assessed to assist in the derivation of EQS values.

Having derived a preliminary EQS from the laboratory data valuable corroboration is sometimes available from field data. Evidence, for example, that a viable fishery is supported in a river containing the substance at concentrations approximating the preliminary EQS provides the most useful validation. The converse does not necessarily invalidate the EQS based on test data since many other factors could be responsible for the absence of fish. Reports recommending EQS values, and including the supporting background data, are subject to peer review and consultation before the DoE decides on values to be adopted as national standards.

Full evaluation of all the relevant data published for those red list substances for which EQS values are not yet available would be time-consuming so that a study is in hand to prepare provisional EQSs to support the evaluation of the screening surveys later this year. These provisional EQSs will be restricted to the EQOs for the support of aquatic life in both fresh and salt waters. It is expected that more detailed work will be required for selected compounds later to substantiate values to be adopted as national EQSs for control purposes.

The provisional EQSs also provide a guide for the required limits of detection of analytical methods since the limit is normally set at one-tenth of the EQS. Although in order to assess loads even lower detection limits might be required.

IMPLEMENTATION OF IPC

IPC will initially be restricted to new plants and to those types of industrial processes that discharge significant quantities of harmful wastes with particular reference to red list substances. HMIP will be responsible for producing the process criteria to achieve IPC by optimising waste disposal to minimise the effect on the different environmental compartments. Such processes could then be described (scheduled) by the Secretary of State for the Environment according to published criteria. At present HMIP is carrying out several studies on different industrial processes to establish the principles for the implementation of the IPC system.

CONTROL OF RED LIST SUBSTANCES

Ministers representing countries bordering the North Sea agreed at a meeting in London in November 1987 to reduce inputs to the North Sea of

Figure 2. Schematic flow chart for the derivation of environmental quality
standards (EQSs)

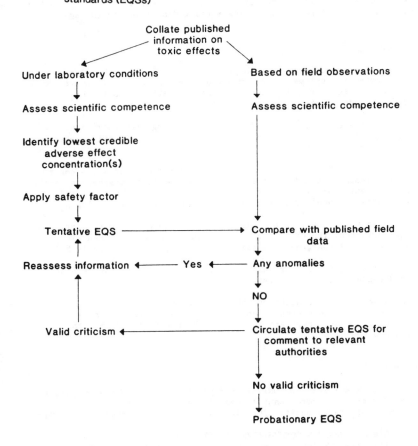

dangerous substances by up to 50% by 1995 (based on 1985 levels)
(Ministerial Declaration 1987). In the UK this commitment is applied to
the red list substances and has been extended to all the seas around the
UK.

The introduction of IPC for process plant is only one aspect of
controlling red list substances. Few of the organic chemicals on the red
list are manufactured in the UK (tributyltin, trifluralin and
1,2-dichloroethane) although most will be imported and occur in
formulations or as process intermediates.

The use of a few substances on the list is already restricted in the
UK (antifouling paints containing tributyltin are banned for use on small
boats) or are banned DDT and proposals have been made for phasing out the
use of others (dieldrin, aldrin and endrin).

For some red list substances reduction in loads can be achieved by
applying batneec but this might not be sufficient to reduce the loads by
50% and some additional restriction of use might be required.

As already mentioned approximately half the compounds on the red list
are pesticides and herbicides. For these some reduction in load might be
achieved by more careful management of the production, formulation and
handling of such substances to avoid significant point-source discharges
direct to watercourse or to sewer, but substantial reductions in inputs to
the environment will require wider considerations. Thus guidelines on

levels of application, frequency of use and timing of applications (seasonal or growth stage) will need to be established and enforced. In addition for some of the compounds a reduction in use might be required to obtain a significant reduction in loads being discharged to the sea.

CONCLUSIONS

1. In setting up Her Majesty's Inspectorate of Pollution (HMIP) the UK is attempting to implement the concept of "best practicable environmental option" (BPEO) for the control of dangerous substances to the different environmental compartments (air, water and land).

2. HMIP is responsible for implementing the "integrated pollution control" concept initially for the control of particularly dangerous substances, the "Red List".

3. The UK is adopting a more precautionary approach towards the control of particularly dangerous substances, the red list, discharged to the aquatic environment by using limit values, based on batneec, and environmental quality standards, whichever is the more stringent.

4. In response to the second Ministerial Conference on the Protection of the North Sea, the UK has agreed to reduce the loads of red list substances discharged to UK coastal waters by up to 50% by 1995.

5. Using a selection scheme based on scientific criteria the UK has produced a 'Red List' containing 23 substances.

6. For some Red List substances the introduction of batneec might be sufficient to achieve an acceptable decrease whereas for others optimisation or a decrease in use will be needed to achieve the desired reduction.

REFERENCES

COUNCIL OF EUROPEAN COMMUNITIES (1976) Council Directive on pollution caused by certain dangerous substances discharged into the aquatic environment of the Community (76/464/EEC). Official Journal L129 18 May.

COUNCIL OF EUROPEAN COMMUNITIES (1980) Council Directive relating to the quality of water intended for human consumption (80/778/EEC). Official Journal L229, 30 August.

COUNCIL OF EUROPEAN COMMUNITIES (1985) Council Directive on the assessment of the effects of certain public and private projects on the environment (85/337/EEC). Official Journal L175, 5 July.

DEPARTMENT OF THE ENVIRONMENT (1988a) Water Facts. HMSO, London.

DEPARTMENT OF THE ENVIRONMENT (1988b) Inputs of dangerous substances to water, proposals for a unified system of control. The Government's consultative proposals for tighter controls over the most dangerous substances entering the aquatic environment "The Red List", July 1988.

DEPARTMENT OF THE ENVIRONMENT (1989) Agreed "Red List" of dangerous substances confirmed by Lord Caithness. News Release 194, 2 Marsham Street, London, UK.

GARDINER J and MANCE G (1984) Proposed environmental quality standards for List II substances in water, Introduction. WRc Technical Report TR 206.

MINISTERIAL DECLARATION (1987) Second International Conference on the protection of the North Sea, London, 24/25 December.

ROYAL COMMISSION ON ENVIRONMENTAL POLLUTION (1976) 5th Report. Air Pollution Control: an Integrated Approach Cmnd 6371, HMSO, London.

Chapter 8

AN INTEGRATED ENVIRONMENTAL MONITORING PROGRAM AT A US NUCLEAR RESEARCH FACILITY

R H Gray (Pacific Northwest Laboratory, Richland, USA)

ABSTRACT

A comprehensive environmental monitoring and surveillance program is conducted on the U.S. Department of Energy's Hanford Site, southeastern Washington, U.S.A. In addition to monitoring radioactivity and chemicals in surface and ground water, the program monitors air, foodstuffs, wildlife, soil and natural vegetation. Population numbers of key wildlife species are also determined. In 1987, measured exposure to penetrating radiation and calculated radiation doses to the public were well below applicable regulatory limits. The calculated effective dose potentially received by an individual using hypothetical worst-case assumptions for all routes of exposure in 1987 was 0.05 mrem/yr.

Fish and wildlife monitoring shows that chinook salmon spawning in the Columbia River at Hanford has increased in recent years with a concomitant increase in the number of bald eagles during winter. Nesting Canada goose and great blue heron, and various plants and other animals, e.g., elk, mule deer, and coyotes are common.

Key words: environmental monitoring, radionuclides, chemicals, fish, wildlife.

INTRODUCTION

The U.S. Department of Energy's (DOE) Hanford Site occupies a land area of about 1,450 km^2 (560 mi^2) in semi-arid southeastern Washington, U.S.A. (Fig. 1). The Columbia River flows through the Site and forms part of its eastern boundary. Flow of the Columbia River is regulated daily according to electric power demands. Although the river was once closed to public access, public use for recreational and barge traffic is again practical. The southwestern portion of the Site includes the southern terminus of the Rattlesnake Hills with elevations exceeding 1000 m. Both unconfined and confined aquifers lie beneath the Site.

Nuclear and non-nuclear industrial and research activities have been conducted at Hanford since 1943. The most environmentally significant activities have involved the production of nuclear materials and the chemical processing and waste management associated with the major product, plutonium. Byproduct wastes have included gamma, beta, and alpha-emitting radionuclides and various nonradioactive chemicals in gaseous, liquid and solid forms.

There are currently four major DOE operations areas on the Hanford Site (Fig. 1). The 100 Areas located along the Columbia River include the dual-purpose N Reactor that produced plutonium for national defense and steam for the Hanford Generating Project (HGP), operated by the Washington Public Power Supply System (WPPSS), and eight, now deactivated single-purpose, plutonium production reactors. The plutonium uranium extraction (PUREX)

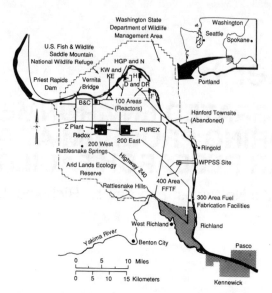

Fig. 1. The Hanford Site. HGP = Hanford Generating Project; REDOX = reduction-oxidation; PUREX = plutonium uranium extraction; WPPSS = Washington Public Power Supply System; FFTF = Fast Flux Test Facility.

plant (reactor fuel reprocessing), plutonium finishing plant (Z Plant), and waste-disposal facilities are located in the 200 Areas on a plateau (elevation 229 m) about 11.3 km west of the Columbia River. The 300 Area, located just north of Richland, Washington contains the uranium fuel manufacturing facilities in support of N Reactor, and research and development laboratories. The Fast Flux Test Facility (FFTF) which has operated intermittently since 1981 to test new fuels and materials for future breeder reactor technology is located in the 400 Area. Nongovernment facilities within Hanford Site boundaries include HGP, the WPPSS nuclear plant (WNP) sites, WNP-1, WNP-2 and WNP-4, including one commercial reactor (WNP-2) that achieved full operation status in the fall of 1984, and a commercial low-level radioactive-waste burial site near the 200 Areas, operated by U.S. Ecology.

Environmental monitoring at Hanford has been ongoing for over 40 years. The program is conducted to assess potential impacts to individuals and populations that may be exposed to radionuclides, ionizing radiation and hazardous chemicals. Environmental monitoring currently includes air, ground and surface water, fish, wildlife, soil, vegetation, and foodstuffs (fruits, vegetables, milk). Fish and wildlife are monitored for radioactivity and to determine the population status of key species.

RADIOLOGICAL MONITORING

Air

Potential airborne transport of stack releases containing radionuclides from Hanford facilities offers a direct pathway for human exposure. Thus, air is sampled continuously for airborne particulates and analyzed for radionuclides at 50 locations onsite, at the Site perimeter, and in nearby and distant cities [PRICE (1986); PNL (1987); JAQUISH and MITCHELL (1988)]. At selected locations, gases and vapors are also collected and analyzed. Many of the longer-lived radionuclides released at Hanford are also present in atmospheric fallout that resulted from nuclear weapons testing in the 1950s and 1960s or from nuclear accidents that occurred elsewhere.

In May and June, 1986, air samples collected onsite as well as those from distant locations showed increases in several long- and short-lived radionuclides (e.g., ^{137}Cs, ^{131}I, ^{103}Ru) that resulted from the reactor accident at Chernobyl, April, 1986, in western Russia. However, even then, no sample exceeded 0.17% of the applicable DOE derived concentration guide (DCG) for areas permanently occupied by the public.

Ground Water

The shallow unconfined (water-table) aquifer has been affected by waste-water disposal practices at Hanford more than the deeper, confined aquifers. Discharge of water from various industrial processes has created ground-water mounds near each of the major waste-water disposal facilities in the 200 Areas, and in the 100 and 300 Areas (Fig. 1). Discharge to ground water in the 200 Areas may contribute ten times more water annually to the unconfined aquifer than natural input from precipitation and irrigation [GRAHAM et al. (1981)]. These ground-water mounds have altered local flow patterns in the aquifer, which are generally from west to east.

Ground water, primarily from the unconfined aquifer, is currently sampled from over 560 wells [JAQUISH and MITCHELL (1988)]. Tritium (^{3}H) occurs at relatively high levels in the unconfined aquifer, is one of the most mobile radionuclides, and thus, reflects the extent of ground-water contamination from onsite operations. Many liquid wastes discharged to the ground at Hanford have contained ^{3}H. The PUREX facility is currently the main source for ^{3}H-containing wastes [DOE (1983)]. Tritium from releases prior to 1983 that passed-downward through the vadose (unsaturated) zone to the unconfined aquifer continues to move with ground-water flow toward the Columbia River. Tritium concentrations in Hanford ground water range from less than 300 pCi/L to over 2,000,000 pCi/L near or within the 200 Areas [PNL (1987); JAQUISH and MITCHELL (1988)].

Ground water from the unconfined aquifer enters the river through subsurface flow and springs that emanate from the riverbank. McCORMACK and CARLILE (1984) identified 115 springs along a 41-mile stretch of river. Tritium concentrations in wells near the springs ranged from 19,000 to 250,000 pCi/L and averaged 176,000 pCi/L in 1985 [PRICE (1986)]. Although the distribution of ^{3}H and other radionuclide concentrations in springs generally reflected those in nearby ground-water wells, the magnitude was generally less in springs due to mixing of ground and surface water. Tritium concentrations in the river were generally less than those in springs. It is noteworthy that ^{3}H also occurs naturally in the Columbia River upstream from Hanford. Tritium concentrations in springs were less than 4% of the DOE DCG (2,000,000 pCi/L). Tritium concentrations in the river were less than 0.5% of the DCG and less than half the regulatory limit for drinking water (20,000 pCi/L) [EPA (1976)].

Surface Water

Columbia River water is used for drinking at downstream cities, for crop irrigation and for recreational activities (fishing, hunting, boating, waterskiing, swimming). Thus, it constitutes the primary environmental pathway to people for radioactivity in liquid effluents. Radionuclides can be delivered to human foodstuffs through crops irrigated with river water and cow's milk through irrigated alfalfa and other cattle forage. Although radionuclides associated with Hanford operations, worldwide fallout and natural phenomena continue to be found in small but measurable quantities in the Columbia River, concentrations are below Washington State and Environmental Protection Agency (EPA) drinking water standards.

Deep sediments in downstream reservoirs still contain low concentrations of some long-lived radionuclides [ROBERTSON and FIX (1977); HAUSHILD et al. (1975); NELSON et al. (1979); SULA (1980); BEASLEY et al. (1981)]. Trace amounts of ^{239}Pu, ^{60}Co, ^{137}Cs, and ^{152}Eu persist in sediments accumulated above the first downstream dam (McNary). In 1977, about 20 to 25% of the total plutonium inventory ($^{239, 240, 241}$Pu) in Lake Wallula sediments, 100 km downstream, was believed to originate from the 1944 through 1971 releases at Hanford [BEASLEY et al. (1981)]. However, only ^{239}Pu was believed to actually reflect earlier reactor operations. Further, this ^{239}Pu was derived from ^{239}Np (produced by neutron capture in natural uranium followed by decay to ^{239}Np), an abundant isotope in Columbia River water. Thus, plutonium may not have been released to the river from reactor operations.

Fish and Wildlife

Fish are collected at various locations along the Columbia River and boneless fillets are analyzed for ^{60}Co, ^{90}Sr, and ^{137}Cs. Carcasses are analyzed to estimate ^{90}Sr in bone. Following shutdown of the last single-purpose, once-through cooling reactor and installation of improved liquid effluent control systems at N Reactor, short-lived radionuclides, including the biologically important ^{32}P and ^{65}Zn, essentially disappeared from the river [CUSHING et al. (1981)] through radioactive decay. Radionuclide concentrations in fish collected from the Hanford Reach of the Columbia River are similar to those in fish from upstream locations.

Deer (Odocoileus sp.), ring-necked pheasants (Phasianus colchicus), mallard ducks (Anas platyrhynchus), Nuttall cottontail rabbits (Sylvilagus nuttallii) and black-tailed jack rabbits (Lepus californicus) are collected and tissues are analyzed for ^{60}Co and ^{137}Cs (muscle), $^{239, 240}Pu$ (liver) and ^{90}Sr (bone). The doses that could be received by consuming wildlife at the maximum radionuclide concentrations measured in 1985-1987 were below applicable DOE standards [PRICE (1986); PNL (1987); JAQUISH and MITCHELL (1988)].

Soil and Vegetation

Airborne radionuclides are eventually deposited on vegetation or soil. Samples of surface soil and rangeland vegetation (sagebrush) are currently collected at 15 onsite and 23 site perimeter and offsite locations [JAQUISH AND MITCHELL (1988)]. Samples are collected from nonagricultural, undisturbed sites so that natural deposition and buildup processes are represented. Sampling and analyses in 1985 through 1987 showed no radionuclide buildup offsite that could be attributed to Hanford operations [PRICE (1986); PNL (1987); JAQUISH and MITCHELL (1988)].

Foodstuffs

The most direct way for deposited radionuclides to enter the foodchain is through consumption of leafy vegetables. Samples of alfalfa and several foodstuffs, including milk, vegetables, fruit, beef, chickens, eggs and wheat, are collected from several locations, primarily downwind (i.e., south and east) of the Site. Samples are also collected from upwind and somewhat distant locations to provide information on radiation levels attributable to worldwide fallout. Foodstuffs from the Riverview Area (across the river and southeast) are irrigated with Columbia River water withdrawn downstream of the Site. Although low levels of ^3H, ^{90}Sr, ^{129}I, and ^{137}Cs have been found in some foodstuffs, concentrations in samples collected near Hanford are similar to those in samples collected away from the Site.

Penetrating Radiation

Penetrating radiation (primarily gamma-rays) is measured in the Hanford environs with thermoluminescent dosimeters to estimate dose rates from external radiation sources. Radiation surveys are routinely conducted at numerous onsite locations including roads, railroads and retired waste-disposal sites located outside of operating areas. Onsite and offsite measurements and survey results for 1985-1987 were similar and comparable to past years. Dose rates near some operating facilities were only slightly higher than natural background rates.

Overall Impact from Hanford Operations

Beginning in 1974 the evaluation of radiation doses has included assessment of the maximum external dose rate at a location accessible to the general public, doses to a hypothetical maximally exposed individual, and doses to the population within 80 km of the Site. Based on these assessments, potential radiation doses to the public from Hanford operations have been consistently below applicable standards and substantially less than doses normally received from common sources of background radiation. The calculated 50-year whole-body cumulative dose received by the maximally exposed individual ranged from 0.5 to 3 mrem during the years 1981 through 1986 [PNL (1987)]. The maximally exposed individual is a hypothetical person who receives the maximum calculated radiation dose when worst case assumptions are used concerning location, inhalation of radioactive emissions, consumption of contaminated food and water and direct exposure to contaminants. Expressed as effective dose equivalents, the calculated dose received by a hypothetical maximally exposed individual was 0.1, 0.09 and 0.05 mrem respectively, in 1985, 1986 and 1987. The average per capita whole-body cumulative (effective) dose for 1985, 1986 and 1987, based on the human population of 340,000 within 80 km of the Site, was 0.02, 0.03 and 0.01 mrem annually, respectively [PRICE (1986); PNL (1987); JAQUISH and

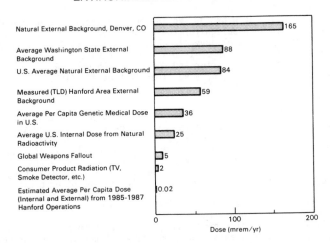

Fig. 2. Annual radiation doses from various sources: external background, Denver, Colorado, Washington State and U.S. average from OAKLEY (1972). Genetic medical dose, U.S. internal dose, weapons fallout and consumer product radiation from KLEMENT et al. (1972); TLD = thermoluminescent dosimeter, does not include neutron component; mrem/yr = millirem per year.

MITCHELL (1988)]. These estimates and the measured Hanford area background radiation can be compared to, and are considerably less than, doses from other routinely encountered sources of radiation, such as natural terrestrial and cosmic background radiation, medical treatment and x-rays, natural internal body radioactivity, worldwide fallout and consumer products (Fig. 2).

CHEMICAL MONITORING

Air Quality

Nitrogen oxides (NO_x) are routinely released onsite from fossil-fueled steam and chemical processing facilities, most notably the PUREX plant. Nitrogen dioxide is sampled at eight onsite and one offsite locations by the Hanford Environmental Health Foundation (HEHF). Nitrogen dioxide concentrations measured in 1984-1987 were well below federal (EPA) and local (Washington State) ambient air quality standards [PRICE (1986); PNL (1987); JAQUISH and MITCHELL (1988)].

Ground Water

In 1987, over 3000 ground-water samples were collected and analyzed for inorganic constituents, and over 290 wells were monitored for potentially hazardous materials including pesticides, herbicides and total organic halogen [JAQUISH and MITCHELL (1988)]. In addition, well samples were analyzed by HEHF for water quality. Detected constituents included several metals, anions, coliform bacteria, radionuclides and total organic carbon. Many of these constituents are expected in natural ground water. Chromium, cyanide, fluoride and carbon tetrachloride were found in wells not used for drinking water near operating areas.

Columbia River

Nonradioactive waste water is discharged at eight locations along the Hanford reach of the Columbia River. Discharges consist of backwash from water intake screens, cooling water, water storage tank overflow and fish laboratory waste water. Effluents from each outfall are monitored by the operating contractors. The Columbia River is also monitored by the United States Geological Survey, upstream and downstream of the Site, to verify compliance with Class A [WSDOE (1977)] water-quality requirements. In addition, numerous studies have evaluated and resolved the potential environmental issues associated with water intake and thermal discharge structures

on the Columbia River at Hanford. For example, retrofitting of the HGP
water intake and a newer design for the intake used at WNP-2 have ensured
safe downstream migration of juvenile chinook salmon [PAGE *et al.* (1977);
WPPSS (1978); GRAY *et al.* (1979, 1986)]. Other studies have concluded that
thermal discharges from N reactor and HGP to the Columbia River were
biologically insignificant [DOE (1982); NEITZEL *et al.* (1982)].

HANFORD FLORA AND FAUNA

Most of the Hanford Site consists of undeveloped land that supports
stands of native vegetation and a few exotic species (e.g., cheatgrass,
<u>Bromus</u> <u>tectorum</u>; Russian thistle, <u>Salsola</u> <u>kali</u>; and tumble mustard,
<u>Sisymbrium</u> <u>altissimum</u>), is free from agricultural practices, and has been
essentially free from livestock grazing and hunting for 45 years. Thus, the
Site serves as a refuge for migratory waterfowl, elk (<u>Cervus</u> <u>elaphus</u>), mule
deer (<u>Odocoileus</u> <u>hemionus</u>), coyote (<u>Canis</u> <u>latrans</u>) and other plants and
animals. Restricted land use has favored native wildlife that frequent
riverine habitats, for example, mule deer, Canada goose (<u>Branta</u> <u>canadensis</u>),
and great blue heron (<u>Ardea</u> <u>herodias</u>).

The Columbia River at Hanford supports up to 48 species of fish [GRAY
et al. (1977)] and serves as a migration route for upriver runs of Chinook
(<u>Oncorhynchus</u> <u>tshawytscha</u>), coho (<u>O.</u> <u>kisutch</u>) and sockeye (<u>O.</u> <u>nerka</u>) salmon,
and steelhead trout (<u>Salmo</u> <u>gairdneri</u>). The Hanford Reach supports the last
remaining mainstem spawning habitat for fall chinook salmon. Steelhead
trout also spawn in the Hanford Reach. The salmon population is maintained
by a combination of natural spawning, artificial propagation and regulated
commercial and sport harvest of returning adults.

Based on redd (nest) counts from the air, fall chinook salmon spawning
in the Hanford Reach of the mainstem Columbia River has increased dramatically since 1980 (Fig. 3). Recent observations by divers [SWAN *et al.*
(1988)] showed salmon redds at depths below those visible by boat or aircraft and suggests that salmon spawning in the Hanford Reach may be even
greater than previously estimated. The increase in salmon spawning has
attracted increasing numbers of wintering bald eagles (<u>Haliaeetus</u> <u>leuco-</u>
<u>cephalus</u>). The bald eagle is listed by the U.S. Fish and Wildlife Service
as "threatened" in the state of Washington [RICKARD and WATSON (1985)].

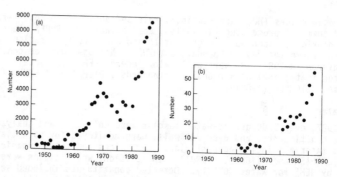

Fig. 3. Numbers of (a) salmon redds (nests) and (b) wintering bald eagles
(there were no counts from 1969-1974) at Hanford [adapted and updated from
RICKARD and WATSON (1985)].

The sparsely vegetated islands in the Columbia River have historically
been used as nesting habitat for great basin Canada goose [HANSON and
EBERHARDT (1971); FITZNER and RICKARD (1982)]. From the mid-1950s to the
mid-1970s the number of goose nests declined from a high of 250-300 to about
100 annually. From the late 1970s to the present, the number of nests has
increased and appears to have stabilized at about 150-200. Initially,
closure of the Hanford Reach was beneficial to the geese by providing freedom from human intrusion. However, the coyote, a natural goose predator,
also benefitted, and is believed to be the major cause of the decline in
numbers of goose nests into the mid-1970s.

Initially there were no nesting great blue heron on the Hanford Site. However, there are now three active colonies consisting of about 35-40 birds each and herons are present year round.

Elk first arrived on the Hanford Site in 1972 [RICKARD *et al.* (1972)]. From a small founding population, the herd size grew to about 80 animals in 1987. The rapid increase in elk is attributed to the lack of predation or human disturbance during calving, absence of onsite hunting, and the lack of competition from sheep and cattle for available forage. For the last three years, offsite hunting has limited further population increases by removing about 15 to 20 animals annually from the herd.

The mule deer population at Hanford is estimated at several hundred animals and appears stable even in the absence of onsite hunting. Coyote predation on fawns is believed to be an important factor that maintains the stable deer population [STEIGERS and FLINDERS (1980)].

SUMMARY

The Pacific Northwest Laboratory (PNL) conducts an environmental monitoring program to assess potential effects of Hanford Operations on the local environs, onsite workers, and the offsite public. Monitoring for radiological emissions at Hanford has been ongoing for 45 years and includes air, ground and surface water, fish, wildlife, soil, vegetation, and foodstuffs. Measured and calculated radiation doses to the public have been consistently below applicable regulatory limits. The Hanford Site now serves as a refuge for key fish and wildlife species.

ACKNOWLEDGEMENTS

Environmental monitoring at Hanford reflects the cooperative efforts of numerous individuals representing the staffs of DOE, PNL, HEHF, and other contractor, state and federal organizations. Environmental monitoring has been conducted by the PNL since 1965, and is supported by DOE under Contract DE-AC06-76RLO 1830 with Battelle Memorial Institute.

BIBLIOGRAPHY

BEASLEY, T.M.; BALL, L.A.; ANDREWS, III, J.E.; HALVERSON, J.E. - "Hanford-Derived Plutonium in Columbia River Sediments." Science, 214, 1981, pp. 913-915.

CUSHING, C.E.; WATSON, D.G.; SCOTT, A.J.; GURTISEN, J.M. - "Decrease of Radionuclides in Columbia River Biota Following Closure of Hanford Reactors." Health Phys., 41, 1981, pp. 59-67.

DOE - 316(a) Demonstration for Test of N Reactor in Plutonium-Only Mode of Operation. U.S. Department of Energy, Richland, WA, 1982.

DOE - Final Environmental Impact Statement: Operation of PUREX and Uranium Oxide Plant Facilities. DOE/EIS-0089, U.S. Department of Energy, Washington, D.C., 1983.

EPA - National Interim Primary Drinking Water Regulations. EPA-570/976-003, U.S. Environmental Protection Agency, Washington, D.C., 1976.

FITZNER, R.E.; RICKARD, W.H. - "Canada Goose Nesting Performance Along the Hanford Reach of the Columbia River, 1971-1981." Northwest Sci., 57, 1982, pp. 267-272.

GRAHAM, M.J.; HALL, M.D.; STRAIT, S.R.; BROWN, W.R. - Hydrology of the Separations Area. RHO-ST-42, Rockwell Hanford Operations, Richland, WA, 1981.

GRAY, R.H.; DAUBLE, D.D. - "Checklist and Relative Abundance of Fish Species from the Hanford Reach of the Columbia River." Northwest Sci., 51, 1977, pp. 218-215.

GRAY, R.H.; NEITZEL, D.A.; PAGE, T.L. - "Water Intake Structures: Engineering Solutions to Biological Problems." The Northern Engineer, 10, 1979, pp. 26-23.

GRAY, R.H.; PAGE, T.L.; NEITZEL, D.A.; DAUBLE, D.D. - "Assessing Population Effects from Entrainment of Fish at a Large Volume Water Intake." Environ. Sci. and Health, A21, 1986, pp. 191-209.

HANSON, W.C.; EBERHARDT, L.L. - "A Columbia River Goose Population, 1950-1970." Wildlife Soc. Monograph No. 28, 1971.

HAUSHILD, W.L.; DEMPSTER, Jr., G.R.; STEVENS, Jr, H.H. - Distribution of Radionuclides in the Columbia River Streambed, Hanford Reservation to Longview, Washington. Geological Survey Prof. Paper, 433-O, U.S. Government Printing Office, Washington, D.C., 1975.

JAQUISH, R.E.; MITCHELL, P.J., eds. - Environmental Monitoring at Hanford for 1987. PNL-6464, Pacific Northwest Laboratory, Richland, WA. National Technical Information Service, Springfield, VA, 1988.

KLEMENT, Jr., A.W.; MILLER, C.R.; MINX, R.P.; SLEIEN, B. - Estimates of Ionizing Radiation Doses in the United States, 1960-2000. ORP/CSD 72-1, U.S. Environmental Protection Agency, Washington, D.C., 1972.

McCORMACK, W.D.; CARLILE, J.M.V. - Investigation of Ground-Water Seepage from the Hanford Shoreline of the Columbia River. PNL-5289, Pacific Northwest Laboratory, Richland, WA. National Technical Information Service Springfield, VA, 1984.

NEITZEL, D.A.; PAGE, T.L.; GRAY, R.H.; Dauble, D.D. - "Once Through Cooling on the Columbia River--the Best Available Technology?" Environ. Impact Assess. Rev., 3, 1982, pp. 43-58.

NELSON, J.L.; HAUSHILD, W.L. - "Accumulation of Radionuclides in Bed Sediments of the Columbia River Between the Hanford Reactors and McNary Dam." Water Resourc. Res., 6, 1979, pp. 130-137.

OAKLEY, D.T. - Natural Radiation Exposure in the United States. ORP/SID 72-1, U.S. Environmental Protection Agency, Washington, D.C., 1972.

PAGE, T.L.; NEITZEL, D.A.; GRAY, R.H. - "Comparative Fish Impingement at Two Adjacent Water Intakes on the Mid-Columbia River." In Proceedings, Fourth National Workshop on Entrainment and Impingement, ed. L. D. Jensen, Ecological Analysts, Melville, NY, 1977, pp. 257-266.

PNL. Environmental Monitoring at Hanford for 1986. PNL-6120, Pacific Northwest Laboratory, Richland, WA. National Technical Information Service, Springfield, VA, 1987.

PRICE, K.R. - Environmental Monitoring at Hanford for 1985. PNL-5817, Pacific Northwest Laboratory, Richland, WA. National Technical Information Service, Springfield, WA, 1986.

RICKARD, W.H.; HEDLUND, J.R.; Fitzner, R.E. - "Elk in the Shrub-Steppe Region of Washington: An Authentic Record." Science, 196, 1977, pp. 1009-1010.

RICKARD, W.H.; WATSON, D.G. - "Four Decades of Environmental Change and Their Influence Upon Native Wildlife and Fish on the Mid-Columbia River, Washington, USA." Environ. Conser., 12, 1985, pp. 241-248.

ROBERTSON, D.E.; FIX, J.J. - Association of Hanford Origin Radionuclides with Columbia River Sediment. BNWL-2305, Pacific Northwest Laboratory, Richland, WA. National Technical Information Service, Springfield, VA, 1977.

STEIGERS, W.D.; FLINDERS, J.T. - "Mortality and Movements of Mule Deer Fawns in Eastern Washington." J. of Wildlife Manage., 44, 1980, pp. 381-388.

SULA, M.J. - Radiological Survey of Exposed Shorelines and Islands of the Columbia River between Vernita and the Snake River Confluence. PNL-3127, Pacific Northwest Laboratory, Richland, WA. National Technical Information Service, Springfield, VA, 1980.

SWAN, G.A.; DAWLEY, E.M.; LEDGERWOOD, R.D.; NORMAN, W.T.; COBB, W.F.; HARTMAN, D.T. - Distribution and Relative Abundance of Deep-Water Redds for Spawning Fall Chinook Salmon at Selected Study Sites in the Hanford Reach of the Columbia River, Final Report. National Marine Fisheries Service, National Oceanic and Atmospheric Administration, Seattle, WA, 1988.

WPPSS - Supplemental Information on the Hanford Generating Project in Support of a 316(a) Demonstration. Washington Public Power Supply System, Richland, WA, 1978.

WSDOE - Washington State Water Quality Standards, Chapter 173-201. Washington State Department of Ecology, Olympia, WA, 1977.

PART II

Policies for pollution control

Chapter 9

WATER POLLUTION CONTROL: LEGAL, ECONOMIC AND INSTITUTIONAL ISSUES

M R Solanes (United Nations, New York, USA)

ABSTRACT

Pollution is an externality resulting from point and nonpoint sources. Its control includes regulatory measures and economic incentives. Regulations can consist of effluent standards, technological requirements, and standards of water quality for receiving bodies.

Economic incentives can consist of subsidies or charges.

All of the above can be combined through permit and charge systems.

Pollution control demands adequate planning and the granting of enforcing powers to implementing agencies.

Legal actions for water control include a variety of public and private measures in administrative and judicial fora.

Authority to act "ex officio", to issue injunctions, and to enforce abatement and cease and desist orders are crucial to pollution control.

Legal remedies include strict liability and accumulative fines. Procedural changes have liberalized the rules to confer standing to sue and the system of proofs, thereby fostering citizen participation and suits.

Yet, it is often argued that economic incentives and financial charges are more effective than regulations.

Institutional arrangements for the enforcement of pollution include national and local authorities. The former set policies, objectives, standards, and basic procedures. They also monitor compliance. Enforcement is predominantly the role of the latter. The river basin is often used as a geographical unit for water management and pollution control.

Finally, groundwater protection is of increasing concern. Protection measures include control of well injection, wellhead protection, underground tanks, and percolation of polluting substances.

Key words: pollution, externality, police power, regulation, charges, point and nonpoint sources, best practical technology, best available control technology, best conventional control technology, effluent standards, water quality standards, enforcement, strict liability, standing to sue, river basin.

1 - EXTERNALITIES

Water protection and conservation are highly correlated with the economic concept of externality. Externalities are those costs of productive activities that are not taken into account by the subjects making economic decisions. Moreover, they do not bear these costs, but transfer them to other persons or to society as a whole. Externalities can affect common, public or private goods. [Barkley and Sekler (1972)]

1.2 - Externalities and Water

Water is a flow resource. Therefore it is particularly vulnerable to externalities. In addition, the qualitative limitations of water supply have only recently been understood.

In the past few people understood that damage inflicted upon water was eventually an injury to their rights and interests. [Barkley and Sekler (1972)] Only highly motivated groups were aware of this fact and acted accordingly.

However, deterioration of water quality can affect private, public, and common goods and rights.

The externality is a transfer of the adverse effects of economic decisions to third parties, who have not been consulted in the decision-making process, and who do not benefit from the decision.

Legally defined, the behaviour of a polluter is an undue appropriation of some socially needed qualities of water.

1.3 - Externalities and Police Power

The State regulates social behaviour to prevent and to ameliorate damages to private, public, and common goods. The legal tool often used is the police power.

It can be exercised in preventive or reparative forms: ex-ante or ex-post. The preventive exercise of police power demands the adequate regulation of certain activities, to prevent the deterioration of common, public, or private goods. Reparative exercise implies acts and orders tending to halt activities injuring common goods (cease and desist orders). The police power is often invoked to control and curb water pollution.

1.4 - Water Quality and Water Pollution

In general water quality refers to the physical, chemical, biological, radiological, and other characteristics of water resources, affecting their usefulness for a particular use.

"Pollution" is the man-made damage or the man-induced alteration of the chemical, physical, biological, and radiological integrity of water.

The sources of pollution can be identified (point source pollution) or not (nonpoint source pollution).

Pollutants have been classified into eight basic categories: organic waste, infectious agents, plant nutrients, synthetic organic chemicals, inorganic chemicals, sediments, radioactive pollution and temperature increases. [Beck and Goplerud III (1988)]

1.5 - Pollution as an Externality: Controlling Mechanism

Two basic mechanisms can be used to curb pollution: direct regulations and economic incentives. Direct regulations can include: 1) setting of standards; 2) prescription of appropriate technologies; 3) determination of allowable concentration of pollutants at the source; 4) allowable concentration of pollutants of various types in the receiving body of water. [Palange and Zavala (1987)]

Economic incentives include: financial incentives to, and charges on, the polluter.

2 - REGULATORY MEASURES

2.1 - Permits and Effluent Regulations

In some systems the principal water pollution control mechanisms were quality control standards for the receiving body of water. They relied upon an assumed capacity of the water body to receive and assimilate a certain amount of pollutants and still be of the quality desired for certain uses. Generally, it was not required to reduce the pollutant content of preexisting discharges.

At present controlling mechanisms have evolved to require a permit and comply with standards of technology based effluent limitations, if they are point sources.

However, standards of quality for the receiving body are not abandoned. They continue and might be the base to require, should water quality fall below accepted levels, further reductions in the content of pollutants of any given source. [Beck and Goplerud III (1988)]

Emphasis is placed on the prohibition to add any pollutant from any point source. The definition is result oriented. It focuses on the consequences of an act, and not on the means by which the act is accomplished.

Therefore, the prohibition includes intentional, unintentional, and accidental addition of pollutants. Every type of discharge is subject to regulation. Therefore, "discharge of a pollutant" is any addition of pollutants to waters. Discharges are to be allowed by a permit. [Beck and Goplerud III (1988)]

The adoption of permit systems to control water quality and pollution is widespread. Thus a recent ECE report signals the requirement of permits in countries such as Canada, the Federal Republic of Germany, the Netherlands, and Spain. [ECE (1987)]

2.1.1 - Effluent Limitations and Guidelines

Permits contain standards and limitations which are the essence of a pollution control programme.

Technology based limitations provide standards that must be achieved before a pollutant is discharged.

The American system has defined three levels of technology that are applicable to existing classes or categories of point sources:

1) Best Practical Control Technology Currently Available (BPT);
2) Best Available Control Technology Economically Achievable (BAT); and
3) Best Conventional Control Technology (BCT).

Specific timetables and circumstances are set out for achieving these technological levels. The starting point is the identification of the industries to be regulated.

BPT

BPT is the initial technological level to be achieved by industry. All industrial sources, unless excused through variances, were required to reach this level of technology within a given term.

Factual criteria to be taken into consideration when assessing BPT include: total cost of application of technology in relation to the effluent reduction benefits to be achieved, age of equipment and facilities involved, the process employed; the engineering aspects; non water quality environmental impact and processed changes.

It is a technology which has a reasonable level of engineering and economic viability at the time of commencement of actual construction of the control facilities.

It is based upon the average of the best existing performance by plants of various sizes, ages, and unit processes within each industrial category.

BPT is not innovative technology.

Variances based on factors different from those considered when issuing the guidelines can be granted. However, economic ability to meet the costs of the treatments will not be considered.

BAT

BAT might require the development of a new technology for a particular industry. BAT is technology forcing.

The determination of BAT is based upon the degree of effluent reduction achievable by the best performance within a class or category, with a final goal of eliminating pollutant discharges.

Economic costs are accorded less weight in the BAT assessment than in the BPT assessment process.

BCT

BCT is an answer to the problems found in implementing BAT. The factors related to the assessment of BCT include: the consideration of the unreasonableness of the relationship between the costs of attaining a reduction in effluent and the effluents reduction benefits derived, and the comparison of the cost and level of reduction of such pollutants from a class or category of industrial sources.

Thus BCT is intended to provide some escape from the more onerous and costly requirements of BAT. [Beck and Coplerud III (1988)]

2.1.2 - New Source Performance Standards (NSPS)

New sources are those which have commenced construction after proposal of regulations prescribing a new source performance standards, for certain designated categories of sources.

NSPS must reflect the greatest degree of effluent reduction which the administration determines to be achievable through the application of the best available demonstrated control technology.

New sources are protected from the imposition of new standards for a ten-year period following completion of construction.

2.1.3 - Publicly Owned Treatment Works (POTW)

POTW are subject to effluent limitations (to be implemented within a given term) which are based upon secondary treatment as defined by the controlling authority or more stringent limitations necessary to meet water quality standards, treatment standards or schedules of compliance established pursuant to state law.

2.1.4 - Toxic pollutants

Toxic pollutants are subject to particularly stringent regulations. In the American system pollutants are listed and published. Hearings must be held before enactment of standards.

Standards are implemented through the permit programme. Permits are automatically revised each time the underlying standards are changed. Furthermore, special, more stringent effluent limitations can be imposed to dischargers where site specific water conditions demand custom-designed standards. At the same time, particular discharges could be adjusted to allow for the toxic pollutants already existing in intake waters.

Certain pollutants like aldrin, dieldrin or D.D.T. were to be eliminated.

Generally, toxic standard effluents shall be at a level which provides an ample margin of security and safety. Therefore, standards should guard against incompletely understood dangers.

Generally, limitations on the discharge of toxic pollutants do not require consideration of the economic and technical factors involved. [Beck and Goplerud III (1988)]

Some toxic substances can be subject to BAT effluent limitations, and plants are given a term for their compliance.

Variations allow for plant-wide limitations based on cost effectiveness, instead of outfall limitations. Qualified plants will have to reach a net reduction in traded pollutants.

An area of concern is the case of the discharges that, although in compliance with effluent limitations according to mandatory BAT, still do not attain the water quality standards required for the receiving body of water. These areas are to be subject to special strategies and controls, achieving the applicable standards no later than three years after the establishment of the strategies.

2.1.5 - Pretreatment Standards

In the USA industry discharges 25% of its waste into publicly owned treatment works. Therefore pretreatment standards can be required. These pretreatment standards should include discharges of all pollutants "that are not susceptible to treatment by the POTW, or which would interfere with the operation of such treatment works".

The standards must prevent the discharge of any pollutant that "interferes with, passes through, or otherwise is incompatible with such works".

Specifically, legal authority must enable the POTW to: 1) deny or condition discharges; 2) require compliance; 3) control the discharges into the POTW; 4) require a compliance schedule and the submission of all notices and self-monitoring reports by industrial users as necessary to assure and assess compliance; 5) carry out inspections, surveillance and monitoring, to determine compliance or non-compliance independent of the industrial user; and 6) obtain remedies for non-compliance including authority and procedures to immediately and effectively halt or prevent any discharge of pollutants, which reasonably appear to present an imminent endangerment of the health and welfare of persons, of the environment, or of the operations of the POTW.

POTW's must also demonstrate that they have enough financial and human resources to carry out the authorities and procedures. [Beck and Goplerud III (1988)]

2.1.6 - Obtaining a Permit

The permit issuer reviews the application, and if complete, prepares a draft permit and issues a public notice of its intent to issue the permit.

Public comments and hearings are provided for. After comments and testimonies have been considered, the permit may be issued for a limited term. It will contain technology requirements, provisions for bypasses and upsets and notice and reporting requirements.

Some permits will have a re-opener clause which allows for the imposition of any subsequently promulgated BAT limits for toxics which are more stringent than the conditions of the permit when issued. It also contains the monitoring requirements imposed on the operator. [Beck and Goplerud III (1988)]

2.2 - The Water Quality Approach

2.2.1 - Quality Standards

Effluent standards are complemented with water quality standards. In the past these standards intended to preserve the water quality of the receiving body according to expected, or existing water uses.

At present water quality standards are linked to environmental concerns, aimed, in some countries, to restore and maintain the chemical, physical and biological integrity of national waters. [Beck and Goplerud III (1988)]

2.2.2 - Anti-Degradation Policy

The Environmental Protection Agency (EPA) of the USA requires each state to develop an anti-degradation policy and to identify methods to implement the policy.

2.2.3 - Application and Use of the Surface Water Quality Standards

Total Maximum Daily Load

Technology based effluent limitations can be insufficient to implement the applicable water quality standard for some waters. Once those waters have been identified, they must be ranked taking into account the severity of the pollution and the uses to be made of the water and establish for those waters "the total maximum daily load" (TMDL) of the pollutants identified as suitable for such waters. The maximum daily load is not to exceed what is necessary to implement the water quality standard "with seasonal variations and a margin of safety which takes into account any lack of knowledge concerning the relationship between effluent limitations and water quality". Such load limitations should be allocated among contributors and discharge limitations incorporated into the permits. [Beck and Goplerud III (1988)]

Toxic Pollutants Control Strategy

Effluent limitations might not suffice to achieve desired water quality standards due to the presence of toxic pollutants.

Water bodies facing this situation must be identified determining the sources and amount of toxic pollutants being discharged. For each of these water segments, a control strategy shall be prepared with a view to achieve the applicable water standard within a given term.

2.2.4 - Drinking Water Quality Standards

Drinking water is normally subject to its own set of primary and secondary regulations.

Enforcement

Enforcement includes: monitoring, inspecting, record keeping, reporting and provision of drinking water in emergencies.

Compliance with standards can also be required through administrative orders or civil suits.

2.3 - Integration of Technological and Quality Approach

In the American system the technological approach is primary and the quality approach is secondary. However, the latter is to be applied when the former is insufficient to meet community needs and legal specifications.

Applicants for permits must provide a certification that compliance with regard to effluent limitation and standards, general effluent limitation, new source performance, and toxic pretreatment requirements will be met. The certification process requires public notice and discretionary public hearings.

In certain critical cases, the administration can establish water quality related effluents whenever technological limitation will not be sufficient to reach, or maintain desired water quality standards. However, there must be a reasonable relationship between the limitations and economic and social costs and benefits. [Beck and Goplerud (1988)]

2.4 - Economic Considerations

Economic considerations explain why pollution occurs, suggest alternatives for their management, and provide criteria to make cost-effective decisions to reduce or eliminate pollution. Three mechanisms can be used to this effect: regulations, subsidies, and taxes. The three can also be combined. [Palange and Zavala (1987)]

Critics of the regulatory approach point out that it is not as cost-effective as a charge system.

Thus, it has been affirmed that failure is inherent to the regulatory process, since regulatory agencies bargain on regulations, and courts seek "reasonable compromises" between public and private interests. Litigation makes the enforcement process long and drawn out and often inconclusive. In addition, regulation does not take into account that different polluters have different costs.

It is therefore argued that user and effluent charges are more useful than regulations in securing water quality objectives. Accordingly, charges based on the amount and kind of pollutants being discharged are a strong incentive to pollution control. [Freeman and Haveman (1973)]

In addition, properly designed charges provide continued incentives for reduction of waste loads and encourages improvement of technologies. [Knesse and Bower (1973)]

It has been suggested that the most appropriate system (in economic terms) is to establish water quality standards for receiving waters coupled with a charge system related to the volume and strength of effluents which adversely affect the quality of water. [Palange and Zavala (1987)]

Three types of charges have been tried in order to control water pollution: a) charges based on water quality objectives; b) charges based on financing of a pollution abatement programme; and c) charges in conjunction with effluent standards.

(a) Charges based on water quality objectives are intended to induce individual sources to take measures to reduce the quantities of pollution indicators.

(b)   Financing charges intend to allocate the costs of common treatment or programmes, according to a selected indicator.

(c)   Finally, charges in conjunction with effluent standards are imposed on effluents in excess of standards.

Effluent charges are used in the Federal Republic of Germany, where the Waste Water Charges Act requires payment for waste water discharges in proportion to their pollution load. In the Netherlands, the Pollution of Surface Water Act provides for a Water Pollution Levy on both direct and indirect discharges. [ECE (1987)]

The Water Law of Spain (1985) and its regulations (1986) also provide for a disposal charge inversely proportional to the quality of the discharged water. [ECE (1987)]

The privatization of the water basin authorities of the UK will provide increased opportunities for the application of the polluter pays principle. [ECE (1987)]

The assessment of the benefits of a pollution control programme are difficult, since many benefits are sometimes of difficult monetary measurement. However, a negative or neutral value in the cost/benefit analysis does not mean that a programme should not be sought. Yet, it might indicate that the benefits should be made explicit. [Palange and Zavala (1987)]

2.5 - Planning

Metropolitan areas and nonpoint pollution sources pose particular problems. Nonpoint sources include: agriculture, construction, mining, and salt-water intrusion. Their control is difficult. Therefore, special legal provisions aim at the development and implementation of management plans on a regional basis, with specific provision for the identification and control of nonpoint sources.

Management plans are part of a continuing planning process, integrating water quality standards and technology, basin planning, and waste treatment facility planning.

The areas with substantial water quality problems must be identified and a single organization (area-wide planning agency) must be responsible for each one area.

A plan must "contain alternatives for waste treatment management, and be applicable to all wastes generated within the area involved". In addition the plan must: a) identify waste treatment needs over a given period; b) designate construction priorities for treatment works; c) develop a regulatory programme to implement management requirements, facility construction, and pretreatment; d) identify the agencies necessary to construct, operate, and maintain facilities and carry out the plan; e) identify the necessary measures, time frame, costs, and impacts for, and of, carrying out the plan; f) develop a process to: (1) identify agricultural- and silvicultural-related, mine-related and construction-related nonpoint sources, and (2) set forth control procedures and methodes for each; g) develop a process to: (1) identify salt-water intrusion, and (2) set forth control procedures and methods; h) develop a process to control disposition of all area "residual waste" that could affect water quality; i) develop control processes for disposal of pollutants on land or in subsurface excavations to protect water quality; and j) confer enough legal authority upon enforcement agencies.

One of the principal substantive requirements is the identification and adoption of best management practices (BMPs) for the control of nonpoint source pollution, which has been defined as:

"Methods, measures or practices selected by an agency to meet its nonpoint source control needs. BMPs include but are not limited to structural and nonstructural controls and operation and maintenance procedures. BMPs can be applied before, during and after pollution producing activities to reduce or eliminate the introduction of pollutants into receiving waters." [Beck and Goplerud III (1988)]

The plan shall describe the regulatory and nonregulatory programmes, activities and Best Management Practices (BMPs) selected as the means to control nonpoint source pollution. Economic, institutional, and technical factors shall be considered in a continuing process of identifying control needs and evaluating and modifying the BMPs as necessary to achieve water quality goals.

2.6 - Groundwater

Concern for groundwater pollution has prompted governments to take special measures for their protection. Czechoslovakia has implemented testing, land use planning, and catchment protection programmes. A similar process is taking place in Finland.

The Federal Republic of Germany requires a certification of "no concern" before issuing groundwater permits. Legislation also provides for protection areas and protection from diffuse pollution. The latter includes regulations of pesticides and fertilizers and their adequate planning.

The Netherlands is currently planning a permit system for groundwater, and the United States has provisions for the protection of groundwater used for drinking water supplies. [ECE (1987); Beck and Goplerud III (1988)]

3 - LEGAL REMEDIES AGAINST WATER POLLUTION

Several legal actions have been used against pollution including nuisance, invasion of riparian rights, interference with prior appropriation rights, trespass, negligence and strict liability.

Development of administrative regulations and statutory law have reduced the resort to these actions. However, to the extent that some aspects of water pollution might not be regulated, remedies of common law, or of common civil legislation, still have a broad scope of application.

Available remedies included: injunction, self-help, or an administrative order to terminate the pollution. Damages to injured parties are also relevant legal elements. [Beck and Goplerud III (1988)]

3.1 - Compliance Monitoring

Controlling agencies can require the maintenance of records, the making of reports, and the use monitoring equipment and sample effluents.

Records must be available to the public, except for trade secrets. Agencies also have authority to enter and inspect the premises of any effluent source.

3.2 - Administrative Enforcement

Enforcement should provide swift and direct sanctions against polluters.

Public domain waters can be protected by several legal subjects and actions in law depending on the facts of each situation and on the legal status of the goods.

It is the general duty of the state to protect the public domain. Some countries have developed the doctrine of "public trust", based on the mission of state stewardship. More specifically, the state has the attributes derived from its condition of titular and possessor of the public domain.

In addition to this, for some cases there are punitive remedies.

The state can issue administrative injunctions of cease and desist with respect to those activities illegally affecting the public domain, general wellbeing, or the environment.

Options include: notice of violations, administrative compliance and extinction orders, civil actions, administrative penalties, civil penalties and criminal actions.

Administrative compliance orders are combined with a deadline for compliance. The "finding" of a violation is made ex parte. Different penalties are provided for, including accumulative daily fines.

The state can also resort to civil suits, seeking appropriate relief, including a permanent or temporary injunction for any violation.

The factors to be considered when setting the amount of a civil penalty include: "the seriousness of the violation; the economic benefit, if any, resulting from it; the history of past violations; good faith efforts to comply; and the economic impact of the penalty on the violator". [Beck and Goplerud III (1988)]

Criminal enforcement

Legislation imposes criminal liability for certain violations, of responsible corporate officers, when negligent or knowing violations happen. They include fines, accumulative daily fines, imprisonment or both. [Beck and Goplerud III (1988)]

Emergency powers

These powers authorize the filing of a suit to restrain polluting activities when they: "present an imminent and substantial endangerment to the health of persons". Some regulations authorize administrative actions and direct administrative enforcement.

3.3 - Citizen suits

These suits provide an alternate enforcement mechanism to supplement inertia in administrative enforcement.

The main issue related to citizen suits is standing to sue.

This is one of the most restrictive issues to private action. In the absence of private affected interests, procedural laws tend to restrict the judicial standing of individuals. Therefore, their actions are limited to administrative complaints, without the possibility to go to court.

Aware of the problem, some countries (mostly industrialized societies) allow a varied degree of flexibility in granting of standing to sue, and in recognizing substantive and procedural rights for the defense of environmental goods. [Anderson (1973)]

As a result, the requirements for standing to sue have been liberalized in a double sense: on the one hand, as regards the value to be protected; and on the other, concerning the particularization required of the injury. The values to be protected are not only economic but also aesthetic, recreational and conservational interests. The particularization required of the injury has also undergone certain changes although injury in fact must be shown and redress must be sought.

This change attempts to strengthen the role of the judiciary and of citizens in protecting common goods.

New interests are admitted as a basis for standing. Conservation, aesthetic, and recreational concerns are given such status as to become autonomous bases for legal actions.

Private persons are allowed to act on behalf of general public interests, and the right to litigate in defense of public environmental policies are granted to them.

The restrictive requirement of exclusive subjective rights is abandoned, and standing to sue is granted on a broader basis.

The process can be summed up by saying that the fact that some interests are shared by many and are not the property of a few does not make them less deserving of legal protection under the principle of due judicial process.

There have also been simultaneous, complementary developments that tend to reinforce the role of private persons in defense of the environment and its common goods. Those developments are reversal of the burden of proof, information rights, measures to assure proofs, expert testimony, etc.

In addition, some systems, like in the USA, provide for judicial review of the activities of administrative agencies. [Beck and Goplerud III (1988)]

3.4 - Liability Matters

In some cases, the owners, operators, or persons in charge of a facility from which pollution results are subject to personal liability.

Penalties are assessed after notice and opportunity for a hearing. These are "strict liability" clauses to be assessed regardless of the cause of the discharge or fault.

In cases where discharges result from wilful negligence or misconduct within the privity and knowledge of the owner, operator or person in charge, the penalties are substantially increased.

Discharges must be immediately notified to the appropriate authorities. Yet the information cannot be used against the notifier (to encourage prompt notification). The duty to inform falls upon the person in charge of the facilities from which the discharge results.

The responsible party is not only liable to civil penalties, but also for the costs of clean-up and associated activities incurred by the government.

The exemptions to liability are: 1) act of God; 2) act of war; 3) negligence on the part of the government; and 4) act or omission of a third party.

Yet, the owner or operator must prove that the discharge was caused, solely, by one of these factors. [Beck and Goplerud III (1988)]

4 - INSTITUTIONAL ARRANGEMENTS

Modern legislation is based on a comprehensive approach, whereby water quality is preserved and improved for all legitimate uses.

Under this system the administering agency is authorized to deal with pollution in all waters within its area of authority. It is empowered to determine limits of waste discharges, to enforce the abatement of existing pollution, and to regulate discharges through a system of permits.

In some American states control is vested in the health agency; in others, it is vested in a special independent agency; finally, others have created agencies within the health sector and charged them with water pollution control. [Beck and Goplerud III (1988)]

The trend to concentrate water protection measures in a single or overseeing national authority is confirmed by national experiences.

However, national authorities work in co-ordination with district, state, and basin organizations. [ECE (1987)]

4.1 - Alternative Arrangements

The institutions controlling water pollution cover a wide range of alternatives; they can be public and private, and appertain to different levels of Government. They can be functionally specialized (like health agencies) or have a wide mandate to monitor environmental quality including water.

They can have policy, planning, and enforcing capabilities or they can mainly deal with field implementation and control.

National level functions should include: 1) development and assessment of programme policy; 2) establishment of programme planning criteria; 3) development and enforcement of criteria for the setting of pollution sources; 4) establishment and enforcement of standards; 5) research; 6) demonstration programmes; 7) criteria and guidelines for the use and disposal of toxics; 8) establishment and maintenance of nation-wide monitoring programme; 9) establish monitor and enforce a permit system; 10) establish criteria for resource allocation, priority projects, and cost recovery; 11) provide assistance; 12) set up criteria for training; 13) collect and disseminate information; 14) co-ordinate with other government agencies; and 15) provide for citizen participation and advisory bodies. [Palange and Zavala (1987)]

State, provincial and district level activities must assure appropriate co-ordination at provincial or state level. They must keep a balance including securing funds, keeping updated information, ensuring a full range of measures for compliance, monitoring of water resources, and technical review and assistance.

Local organizations usually have disposal functions like operation and maintenance of the system; controlling the pollutants entering the system; establishing local standards and cost recovery measures and pricing policies; monitoring and surveilling facilities, effluents and water quality; management of storm flows and overflows; and co-ordination with other levels of government.

Functional organizations

Some countries have created functional organizations that deal with water supply, sanitation and disposal problems. These organizations have specific functional duties and cut across political and jurisdictional boundaries.

River Basin Organization

There has been a decided trend in the developed countries towards creation of a single authority to manage the water resources in an entire river basin.

The basin authority should be given sufficient powers to administer broad policies and enforce legislation. In federal countries the basin organization may also be formed by mutual agreement among the states or provinces lying within its boundaries.

One of the more successful organizations of this type has been that for the Ruhr District in Germany, organized in 1904 and known as the Ruhrverband. This entity has restored recreational water use within one of the most highly industrialized basins in the world.

Expenses for operating the system are allocated to each discharger - domestic and industrial - according to the effluent loads. For an industry the charge is based on the level of production, number of employees, waste flow volume, and character of wastes - all according to a complex formula. Similar considerations govern the charges for municipalities. The charges provide the funds for operating the treatment facilities, interceptors, pumping stations, and other related works. A network of sampling stations is also maintained to monitor the quality of the water and assure standards compliance for each section of the river in accordance with established uses. [Palange and Zavala (1987)]

In France, the Ministries of Agriculture and the Environment launched a pollution control action programme (1985). It is to be implemented in co-operation with the basin agencies. [ECE (1987)]

The public basin authorities of Spain have authority to grant permits and concessions concerning water resources. [ECE (1987)]

5 - CONCLUSIONS

5.1 - Evolution

Pollution control has evolved from case based, ex-post, reparative remedies of common and civil law, to comprehensive legal systems for pollution prevention and control. They are based on statutory legislation expressely aimed at protecting and improving water quality.

The modern approach is systemic, including source and nonsource pollution. A variety of legal, economic, planning, and institutional measures are used to protect surface and groundwater resources.

5.2 - Key legal tools

The key legal tools are standards, permits, and procedural rules expediting administrative control, monitoring, inspection, decision-making and enforcement.

The liberalization of the requirements to grant standing to sue, the enforcement of strict liability, the use of accumulative fines, and the procedural changes occurred in the rules on evidence and proof, are also important elements in present legal systems for pollution control. They have fostered citizen suits and participation, which have become important factors in supporting - and controlling - administrative compliance with - and enforcement of - laws and regulations.

In many systems, legal rules are applied in conjunction with financial charges, within a plan for pollution control and management.

5.3 - Institutional arrangements

Institutional arrangements are based on a tier system where national organizations set policies, basic criteria for standards, monitor compliance, and provide assistance and support to state, provincial, and local authorities. Direct enforcement and application of standards and measures are mainly a responsibility of the latter.

The river basin has proven to be an appropriate unit for pollution control.

Bibliography

ANDERSON, F. - "Nepa in the Courts", in A Legal Analysis of the National Environmental Policy Act, USA, Environmental Law Institute, Resources for the Future, 1973.

BARKLEY, P.; SEKLER D. - Economic Growth and Environmental Decay: The Solution Becomes the Problem, New York (USA), Harcourt Brace Jovanovich, Inc., 1972.

BECK, R.; GOPLERUD, C. - "Water Pollution and Water Quality: Legal Controls" in Water and Water Rights, Vol. III, Editor in Chief Clark, R., Charlotsville, Virginia (USA), The Michie Company, 1988.

Economic Commission for Europe (ECE) - Current Trends and Policies and Future Prospects regarding the Use of Water Resources and Water Pollution Control [ENVWA/R.2/add 1], Geneva (Switzerland), 1987.

PALANGE, R.; ZAVALA, A. - Water Pollution Control Guidelines for Project Planning and Financing, World Bank Technical Paper No. 73, Washington D.C. (USA), The International Bank for Reconstruction and Development, 1987.

Chapters in Books

FREEMAN, M.; HAVEMAN, R. - "Clean Rhetoric and Dirty Water", in Pollution, Resources and the Environment, edited by Enthoven, A. and Freeman, M., New York (USA), W.W. Morton and Company, Inc., 1973, pp. 122-137.

KNEESE, A.; Bower, B. - "The Delaware Estuary Study: Effluent Charges, Least Cost Treatment, and Efficiency", in Pollution, Resources and the Environment (see above), pp. 112-121.

Chapter 10

'SHERLOCK': AN INTEGRATED APPROACH TO POLLUTION MONITORING

C J Swinnerton, D J Palmer (Wessex Rivers, Bridgwater, UK),
P N Williams (Wessex Rivers, Bath, UK)

ABSTRACT

This paper describes the development of a new monitoring system designed to detect pollution incidents and changes in river water quality.

When, for example there is an increase in ammonia concentration or a decrease in dissolved oxygen, several things automatically happen.

Firstly, a signal is sent to alert the pollution control scientist that something is amiss. The system can then be interrogated by telephone and any results examined using a portable computer.

Secondly, a sampling machine can be automatically triggered when a predetermined level has been reached. This could be when the dissolved oxygen level falls below 50% saturation or when the ammonia concentration rises above 2 mg/l for example. It is possible to set such threshold values for any of the other parameters measurable by the system. These samples can then be further examined in the laboratory.

Finally, a radio signal will alert other sampling units strategically placed upstream or downstream of the discharge, at distances of up to 4 km, automatically starting those machines as well, so that data can be obtained at these locations in an integrated manner.

Key words: pollution, monitoring, rivers.

1 - INTRODUCTION

1.1 -

At present, the responsibility for pollution control in England and Wales rests with the Water Authorities but when Water Supply, Sewerage and Sewage Treatment are privatised, likely to be September 1989, Pollution Control will pass to the newly formed National Rivers Authority (NRA). In preparation for the change each Water Authority has had to establish a shadow NRA unit by the 1st April 1989.

1.2 -

This paper describes the development of a new monitoring system designed to detect pollution incidents and changes in river water quality.

1.3 -

To put the work in context a brief description of Wessex Rivers and the structure that has been adopted to carry out the various responsibilities is presented.

2 - WESSEX RIVERS

2.1 -

The Wessex Region covers an area of almost 10,000 km^2 and comprises the counties of Avon, Dorset, Somerset, most of Wiltshire and small parts of Devon, Gloucestershire and Hampshire.

2.2 -

Along with the other shadow NRA units Wessex Rivers is responsible for:-

(i)   Pollution control and the protection of the Water Environment,
(ii)   Water resources
(iii) Fisheries, Recreation and Conservation
(iv)   Flood Defences

2.3 -

The structure of the organisation is:

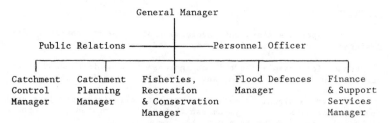

2.4 -

The Catchment Control Manager is responsible for Pollution Control and monitoring water quality and water resources. This involves monitoring compliance with discharge consents and abstraction licences and the biological quality of rivers.

2.5 -

The development of Sherlock has been carried out by staff within the Catchment Control function.

3 - THE NEED

3.1 -

In general, pollution control scientists respond to pollutions by reacting to visual signs such as dead fish or a discolouration of the water. To detect when these events have occurred reliance is placed upon someone noticing the effects of the pollution and reporting it. Obviously there is a time lag inherent in this system and frequently by the time an investigation is underway the discharge causing the problem has stopped. This makes it difficult to trace the source of the pollution which can by this time be a considerable distance from the observed effect.

3.2 -

To be successful, remedial measures often rely upon quick action being taken at the time the pollution is occurring. A quick response can lead to more efficient preventative measures being taken to stop a discharge and prevent considerable amounts of later work to clear up the effects of pollution.

4 - DEVELOPMENT OF THE MONITORING SYSTEM

4.1 -

The monitoring system was developed in a number of stages over a period of approximately two years. This allowed for a degree of evolution in the development in addition to making the best use of new developments, especially in the field of communications, as they became available.

4.2 -

The first part of this development was to input data from field instruments that were already in use to a data logger and develop the interface and software packages to enable the data to be downloaded directly to a microcomputer. The data loggers used are Squirrel 8 and 12 bit instruments. The 8 bit logger has an 8k memory with 4 channels and similarly the 12 bit logger has 42k memory with up to 13 channels and comes with built-in alarms. Both have a variable logging frequency. Information can be downloaded to either an office-based microcomputer or a portable microcomputer carried in a car. Graphical interpretation of the recorded data can therefore be carried out equally well in the field or in the office.

Dissolved oxygen, ammonia, pH, electrical conductivity, turbidity and temperature were amongst the first determinands to be included relating to the quality of the water. Later it was realised that flow or changes in flow also provided useful additional information.

Flow measurement is carried out using an ultrasonic instrument and can be used in one of two ways. If a flume is available at the measuring site then the characteristics of the flume and the measured current depth can be programmed into the Ranger and the instrument will calculate the depth and rate of flow, which can then be logged. If no flume is available and the watercourse is too large to facilitate the temporary installation of a flume then the instrument can be set up to measure any rise and fall in the level of the watercourse by measuring the distance between the ultrasonic head and the water level.

The system was further expanded to include river flow, rainfall and flow or no flow from piped discharges. Additional battery power is supplied to operate the instruments over the required time period.

Information from any instrument which delivers a millivolt signal can be captured by the data logger for later examination.

4.3 -

The next phase of development was to use the information gathered by any of these field instruments to activate an automatic sampling machine. Should the flow or chemical information exceed pre-set limits, used as a high or low level alarm, then a trip amp monitoring the millivolt output from any of the instruments sends a signal to start the sampling machine. The automatic sampling machine can be pre-programmed to take samples over various pre-determined time intervals or take composite samples on a time average basis. Samples taken in this manner are submitted for laboratory analysis to supplement the information captured on the data logger. When the selected channel or channels drop out of alarm, the sampling machine can continue to sample or can be switched off until the next alarm occurs.

4.4 -

Having obtained information and samples at a particular location very soon it was realised that more comprehensive data could be obtained if many

units were interlinked at different locations. This can provide useful
information both upstream and downstream of the discharge being monitored,
so enabling an accurate assessment of the effects of the discharge.
Similarly if interlinked, units spaced out along a river system can generate
survey data without the intensive use of manpower. The communication link
between units is achieved by the use of a command radio sending signals to
receiving radios attached to sampling machines at the other sites. These
sites can be up to 4 km apart.

4.5 -

The ability to interrogate the system from a car, the office or at home
without having to visit the site gives further flexibility. The need to
know when something unusual has happened at a river site also requires an
alarm system if we are to act quickly to take any remedial action or stop a
discharge from continuing. The use of cellular telephones connected to
the data loggers via modems and modem plates allows the transfer of data
from the monitoring site to any other location, such as a control centre,
office, car or home, with the aid of a computer modem and telephone line.
Two modems are needed for the remote installation, one for the cellular
telephone system and the other for land-line use. Both modems are
intelligent and error checking. To notify the occurrence of a pollution or
alarm situation the same cellular telephone system is used. When the trip
amp monitoring the millivolt signal from the field instruments receives an
alarm signal a relay activates a dial on the alarm mechanism, triggering the
cellular telephone to dial a pre-programmed telephone number. This can also
activate a radio-pager carried by pollution control staff.

The system is shown schematically in fig 1.

4.6 -

Once the system is set up on the river bank, or at a discharge, it can
be used to monitor the normal or background quality of the water. When a
polluting material enters the watercourse and is identified by any one of
the field instruments then the following happens:-

i)   the sampling machine automatically takes samples for laboratory analysis

ii)   a radio signal alerts similar units deployed up to 4 km away and
       associated samples will be taken, for example, upstream and downstream.

iii)   "Sherlock" telephones a pre-programmed number to alert a pollution
       control officer that there is a problem. They can then interrogate the
       control data logger which stores all the information, download the data
       onto a portable computer and assess the gravity of the problem, aided
       by suitable computer programmes to interpret the data and present it in
       a graphical form.

4.7 -

These developments have led to a flexible, mobile monitoring system for
use in the field. Because all of the electrical power is provided by
battery it can be deployed wherever the need for such monitoring has been
identified. The system can also be used to provide survey information in a
more flexible manner than stations at fixed locations - problems rarely
occur where they are expected. Sherlock can also provide more detailed
information than can be gained from spot sampling.

5 - TYPICAL EXAMPLES OF SHERLOCK'S USE

5.1 -

Because of the modular approach, the flexibility of the system allows
for only those instruments of interest to be installed on site. The system
can be adapted to suit particular problems that might be anticipated or a
wide range of probes used, such as ion selective electrodes, to cover a wide
range of situations.

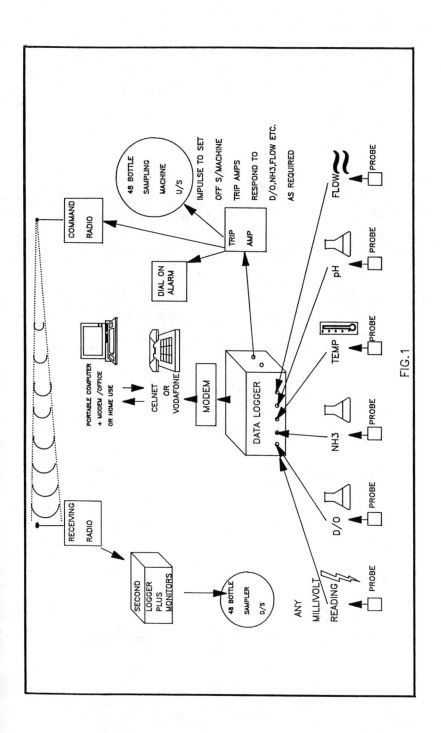

FIG.1

In the last 12 months this monitoring system has been used on a diverse range of problems. Some of these are briefly described below:-

(1) The river Avon, flowing through Hampshire has many commercial fish farms rearing trout predominantly for the table. One particular fish farm, one of the largest in Europe, was thought to have an effect on water quality detrimental to the fishery downstream. Monitoring of the water-course confirmed that there was no significant increase in the ammonia concentration but a significant decrease and diurnal variation in dissolved oxygen was confirmed. The fish farm has now installed an aeration system to overcome these problems.

5.2 -

Wessex is a predominantly intensive agricultural area. This gives rise to many problems associated with the disposal of farm effluents. Pollution incidents from silage liquor and farm slurry are a major problem in this area, not least because they exhibit a very high biochemical oxygen demand on the water causing serious depletion of the dissolved oxygen concentration.

By monitoring the depletion in dissolved oxygen and the increase in ammonia, Sherlock has been particularly successful in alerting pollution control officers to these problems. Two examples are outlined below:-

1 A known problem on the Erlestoke stream in Wiltshire remained unresolved over a long period of time. Monitors were set up at likely locations where pollutions could occur. 1 hour after setting up the equipment and leaving the site a pollution was detected. The decrease in dissolved oxygen and the increase in ammonia concentrations are shown in Fig. 2. When the farmer was confronted with this evidence he admitted the problem and has taken action to prevent a recurrence.

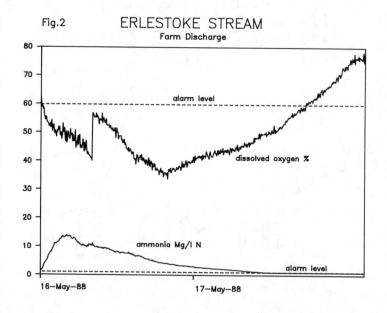

Fig.2   ERLESTOKE STREAM
Farm Discharge

2 A large farm on the Wiltshire Downs prepared their own bulk liquid fertilizer, buying in the necessary chemicals, mixing them and storing the concentrated liquid in tanks. The farm was suspected of being responsible for pollution of the Worton Stream, causing a prolific growth of sewage fungus. After monitoring for approximately a week, following heavy rainfall there was a depletion of dissolved oxygen in the watercourse. This alarm condition activated the sampling machine.

Fig.3

WORTON STREAM
URCHFONT BOTTOM

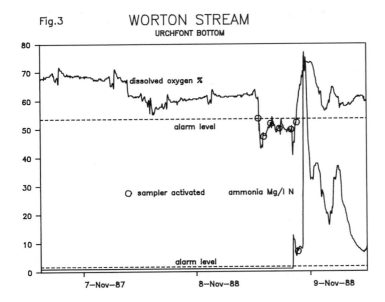

Seven hours later the ammonia concentration in the watercourse increased from less than 1 mg NH_3-Nl^{-1} to in excess of 70 mg $N_{H3}-Nl^{-1}$ due to fertilizer being washed from yards and gullies. The results are shown in Fig. 3.

5.3 -

Since Sherlock has been developed Wessex has carried out more than 40 investigations using this equipment. (Similar examples can be illustrated from a wide range of activities). This system has been particularly useful in monitoring discharges from sewage treatment works and their intermittent storm overflows. The flexibility and modular system has enabled many types of industrial discharge to be examined. This has included the effect on the watercourse of varying pH and soluble iron from a sand extraction; the efficiency of industrial pretreatment plants in meeting compliance with their discharge consent; and in one case, illustrating how changes in the working pattern through the week at a company affected the operation of an industrial treatment plant and the concomitant effect on the watercourse.

6 -

Sherlock is a monitoring system which is flexible, completely self contained and easily transportable by car or van. It does not rely on external power sources allowing it to be operated in remote locations. The ability to alert pollution control staff to problems enables the cause of the problem to be dealt with much more quickly than if we respond to the effects caused by the pollutant. This will represent an important advance in the prompt and effective investigation of pollution incidents by environmental protection organisations.

Chapter 11

THE UK APPROACH TO CATCHMENT-WIDE POLLUTION MANAGEMENT

D Fiddes (Water Research Centre, Swindon, UK)

ABSTRACT

The organisation of the UK Water Industry into large regional authorities
with responsibility for the total water cycle over complete river basins has
resulted in the development of approaches to pollution management which are
very cost-effective but are sometimes represented as being in conflict with
approaches evolving elsewhere in Europe. The paper describes the evolution
of the methodology and its associated technology, together with the benefits
resulting. It is suggested that there are significant benefits to both the
utility and regulatory authorities through following this integrated·
approach. There is therefore no conflict of interest.

Key words: pollution control, river basin management, environmental
quality, sewerage, receiving water, capital works.

INTRODUCTION

In the UK, responsibility for water supply, waste water and rivers has
been with large regional authorities for about 15 years. This fortunate
position has facilitated the development of integrated, cost effective,
pollution management techniques. The UK approach has sometimes been
represented as being in conflict with approaches evolving elsewhere in
Europe. The paper describes the development of the UK methodology and shows
that there are significant benefits to both the utility and the regulatory
authorities, avoiding any conflict of interest.

HISTORICAL BACKGROUND

Many pollution problems can be traced back to the Industrial Revolution
and the associated growth in urban populations. This transformed many
previously clean and healthy rivers into open sewers. The public disquiet
resulting caused a number of Royal Commissions to be set up in the 1860s
which resulted in the 1875 Public Health Act and the 1876 River Pollution
Prevention Act. These controlled discharges to and from sewers and direct
discharges to rivers. Unfortunately enforcement was not rigorous due to the
fragmentation of the responsible local sanitary authorities and a weakening
amendment making damage to industrial interests a defence.

It is interesting that the Royal Commission of 1865 had recognised
that:

"... it was essential for the whole watershed of the river to be
placed under the supervision of a single authority irrespective
of local government boundaries ..."

It was over 100 years before this was achieved when in 1974 the current 10 Regional Water Authorities took over the responsibilities for water, waste water and pollution matters from the previous 29 River Authorities, 1,300 local councils and 198 water supply undertakings. There are proposals at present to privatise the utility function of water supply and wastewater treatment but pollution control will stay in the public sector in a single National Rivers Authority.

CURRENT ENVIRONMENTAL SITUATION IN ENGLAND AND WALES

Water quality surveys are undertaken at 5 yearly intervals. River and estuary reaches are put into one of four classes depending on the ability to support fish life and the water's suitability for treatment for potable supply or agricultural use. The results of the most recent survey are shown in Table 1.

Table 1 – summarised results of 1985 river quality survey

WATER AUTHORITY	CLASSIFICATION OF RIVER REACHES			CLASSIFICATION OF ESTUARY REACHES		
	GOOD/FAIR (%)	POOR/BAD (%)	TOTAL (km)	GOOD/FAIR (%)	POOR/BAD (%)	TOTAL (km)
Anglian	91	9	4,328	98	2	547
Northumbrian	97	3	2,784	70	30	135
North West	78	22	5,323	72	28	451
Severn-Trent	87	14	5,150	100	0	61
Southern	97	3	1,992	97	3	380
South West	94	6	2,941	100	0	355
Thames	93	7	3,546	100	0	138
Welsh	94	6	4,600	98	2	44
Wessex	95	7	2,467	97	3	145
Yorkshire	88	12	5,767	74	26	70
England and Wales	90	10	38,896	92	8	2,730

From the table it will be seen that 10% of river reaches are in the poor or bad classes. That is, they are not good enough to support viable fisheries or to be treatable for drinking water supply economically. Similarly 8% of estuary reaches fall into the poor or bad classes.

These results, although far from ideal, compare favourably with other European countries as will be seen in Table 2 where published results are interpreted using the UK criteria.

In many cases it is not possible to give a single cause for failure but the main contributory factors are inadequately treated sewage treatment works effluents, combined sewer overflow discharges, industrial discharges and farm runoff. The latter is a particular concern as it is a growing problem resulting from the increase in intensive livestock farming. In coastal areas a major concern is the quality of bathing waters. There are about 350 bathing beaches around the UK which are required to comply with the EC Bathing Water Directive. At present only about two-thirds do.

APPROACHES TO POLLUTION CONTROL

There is a general acceptance of the desirability of protecting the aquatic environment from anthropogenic contaminants which either have a detrimental effect on the ecosystem or inhibit the desired use for the water body. There is also agreement that control should be by consenting discharges to the sewerage, river or marine system. Any divergence of opinion comes in the approach to setting and monitoring of consents. Thus the general concepts and ultimate goals are not contentious, only the means of moving towards them.

Table 2 – River quality in Europe

| | PERCENTAGE OF RIVER LENGTHS IN UK QUALITY CLASSES | | | |
	UNPOLLUTED	SATISFACTORY	POOR	GROSSLY POLLUTED
UK:				
England and Wales	67	24	9	2
Scotland	95	4	0.5	0.5
Northern Ireland	84	11	5	0
CONTINENTAL EUROPE:				
Belgium	56	17	16	11
West Germany	45	40	14	1
Luxembourg	72	13	11	4
Netherlands	77	18	4	1
EC Overall	39	35	22	5

Fixed emission standards can be set. For toxic substances this is a very attractive approach with a goal of moving towards zero discharge. However, consideration has also to be given to any alternative risks. For example limiting the discharge of List 1 substances to protect sensitive species of fresh or marine aquatic life requires an alternative "safe" disposal route to land. On balance there should be no net loss of environmental benefit. Fixed emission standards cannot be applied to diffuse sources such as agricultural runoff and may be inappropriate for point-source non-conservative pollutants such as organic loadings in sewage treatment works effluents or storm sewage overflows.

It can be argued that the method of pollution control should be specified rather than performance standards. If the aim is to minimise pollutant loadings, the Best Available Technology (BAT) could be specified. Frequently this will be acceptable to both the polluter and the regulating agency as it simplifies the monitoring requirements. Exceptions will be where either the cost to be incurred by the discharger is considered to be excessive or, despite BAT, the water quality objectives of the receiving water may not be achievable.

An alternative approach is to define "fitness for use" standards (EQSs) for the receiving water body and consent discharges which will allow the water quality objectives (EQOs) to be achieved. This is the approach currently favoured in the UK. The EQO/EQS approach will give the most cost effective upgrading solutions. However, there is a danger that advantage is not taken of readily available pollution reduction options because the quality standard is easily met.

The ideal pollution control policy will be a combination of the EQO/EQS approach with selective use of either fixed emission standards or BAT.

STEPS TOWARDS INTEGRATED POLLUTION MANAGEMENT

The bulk of the Water Industry's assets are its underground pipes – the sewers and water mains. Upgrading the performance of these also accounts for the largest part of the capital expenditure. Cost effective pollution management requires a level of understanding of system performance which has been unavailable in the past. The solution to this dilemma is to build pollution management studies onto asset management studies. This is the approach being followed in the UK.

In the early 1970s, following the formation of the Regional Water Authorities, there was great concern about the condition of the sewerage network. Parts were getting quite old and a series of expensive collapses raised fears that a major programme of reconstruction was going to be required in the near future. Following an extensive programme of research at WRc a procedure was agreed for comprehensive sewerage investigation and set out in the Sewerage Rehabilitation Manual (SRM).

Experience using the SRM has shown that:

o a large proportion of current Water Authority capital and revenue expenditure is spent on dealing with sewerage structural failure, flooding and pollution problems.

o the bulk of such expenditure is on a few large schemes in the core of systems which have wide impact and hence require system-wide studies to generate solutions.

o flooding, structural and water quality problems interact so intimately that cost effective solutions can only be identified following in-depth investigations and a detailed appreciation of system performance.

o survey, analysis and planning costs will go up but are more than offset by capital cost savings and there are further valuable spinoffs in improved measurement of levels of service and better planning.

These findings were proved in a major study of sewerage problems in the north west of England[2].

Comprehensive studies were undertaken in 20% of the North West Water's drainage basins producing outline plans for £325 million of capital works and confirming the need for nearly £2,000 million of sewerage rehabilitation across the region. The studies are now being expanded to cover the required sewage treatment and sea outfall works. The total programme is estimated to require a total expenditure in the order of £3,000 million over the next 20 years.

The concept of undertaking Drainage Area Studies using the SRM procedures in order to produce planned sewerage rehabilitation programmes, is now well established in the UK. The comprehensive system and performance data once obtained can have many uses:

o effective planning of the capital works programme;
o quantification of future funding needs;
o demonstration of levels of service being achieved.

These are all useful for the efficient running of a public utility. If, as is proposed, the UK Water Authorities are privatised, the information contained in what has become known as the Asset Management Plan (AMP) will also be available to satisfy the Government regulating body and the shareholders that the underground assets are being managed efficiently and in the public interest.

To obtain the same benefits in the environmental area, it would be necessary to produce a complementary Catchment Management Plan (CMP) built around hydro-dynamic and water quality models of the receiving waters. Such an approach is set out in a rather idealised way in Figure 1.

Four phases are involved. In the first phase, the region is divided up into catchments and the Uses defined for the individual river reaches and other water bodies. Appropriate standards are defined for these Uses and the current performance monitored against these standards to identify where adequate performance is not being achieved.

In the second phase, flow and water quality computer models are built for the urban drainage, river and marine systems. The urban drainage system will discharge pollutant loads into the river. These will be both continuous, from treatment works, and intermittent discharges of storm sewage. The river model is used to check on the impact of the urban drainage and agricultural discharges and to quantify the pollutant loads passing out to sea. The marine model can then be used to check on the impact on marine Uses, such as bathing and shellfisheries, of the river flows and direct sewage discharges from long sea outfalls and storm overflows.

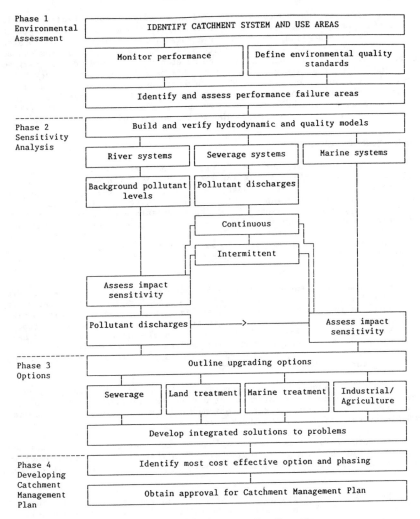

Figure 1 – Catchment management plan flow diagram

In the third phase, the various options that are open to solve the problems identified can be tested, again using the computer models. These can take account of such things as increase in water consumption and changes in population for example. The sort of options available are:

o improving the sewerage network;
o upgrading the treatment works;
o building new long sea outfalls; or
o requiring on-site pre-treatment by industrialists and farmers.

The important point to note is that we are looking for integrated solutions that will ensure the achievement of the required Environmental Quality Standards for the foreseeable future.

The final phase is producing the plan, including details of how the works will be phased.

Figure 1 as noted is a rather idealised plan. In practice, even in the best run Authorities, all of the required information will not be available right at the start.

So initially the procedures are applied to only a representative sample of systems and the results from these aggregated and extrapolated to get an estimate of the Regional costs. It is assumed that the plan will be updated every 5 years or so, when the results of any further investigations can be considered. This updating will also be necessary when standards are revised – say as a result of a new EC Directive. This sampling approach is now being applied by a number of Regional Water Authorities. Typically, the region is divided up into catchments and all available data brought together so that the catchments can be put into groups likely to have broadly similar problems. About 15 catchments are then selected for full detailed study.

DEVELOPMENT OF TECHNOLOGY

The strategy set out in Figure 1 requires, for success, reliable hydrodynamic and quality models for the sewer, river and marine systems and appropriate environmental standards to test the models against. Many of these are already developed but it has been necessary to set up a major research programme to ensure all of the components are available and to an adequate level of accuracy. The research programme has been described elsewhere[3]. Here, it is only necessary to highlight the recent developments in modelling simultaneously the impact of continuous and intermittent discharges from sewerage systems on the marine and river receiving waters.

Many two and three dimensional hydrodynamic/dispersion models are available for simulating the impact of discharges from long sea outfalls. On many beaches the cause of failure is at least in part the bacterial loading coming from the intermittent discharges from Combined Sewer Overflows (CSOs) and Surface Water Outfalls (SWOs). In recent applications these models have also been used in conjunction with sewer flow and quality models and rainfall-time-series to prove the effectiveness of proposed coastal sewerage upgrading works on bathing water compliance.

A similar approach has been adopted for river pollution where again there are continuous and intermittent discharges. In this case the emphasis has been on developing a river impact model which can simulate the chemical and biological effects of the shock loadings.

In both cases it has been possible to show that substantially cheaper solutions can be designed that have a higher level of confidence in delivering the required levels of service than has been possible using traditional methods.

The research programme is still continuing so that the interim procedures currently available can be refined. The comprehensive modelling capability is expected to be available within two years.

CONCLUSIONS

(a) Effective pollution control requires a detailed basin-wide appreciation of the impacts of current pollutant loadings on the total receiving water system and the implications for the associated sewerage systems of any proposed modifications to discharge consents.

(b) The sewerage aspects are adequately covered by procedures and methodologies developed for Drainage Area Studies as part of the preparation of Asset Management Plans.

(c) Interim procedures are available to prepare, in association with Asset Management Plans, Catchment Management Plans for the receiving waters. Research currently in hand will allow these to be enhanced, incorporating comprehensive river and marine impact models within two years.

(d) The integrated approach described provides greater confidence that environmental quality standards will be achieved than rigid adherence to uniform emission standards or Best Available Technology and will show substantial savings in capital cost of the upgrading works.

REFERENCES

(1) WRc/WAA. Sewerage Rehabilitation Manual, Second Edition. WRc. 1986.

(2) BARNWELL F and FIDDES D. The North West Water Sewerage Rehabilitation
     Review. Journal Institution of Water and Environmental Management,
     Volume 2, No. 2, pp124-134. April 1988.

(3) CLIFFORDE I T, SAUL A J and TYSON J M. Urban pollution of rivers - The
     UK Water Industry Research Programme. Proc International Conference
     Water Quality Modelling in the Inland Natural Environment, Bournemouth,
     pp485-491, BHRA. June 1986.

Chapter 12

REGULATIONS AND RESPONSIBILITIES IN WATER POLLUTION CONTROL IN THE FEDERAL REPUBLIC OF GERMANY

H-P Lühr (IWS, Fed. Rep. Germany)

ABSTRACT

The philosophy of water management in the Federal Republic of Germany is described. It covers the principles of the environmental policy, the central laws and regulations related to water as well as the monitoring principles.

Key words: Water Act of the FRG; principles of environmental policy; effluent standards; quality standards; monitoring programs.

1   INTRODUCTION

In recent decades - and especially since the beginning of the 70th - many waters of the Federal Republic of Germany have been subjected to increasing strains as a result of hazardous pollutants and waste heat. The first and initially the most important counter measures for the protection of waters are concentrated on eliminating the worst pollution, particularly that caused by easily degradable substances, which provide an immediate danger to aquatic life due to the drain they make on oxygen. Now it is more essential than ever to combat also other kinds of water pollution, caused by different hazardous substances, particularly those which are harder to degrade and reveal long-term effects, such as heavy metals and organic micro-impurities, including chloro-organic compounds, some of which have carcinogenic effects. To deal with this range of problems, especially in the last 10 years the existing regulations have been extended and new ones have been set up.

Water legislation in the Federal Republic of Germany is aimed at:

- preserving or restoring the ecological balance of waters;
- preventing avoidable impairments of water quality;
- safeguarding the supply of drinking and process water of adequate quantity and quality and
- ensuring that all uses of water that are in the public interest will remain possible.

In 1971 the Government of the Federal Republic of Germany published an environmental program which included the following three principles of its environmental policy:

The principle of precaution, the polluter-pays-principle, the cooperation-principle.

The principle of precaution means: environmental policy is not limited merely to resisting menacing dangers and eliminating damage that has already occured. The balance of nature as a whole must be protected and any claims made upon it should be as sparing as possible in order to maintain the ability of such systems to function over a long-term period. The principle of precaution imposes particular requirements on all levels of planning, and all plan-

131

ning decisions should take into account their possible effects upon the environment. All those with a social and governmental responsibility are obliged to act so as to protect the environment. Moreover, relief in one area should not lead to added pressure in other areas.

According to the polluter-pays-principle the costs of avoiding, eliminating or compensating for environmental pressures are to be borne by those causing them. The polluter-pays-principle represents the application of free-market principles to environmental problems.

The cooperation-principle implies that the shaping and implementation of any environmental policy must be based on the common responsibility of all those involved.

Priority is given to the reduction of the dangerous pollutants listed in the supranational provisions and international conventions. The following measures are of prime importance to reach this aim:

1. Prevention of the use of dangerous substances or their substitution by environmentally compatible substances;
2. modification of manufacturing processes (low-waste or no-waste technologies) and
3. introduction of new waste water treatment methods and improved operation.

2   ORGANIZATION OF WATER MANAGEMENT

To understand the relatively complicated water protection-organization in the Federal Republic of Germany it is essential to know that pursuant to Article 75 para. 4 of the Basic Law (Grundgesetz) the Federation has only the right to enact general provisions concerning water management. The remaining legislative competence concerning water management in particular as regards fulfilment of the framework specified by the Federation and organization of execution, is in the hands of the Federal states.

In this way the Federal states and municipal authorities competent for water management and water rights and whose competence varies from one Federal state to the other attend to the licensing procedure in accordance with the Federal state Water Laws and issue the appropriate notices on water rights.

The Federal state authorities are also responsible for fulfilment of obligations and ensuring that they are complied with. Several regional water boards have, in accordance with the Federal state water laws, been assigned competence governed in the various Water Board Laws. They are of considerable significance, for example the Ruhr authority.

At national level the "Federal Water Act" enacted in 1960 is the central law. Waters pursuant to Article 1 Federal Water Act are:
- surface waters, coastal waters and groundwater -.

The areas of the High seas of the Baltic Sea and the North Sea which are outside of the national territory are not covered by these classic water laws. Their protection is mostly regulated by international conventions. The fulfilment of obligations of these conventions are in the responsibility of the Federation.

Measures in the fields of air pollution, waste management or chemicals, which are also partly very important for water protection, are regulated by federal laws. The execution of these laws, however, is also in these fields largely in the hands of the Federal states authorities.

2.1   National Regulations

The most important regulations within the field of water protection are the following three laws:
- Federal Water Act in the version of 1986,
- Waste Water Charges Act in the version of 1986,
- Detergent Act in the version of 1986.

The 1986 version of the Federal Water Act, together with the water legislation of the various Federal states, covers the entire field of water management. It imposes the following basic regulations on the discharge of waste water into waters:
- any use of water, so any release of waste water into waters is subject to the approval of an appropriate authority;
- permission to release waste water may only be issued if certain minimum

requirements are fulfilled, which comply with the best available technology, for dangerous substances. For all other substances, the minimum requirements have to comply with the generally acknowledged rules of technology. This means, for example, that municipal sewage must be purified in a fully biological way after the elimination of nitrates and phosphorus;

- management plans are intended to regulate the various uses of waters. Priority is given to obtaining drinking water. Particularly strict requirements can be imposed on the treatment of sewage and waste water.

The 1986 Waste Water Charges Act concerns the following basic regulations:

- a levy is payable on the release of waste water into waters, and this is assessed according to the hazardous nature of the waste water. It applies regardless of whether the receiving waters are of good or poor quality;
- the levy is used for cleaning up the receiving waters.

It exerts an economic stimulus for the construction of more water purificating plants, for improving the available sewage treatment, for the development of manufacturing processes producing little or no waste water and for the introduction of such processes as well as a reduction in the amount of products that can only be manufactured with large resulting quantities of waste water, and for the more limited use of such products.

On the basis of the Detergent Act of 1986 two Administrative Regulations have been enacted, setting minimum requirements for the degradability of anionic and non-ionic tensids (at least 80%) and the concentration of phosphate in washing and cleansing agents.

2.2   EEC - Regulations

When the Federal Government started with its environmental program of 1971, the European Economic Communities began also a large action-program concerning environmental protection by a corresponding declaration of the council in 1973. EEC-regulations concerning environmental protection are enacted in form of directives pursuant to Article 100 and Article 235 of the Agreement on the foundation of the European Economic Community. National legislation is being increasingly supplemented by the lawgiving activities of the European Communities. The main EEC-guideline "On Pollution Caused by Certain Dangerous Substances into the Aquatic Environment to the Community" of May 4th, 1976 contains two lists of substances, for which several directives are enacted. This applies to Mercury, Cadmium, Hexachlorocyclohexane, Carbon tetrachloride, DDT, Hexachlorobenzene, Aldrin, Dieldrin, Endrin, Isodrin, Hexachlorobutadiene, Chloroform.
Directives for other substances are in preparation.

2.3   International River Basin Conventions

Internationally agreed regulations and programs - especially for rivers belonging to several countries - are very important for a successful water protection. Effective instruments are international river basin commissions as were set in force, e.g. in 1961 for the rivers Mosel and Saar and in 1963 for the river Rhine. Within the scope of the International Commission for the Protection of the Rhine against Pollution there have been conventions since 1976 governing the reduction of salinity which is unfortunately not yet in force - and the avoidance of chemical pollution. A convention on limiting the thermal loading of the Rhine is being drawn up.

Extensive water protection measures were agreed on in the International Commissions for the Protection of the Mosel and Saar against Pollution and also in the Water Protection Commission for Lake Constance.

On the basis of bilateral agreements the Federal Republic of Germany looks after the interests of boundary waters together with its neighbouring countries.

2.4   Conventions on Protection of the Marine Environment

Appropriate conventions exist in regard to the prevention of pollution of the marine environment; these conventions also include guide-lines on the prevention of pollution of the marine environment outside the EEC. The most important international regulations are

- Convention for the Prevention of Marine Pollution by Dumping from Ships and Aircraft (Oslo 1972),
- Convention for the Discharge of Waste into the Sea (London 1972),
- Convention on the Protection of the Marine Environment of the Baltic Sea Area (Helsinki 1974) and
- Convention for the Prevention of Marine Pollution from Land-Based Sources (Paris 1974).

The Federal Republic of Germany participates in the work of the member country commissions set up by the Conventions.

2.5   Regulations with Indirect Effects for Water Protection

More and more, the interactions between the various environmental media are realized, like air, water and soil, which also serve as living space for man, animals and plants. The following regulations apply indirectly to water protection and are partly of great importance:

- Chemicals Act (ChemG) on protection against dangerous substances,
- Federal Immissions Control Law (BlmSchG) for the prevention of harmful effects on the environment caused by air pollution, noise, vibration and similar phenomena,
- Ordinance on limiting PCB, PCT and VC pursuant to BlmSchG,
- First general administrative regulation pursuant to BlmSchG: Technical instructions for maintaining air purity,
- DDT-Law,
- Waste disposal Act (AbfG),
- Federal Law on epidemic diseases,
- Ordinance on drinking water,
- Foodstuff Law.

3   PHILOSOPHY OF WATER MANAGEMENT POLICY

The FRG persues the application of the precaution principle consequently. From that the emission-minimizing-rule or emission principle is drawn with the sequenz of priorities, that emission avoidance has to come first before emission reduction. Since the emission principle alone cannot guarantee an efficient water protection, the use of the immission principle, that means water quality goals, has to be prooved in each situation to establish stronger and additional requirements for the discharge of waste water. This means, that quality objectives complement the precaution measures. They are not part of them and they do not substitute them.

The principle of the precautionary emission limiting the water protection bases on the following understanding:

Each ecological system reveals a complex relationship of quantities depending on each other, which influence each other and change the system continously. From that and by the input of natural unknown substance as by the input of large quantities of natural substances from human activities the eco-systems will become so undeterminal, that the determination of the processes in the ecological system lies in principle outside of human intellectual capacity, especially the extensive assesment of the cause-effect-relations.

On the other hand, the control of technical processes for producing, manufacturing, use and disposal of substances is a technically solved task. The capability to master greatest problems avoiding emissions of substances has been prooved.

The precaution principle takes into account the scientific-technical capability to minimize the non-controllable problems of the substance-input to the environment by avoiding emissions.

It is obvious, that the emission-minimizing-rule resulting from the precaution principle does not lead to the radical point of zero-emission. Yet, where it may be possible to realize zero, it should be tried.

4   FEDERAL WATER ACT

4.1   Main Articles

The Federal Water Act contains some common regulations, i.e.

"§ 1a: (1)   The waters are to be managed in such a manner, that they serve the public interest and that, in unison with this, they are all to the advantage of individuals and that every avoidable impairment is prevented.

(2)   In the case of measures which can be associated with repercussions on a water, it is the duty of everybody to take the care necessary under the circumstances in order to prevent contamination of the water or any other prejudicial change in its properties."

§ 3   The para. contains the definition about the uses which need a permission by the government.

The regulation to control the effluents is set up in § 7a for direct discharges and for indirect discharges. In this connection requirements for construction and operation are set up in § 18b. Plans for the waste water management are regulated in § 18a.

To protect groundwater, several regulations are to be met:

-    Determination of Water Protection Areas for drinking water supply (§19),
-    Handling of hazardous substances in tanks, pipelines, due to producing and usage etc. (§ 19a-l).

But the main point in groundwater protection is to take care that no danger of any harmful pollution to the groundwater can take place (§ 34).

Water quality management plans can be established after § 36b for water bodies or parts of them. In these detailed plans the different use requirements should be opposed to the necessary measures for the protection of water bodies.

4.2   Minimum Requirements for the Different Industrial Branches and the Communities

Article 7a of the Federal Water Act prescribes minimum requirements (effluent standards) to be met at federal level by all waste water discharges into a water body, independent of the quality of the receiving water. Thus effluent discharges can only be licensed if the minimum requirements are met. This is only possible after adequate waste water treatment.

Minimum Requirements under Article 7a FWA are Effluent Standards on a technical-related base which have no connection to quality objectives in regard to the receiving water body. More stringent requirements based on Quality Objectives may be enforced by the Federal States under Article 36b FWA.

The minimum requirements to be met by effluent discharges are applicable to direct discharges of effluents and to indirect discharges into the municipal sewerage system if those discharges contain dangerous substances. Otherwise the dischargers have to meet the generally acknowlegded rules of technology.

Article 7a in the version of the 1986 law includes three important sentences:

Sentence 1:

"A licence for the discharge of waste water is only to be granted if the quantity and the noxiousness of the waste water are kept as low as it is possible when aplicating procedures that are based on the 'generally recognized rules of technology'".

Sentence 2:

"The Federal Government, with the approval of the Bundesrat (Federal States Chamber), enacts general administrative regulations concerning **Minimum Requirements** for the discharge of effluents, meeting the generally recognized rules of technology in the sense of sentence 1".

Sentence 3:

"The Federal Government enacts general administrative regulations concerning requirements meeting the best available technology, if a waste water of a certain discharger contains substances or substance groups, which are dangerous owing to toxicity, persistence, bioaccumulation or which have mutagenic, cancerogenic or teratogenic effects".

The consequence is that a split niveau of technology exists and that those dischargers which have dangerous substances in their waste water have to be determined. These

Minimum Requirements are compulsory everywhere in the Federal Republic of Germany and are independent of the state of the receiving waters.

These Minimum Requirements are, as a rule, effluent standards in the form of concentrations and/or load of pollutants specific of products.

The water authorities are now in the position to tighten up, in case of need, the local discharge requirements but they may not reduce them below the Minimum Requirements.

On the basis of Article 7a Federal Water Act, nearly 50 General Administrative Regulations have been enacted or are ready to be issued in the near future that contain minimum requirements for the discharge of waste water from various industry branches (Table 1) into waters.

The administrative regulations are structured according to the following uniform scheme:

1.   Scope of application
2.   Minimum requirements
2.1 Individual values
2.2 Analytical methods
2.3 Method of evaluation

A regulation according to Article 7a, Federal Water Act consists of two parts - the scope of application and the minimum requirements.

The scope of application is defined by a positive and by a negative statement. The positive statement defines the scope of application, the negative statement has to delimit the regulation against others.

The second part of an administrative regulation is entitled "Minimum Requirements". In this second part, first (2.1) the minimum requirements are quantitatively listed, that is to say - parameters, limit values and the modalities of the sampling are assigned to each other. The limit values are fixed in terms of concentrations and/or load of pollutants specific of products and in special cases additionally as amount of waste water specific of products. The technological description is integrated in the limit value.

Secondly (2.2) the analysis procedures underlying to the minimum requirement are stated; thirdly (2.3) the monitoring method concerning the control of the minimum requirement follows.

In cases where industrial waste water of different origin is treated jointly in one common waste water treatment plant there may be more than one minimum requirement competent. In those cases a proportional addition of the competent minimum requirements will be applied (mixing calculation). The Federal Government sets only minimum requirements in terms of reduction rates for COD, for the end of the pipe technology using biological treatment plants. At the same time requirements are to be met for special producing plants as pretreatment-requirements.

The most important pollutants/parameters for which minimum requirements are to be fixed are those for which waste water charge has to be paid.

-    COD, Cd, Cr, Ni, Pb, Cu and its compounds, Halogenated Hydrocarbons, Hg, Toxicity for fish -.

Besides these parameters in certain branches minimum requirements will be set for the following pollutants/parameters: BOD, Hydrocarbons, Phenols, Cyanide, Heavy Metals, Halogenated Hydrocarbons, Sulfide, Amonia, Fluoride, Phosphorus.

On the other hand there will be no minimum requirement fixed for pH-value and temperature because limitations for those parameters will be set by the state authorities in accordance with the need of the receiving water.

4.3   Exploitation Plans

In the Exploitation Plans (§ 36b Federal Water Act) for surface waters and groundwater the features shall be laid down which the water shall have in its course and the measures necessary for reaching or maintaining the determined features, as well as the time limits which have to be observed. The Federal Government has issued a General Administrative Regulation concerning the characterization of the features for the condition of the water, and determined the features which shall be incorporated obligatory into the Exploitation Plans and how these features are to be ascertained. By this, plans to the requirements for discharges can be tightened up.

TABLE 1

Minimum Requirements for Different
Waste Water Types

Lignite briquet	Dairy
Oil and fat	Gelatine
Fruit and vegetables	Wood
Soft drinks	Sugar
Fish	Yeast
Potato products	Wine
Meat	Leather tanning
Brewery	Cement
Spirits	Raw pottery
Wood	Glass
Food stuff drying	Textile
Mining	Coke
Iron and steel	Car
Non-ferrous metals	Fibres
Electroplating	Refineries
Pigment	Petrochemicals
Fertilizer	Chlorinated hydrocarbons
Inorganic acids,	
oxydes salts	Pesticide
Sulfite pulp mills	Pharmaceuticals
Paper	Citric acid
Soda	Used oil refineries
Utilities	Offal processing
Mixed water from	
chemical industry	

5   WASTE WATER CHARGES ACT

The basic content of the Waste Water Charges Act may be summarized as follows:
1. Pursuant to Article 2, para. 2, discharging within the meaning of this Act shall
    be deemed to be the immidiate and direct conveyance of waste water into a
    water body. Accordingly, only direct discharging is covered.
2. Pursuant to Article 4, para. 1, the values to be applied for determining the
    number of units of noxiousness shall be taken from the official notice
    licensing the waste water discharge.

3.   Pursuant to Article 9, para. 5, of the Waste Water Charges Act reduction up to zero shall be granted provided the minimum requirement under article 7a of the Federal Water Act is complied with and the reference values specified are observed in assessing the charges.

4.   Dischargers which cannot meet the minimum requirements after § 7a (fixed in the licence) will have to pay more charge depending on the actual value.

5.   The revenue derived from waste water charges can only be used for specific purposes connected with measures for maintaining or improving water quality.

6.   Fines and Penalties are not provided for by this Act for cases in which the reference values specified for the assessment of the charges are exceeded. Here the consequent application of the "polluter-pays-principle", that means, "who pollutes more, pays more", becomes apparent.

The charge is calculated by units of noxiousness. The annual rate levied per unit of noxiousness started with 12 Deutschmarks in 1981 and steadily rises to 40 Deutschmarks in 1986 (Table 2):

TABLE 2

Rate per Unit of Noxiousness

January 1st, 1981	12 DM
January 1st, 1982	18 DM
January 1st, 1983	24 DM
January 1st, 1984	30 DM
January 1st, 1985	36 DM
January 1st, 1986	40 DM

The total charge (TC) is calculated as a sum of the product of the units of noxiousness (UN) and the anual waste water amount (AWA) and the specific anual rate (AR) parameter by parameter:

$$TC = \Sigma (UN \cdot AWA \cdot AR)$$
parameter

Today new rates of 50 and 60 Deutschmarks are under discussion.

6   MONITORING TO MEET REQUIREMENTS

Any system of charges, standards or impact assessment presupposes the existence of a system for monitoring changes in effluent or stream continuously. Monitoring and information-gathering are essential elements in any pollution-control system and should be given priority by governments.

Efficient management of wastes and water quality requires adequate systems of data acquisition and a scientific understanding of the assimilative capacity of receiving waters. The development of such an information base is one of the first steps to be taken in the effort to control water pollution by developing countries.

Two types of standards are typically involved in water pollution control regulations. First, ambient or stream standards are the legal specifications of the minimum conditions which must be met for a given indicator of water quality at a specified location along the stream. For example, a stream standard may require that dissolved oxygen, averaged over a 24-hour period at a selected river mile point, must not fall below 4 parts per million (ppm) more than one day per year. Second, effluent standards are those which specify the mean or maximum permissible discharge of a pollutant, such as Hg or COD, from one particular source. Effluent standards are requirements set on the quality characteristics of actual discharges, while stream standards refer to the quality characteristics of actual discharges, while stream standards refer to the quality requirements for the receiving watercourse.

Stream and effluent standards coexist in control programs today and must be viewed as potential complements in a rational program of management. In a situation where there are numerous waste dischargers, achieving a stream standard through several independent decisions will be impossible. Therefore, a central agency must provide information and incentives which will produce coordinated behaviour. Effluent standards will be meaningful only in the context of water quality goals or standards in the water course. A combined approach of stream and effluent standards may be used by setting individual effluent standards which reflect the size and location of the discharge relative to the waste assimilative capacity of the river stretch.

The success of any measure concerning water protection depends on the efficiency of its control. For this purpose every waste water discharge and all important waters are monitored in the frame of individual monitoring programs.

Relating to effluents, in the case of waste water discharges it is aimed at a sensible combination between a regular self-control of the manager of a waste water treatment plant and a more seldom practised control of the authorities. The motive powers for this monitoring system are the Waste Water Charges Act.

In the Federal Republic of Germany the parameters of effluents are supervised pursuant to the Waste Water Charges Act and to the Federal Water Act in connection to the parameters mentioned in the administrative regulation relating to Article 7a and in addition, parameters defined in the licences. There is no defined system because the licences represent individual regulations based on individual cases.

Self-control of the user of waters serves primarily the purpose that the user may recognize whether, when and to what extent he has to take additional measures to reduce pollution in the waste water or what additional measures of treatment he has to take. Self-control does not replace government supervision. It may at the utmost complement it.

Official supervision of waste water discharge implies examination whether the discharger observes the limit values set for him in the licence or whether there are irregularities. This can be done, however, only by taking random samples.

Chapter 13

FORMULATING AND IMPLEMENTING A POLLUTION CONTROL STRATEGY TO COPE WITH FIRST AND THIRD WORLD POLLUTION IN AFRICA

H D Furness, W N Richards (Umgeni Water, Republic of South Africa)

ABSTRACT

Umgeni Water, a regional water authority in Natal, South Africa, has jurisdiction over a 7092 km^2 area.   Within this area both first and third world conditions are encountered.   Results from a monitoring programme, which were used to assess the different adverse impacts of first and third world developments, on water quality, are briefly discussed.   The formulation and implementation of a pollution control strategy to overcome identified pollution problems and thereby maintain satisfactory raw water quality for both socio-economic groups, is presented.   Essentially this necessitates providing potable water and sanitation facilities to third world communities while formulating and applying appropriate discharge standards,   based   on   water   quality   objectives,   to   highly urbanised/industrialised communities.

Key Words :   Pollution Control, First World Pollution, Third World
               Pollution, Pollution Strategy

INTRODUCTION

Umgeni Water, one of 13 regional Water Authorities in South Africa, has jurisdiction over a 7092 km^2 area.   This includes the industrial and metropolitan areas of Durban and Pietermaritzburg as well as extensive rural areas of KwaZulu (Figure 1).   The industrial-metropolitan areas are typically first world, with a sophisticated water treatment, water reticulation, sewer and wastewater treatment infrastructure.   In the rural areas, presently accommodating 60% of the 5 million people living in Umgeni Water's area, typically third world conditions prevail.   Potable water supplies and sewage treatment facilities are absent.   This has tended to cause extensive informal settlements to develop along river courses and in the vicinity of impoundments.

141

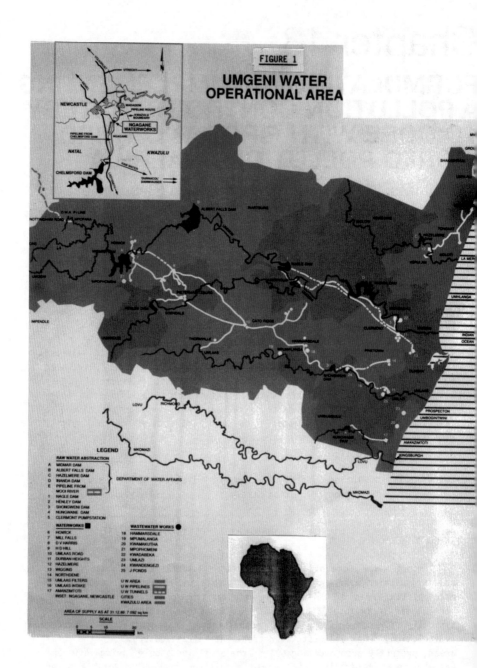

FIGURE 1

UMGENI WATER
OPERATIONAL AREA

REGIONAL WATER MANAGEMENT

In the past management of water resources was regarded as a function of Central Government. As far as pollution control was concerned the philosophy of "the polluter pays" was applied. Point source discharges were regarded as the main source of pollution. As a result all dischargers had to comply with discharge standards, applied on a national basis and based on concentrations. This approach had only limited success because of regional differences and increasing pollution from mushrooming informal settlements.

In response to these developments the Department of Water Affairs (DWA) has adopted a long term strategy that leads logically to regional management of water services including, where appropriate, river basin management. "As much autonomy as practicable is being devolved upon the non-central government sectors, particularly where regional rather than national interests are served. The Department favours the formation of Water Boards because of the highly specialised nature of water management." (Department of Water Affairs, 1986). Additional authority has been given to Water Boards "...to establish, construct, purchase or otherwise acquire and to manage, maintain and control any scheme for the purification or disposal of wastewater, effluent or waste resulting from the use of water ..." (Water Act, 1987). This broadening of responsibility acknowledges the benefits of managing water supply and wastewater treatment facilities on an integrated basis.

The integration of water supply, sewerage and pollution control services is one of Okun's key principles for sound water management. "As wastewaters are treated to higher and higher degrees, the effluent products become important water resources, particularly where natural water resources are fully exploited. If water resources are to be managed efficiently, all elements of the hydrological cycle should fall within the purview of a single agency so that the wastewater of one community on a river may be an asset rather than a problem to its downstream neighbour." (Okun, 1977). In regard to water pollution control the DWA has stated ".... since the effects of water pollution usually extend beyond local, regional, provincial or even international boundaries, control of pollution is the responsibility of the Central Government. However, some duties and responsibilities within an overall policy framework can be delegated to Water Boards." (Department of Water Affairs, 1986).

Recognising the necessity for quality management the DWA has also pointed out that ".... in time quality may become a more important factor than quantity in determining the availability of water in some areas" (Department of Water Affairs, 1986) and that ".... it has already become evident that the efficiency of pollution control measures can be enhanced by a policy of integrated river basin management" (Braune, 1987).

Thus the DWA is gradually devolving responsibility for the management of water services to regional bodies, particularly Water Boards, but currently retains the responsibility for administering pollution control and prosecution in accordance with the Water Act. Although Water Boards have been given the carrot of increased responsibility to facilitate better localised management of water services, the DWA retains the stick of prosecution to ensure that there is public accountability and that the delegated responsibilities are not abused.   In accepting these wider responsibilities Umgeni Water has adopted a policy of seeking to improve the quality of water in its rivers and dams by monitoring discharges and by being actively involved in pollution prevention, sanitation and wastewater treatment.   Water quality management has become essential for the efficient and effective supply of drinking water, particularly in river basins where there is increasing re-use of water.

IMPACT OF DEMOGRAPHIC TRENDS

The management of water resources on a catchment basis involves decision making founded on an accurate data base incorporating information on water quality and quantity, demographic trends and land use.

It is estimated (Horne et al., 1989) that currently approximately five million people reside in the 7092 km^2 making up Umgeni Water's operational area.   By the year 2025 this population will have increased to between ten and sixteen million of which 85% will be urbanised compared to the current 40%.   Faced with such dramatic and dynamic changes in demography Umgeni Water commissioned consultants to produce a Regional Water Plan, an incremental data base predicting the growth and location of population, land use, water demand by area and other pertinent factors. (Horne et al., 1989).

The Regional Water Plan will assist management in making strategic water planning decisions on a macro scale.   It will identify new areas of development, indicating water demand growth and associated sewage loadings, information essential for sound water quality planning.   Data obtained for the water plan correlates water consumption and family income (Figure 2). As the population base shifts from a rural, informal, subsistence lifestyle to an urban, formal, more affluent lifestyle, so the daily per capita consumption of water increases.   At subsistence level, per capita consumption is only ± 15 litres/day but with increasing monthly income, consumption rises to a maximum of ± 370 litres/day, a twenty-five fold increase. Thus not only is it essential to plan for a doubling or trebling of population, but also to plan to meet the increased water supply expectations of these people, as they become urbanised and more affluent. Although Natal is relatively water rich it is not free from water pollution and a substantial amount of water re-use is brought about by population pressure.   Water discharging into the Indian Ocean from the Umgeni and Umlaas Rivers may have been used three or four times, passing sequentially through as many as eight waterworks and sewage works.

This indirect re-use of water increases mineralisation, changes composition and impairs water quality.   Thus, for example, upstream of Midmar Dam the conductivity of the river water is 7 mS/m and its soluble orthophosphate concentration 15 ug/l but approximately 100 km downstream at Clermont Pumping Station the conductivity averages   25 mS/m and the orthophosphate concentration 130 ug/l.   Impoundments in the lower reaches of these rivers are eutrophic and the treatment required to produce potable

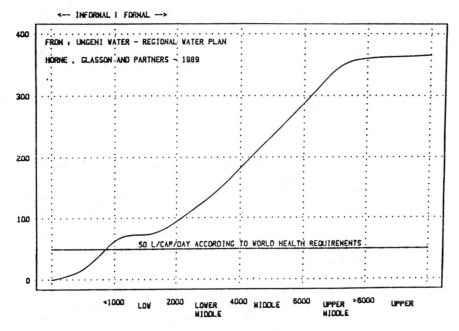

FIGURE 2

PER CAPITA CONSUMPTION V/S AFFORDABILITY

water from these sources is more complex and costly than for treating water from the upland sources. This derogation in water quality arises both from point and diffuse source discharges. The former, predominantly from sewage works and industrial outlets, are regulated and subject to the pollution control provisions of the Water Act. The latter, often a consequence of informal settlements, are uncontrolled and can pose a serious public health threat to downstream users, requiring a very different approach in respect of management and control.

It is against this background of population increase, rapid urbanisation, huge growth in water demand and concommitant effluent loading on its rivers, that the principles of river basin management must be adopted if Umgeni Water is to meet its long term objective of satisfying the needs of the community.

ESTABLISHING A WATER QUALITY DATA BASE

The increasing awareness of potential water quality problems prompted Umgeni Water to look closely at water quality within its area. Although water quality data had been collected on an ad hoc basis by a variety of organisations only the Department of Water Affairs undertook routine sampling. The data obtained from these sources, although valuable for indicating general trends in water quality, did not provide for the day-to-day operational or long-term water quality planning needs of Umgeni Water.

In response to this situation a routine monitoring programme was implemented in March 1987.   This programme is multi-faceted, providing information for day-to-day operational requirements, ensuring compliance with standards, checking process performance, assessing the impact of catchment utilisation on water quality of rivers and impoundments and providing data for water quality modelling purposes.   To meet these objectives approximately 500 sites are sampled on a routine basis, the frequency varying between daily and quarterly.   The number of water quality variables analysed for varies between 11, for daily samples from water treatment plants, and 55, on a quarterly basis, for abstraction points.

In the case of sampling for water quality purposes it was difficult to apply a statistical basis for sample point location, frequency of sampling and frequency of analysis for different water quality variables, due to the lack of a comprehensive water quality data base.   Sampling points were located according to perceived needs e.g. below point sources, informal settlements, major urban or rural developments, at abstraction sites and impoundments.   Sampling frequency and specified water quality variables were determined according to need, or requirements in terms of establishing a suitable data base.

Information from the data base reveals a general trend of deteriorating water quality from the upper to the lower catchments with localised areas of very poor quality scattered throughout. These locations are not restricted to industrial and urban areas or associated solely with point source discharges. It is obvious that deteriorating water quality is being caused by factors such as unplanned, unco-ordinated development (e.g. squatter and informal settlements), leading to diffuse inputs, as well as illegal point source discharges (mainly trade effluents) and legal point source discharges which are not complying with legal discharge standards.

POLLUTION PREVENTION AND WATER QUALITY PLANNING

Recognising the need to maintain and improve the quality of water in its sources, Umgeni Water established a Water Quality Branch within Scientific Services in March 1988. It consists of two sections - Pollution Prevention and Water Quality Planning.   Water Quality Planning staff are responsible for data analysis, assessing the feasibility of water quality standards based on water quality objectives, determining the impact of catchment utilisation on water quality and applying modelling techniques to assist decision making.

Pollution Prevention staff provide the means to ensure that water quality objectives are achieved, providing both a reactive and a proactive response to pollution. The former involves the investigation of reported or perceived pollution incidents while the latter includes routine surveys and establishing background data on pollution occurrences.

Water Quality Officers within this section respond to pollution incidents by carrying out on-site tests, investigating illegal discharges, providing advice and information to dischargers on how to meet discharge standards and, if necessary, instigating formal proceedings with Police Officers and staff of the Department of Water Affairs against polluters.

To help protect water quality these Officers inspect river courses and report on potential pollution problems as well as educating industrialists, riparian residents and owners in measures to prevent pollution.   In addition, the proactive response involves the monitoring of point source discharges and the establishment of a suitable data base. This data base will allow the impact of discharges on receiving waters to be assessed and provide a means of quantifying the effectiveness of pollution prevention measures.

In the case of diffuse inputs, particularly from squatter and informal settlement areas, combating adverse effects on water quality is more difficult than in the case of formal developments. In the long-term the only solution will be the provision of potable water and suitable sanitation. The role of the Water Quality Branch will be to identify the most serious problem areas and demonstrate the potential benefits arising from formalising water supply and sanitation. The planning for and provision of potable water and sanitation, will then be undertaken by other Divisions of Umgeni Water.

This latter aspect is of particular concern to Umgeni Water and is reflected in its objective "To manage water services in response to the socio-economic needs of the population in its region". To achieve this the Board of Umgeni Water has approved the spending of R1,000,000 per year on providing potable water supplies and sanitation facilities to informal settlement areas.   Due to difficulties such as rugged topograph, absence of electricity, the general lack of an adequate infrastructure (roads, drainage etc) as well as a shortage of suitably trained people in informal areas, the provision of a sophisticated water supply/waste water treatment system in these areas is not a viable option. Assistance (financial and technical) in obtaining potable water is being provided for spring protection, location and construction of rain water tanks and in some instances establishing small package plants.   A similar approach to sanitation is also being adopted, where assistance with systems such as improved pit latrines, composting latrines, septic tanks, conservancy tanks etc, is being provided.

CONCLUSION

Inevitably population growth and increasing family affluence will lead to increasing re-use of water. In river basins where water is repeatedly abstracted and returned as treated wastewater by first world communities and where third world communities lack potable water supplies and basic sanitation facilities, the adoption of a river basin management approach is essential. Emphasis will have to be given to formulating standards according to river quality objectives, based on user requirements, which will protect water quality from point source discharges. In the case of informal settlements the provision of potable water and sanitation facilities will not only upgrade living standards but is also the only means of protecting water quality from the diffuse inputs of these communities. River basin management will prove to be in the best interests of the community, not only will it ensure a greater security of supply but also it will enable consumers of previously used water to do so with impunity - their drinking water will remain wholesome with no risk to health.

ACKNOWLEDGEMENTS

The authors wish to express their appreciation to Mr G D J Atkinson, Chief Executive of Umgeni Water, for allowing them to present this paper.

BIBLIOGRAPHY

BRAUNE, E "Catchment Management for Water Quality Objectives". Symposium on Fifty Years of Catchment Research in South Africa. Stellenbosch, 11-12 November 1987.

DEPARTMENT OF WATER AFFAIRS. Management of the Water Resources of the Republic of South Africa, Department of Water Affairs, Pretoria, South Africa, 1986.

HORNE, GLASSON AND PARTNERS. Regional Water Plan for Umgeni Water. Confidential Report to Umgeni Water, Pietermaritzburg, South Africa 1989

OKUN, D.A. Regionalisation of Water Management.   Applied Science Publishers, London (1977)

WATER ACT. Act No. 54 of 1956 of the Republic of South Africa (1977) as subsequently amended (1987).

Chapter 14

PRICING ALTERNATIVES FOR POLLUTION CONTROL SERVICES

R M North (University of Georgia, USA)

ABSTRACT

In economic analyses of costs and benefits, each alternative solution for pollution control will be unique. It is rare to find physical solutions in which prices and fees (or the burden of these prices and fees on various groups) are attached to the proposed solutions. The author suggests some approaches to the problem of "pricing out" pollution control services. Correct pricing of pollution control services can be instrumental in securing the adoption of efficient pollution control practices by directing behavior of all parties. Correct pricing will provide revenues for maintenance and expansion. The economist must construct a capacity charge and a commodity charge into a rate structure that will achieve the desired level of water quality. This rate structure must provide for an equitable distribution of the costs for different classes of users and generate sufficien: :evenues to at least amortize the cost of the system. This paper provides examples for a few pricing systems and rate structures, including recently developed impact fees, that acknowledge the right of publicly owned water utilities to earn returns on invested capital.

Keywords: Pricing, tariffs, rate structures, economics.

INTRODUCTION

Given the objective for pollution control is that of achieving an acceptable standard of water quality, there will be several sets of alternative solutions available to achieve the water quality standard. Each solution will have its own particular aspects regarding the location, type and quantity of pollution, source of pollution, intended use of the water resource and other factors. In turn, each unique physical solution will incur financial and economic costs for each level of water quality attained. In the economic analysis, each alternative solution for each case will be unique, even though several combinations of alternatives may be presented. It is rare, even in our modern society, to integrate the physical solutions for water quality improvement with the economic benefits and costs. It is more rare to find physical solutions suggested in which prices and fees for various solutions (and in turn the burden of these prices and fees on various groups) are attached to the proposed solutions. This makes the adoption of a proposed solution more difficult since the distribution of benefit and cost burdens are not known. The purpose is to suggest some generic approaches to the problem of "pricing out" the pollution control services that would be acceptable to regulators, users and/or beneficiaries as well as to those who will bear the cost burdens. Correct pricing of pollution control services can be instrumental in securing quickly the adoption of pollution control practices and thus in achieving more efficiently the objective of clean water.

In pollution control services we face the traditional situation of a high fixed cost that must be recovered over large spatial and temporal horizons. This situation requires a two part tariff to attain a reasonable level of effectiveness. The basic infrastructure (the capital costs) must be based on some average use criteria associated with capacity requirements. This part of the tariff is often called a capacity charge in the public utility sectors. Capacity charges have been widely recognized and used in the electrical, gas, telephone and water supply sectors. Even though capacity charges are used in the water quality control business they are not so clearly defined nor readily accepted as capacity charges applied in water supply or electricity services. The second type of charge is a commodity charge or a charge per unit/time of use that is directly related to current demand. Commodity charges that one will find in the water pollution control services area will be related to volume of effluent discharge as measured grossly by the quantity of water discharged, or as measured more accurately by the quantity and/or types of pollutants included in the waste discharge.

The appropriate role of the economists in integrating water pollution control practices is to properly construct the capacity charge and the commodity charge into a rate structure that will achieve the desired level of water quality. This rate structure must provide for an equitable distribution of the costs for different classes of users and generate sufficient revenues to at least amortize the cost of the system. In the USA there is now developing a willingness to provide a level of revenues that might be achieved based on the values of services provided. These rate structures, (combinations of capacity and commodity charges) will yield a rate of return on investment or a capital recovery rate that will serve as an upper limit on the total amount of revenue collected from the users. These return on investment (ROI) concepts will provide revenues in excess of historical costs that may be used to offset inflation, obsolescence, mistakes and other results of uncertain outcomes.

There are other types of prices for water pollution control services such as the penalty prices or fines imposed on polluters. These have economic justification in limited cases where downstream or subsequent users are adversely affected. Otherwise, such prices serve only as enforcement tools and do not lend themselves well to economic justification. If we wish to integrate water pollution control services to achieve the optimum level of water quality at the least total cost to society, then we must spend more time developing optimum rate or tariff structures to achieve the objectives and equities described herein.

POLLUTION AND PRICING

Pollution of water normally results from a discharge of waste to the river, lake, ocean or ground. Such discharges are negative values or costs. Furthermore, discharges of waste seek to avoid costs of waste disposal since wastes add nothing to the value of the marketable product or service. Fortunately, those who discharge wastes into water are quite comfortable paying for the water supply service used for waste disposal. The objective then, is to connect the water use to the waste disposal in order to price the combined water use and waste disposal in an acceptable manner to account for both services. This is often, but not always, the case for domestic and industrial users. A surcharge is added to the commodity price of water supply but the capacity charge may be levied separately for water supply and sewage services. There is much greater difficulty in structuring the waste treatment prices than in structuring the water supply prices.

The addition of a surcharge for sewerage to water use may not address the efficiency nor the equity issue of waste disposal costs. For example, a household using water for landscape irrigation will be overcharged for sewerage services relative to a household that does not irrigate. Also an industry that recycles its water or discharges high concentrations of pollutants may not pay an equitable share of costs through the surcharge method. For these reasons some water/sewerage service vendors select an average or non-irrigating period (usually winter) as the basis for establishing the annual commodity surcharge for sewerage services. For industries, pretreatment of water or added fees for waste loads may be required. There are many other physical aspects of waste treatment that affect its pricing. The essential criterion is that we structure the prices of sewerage to direct behavior (use the least service) and to generate revenue sufficient to recover costs, replace/expand facilities and absorb risks.

CAPACITY PRICES

Capacity charges (prices) are levied to recover all or part of the capital costs of providing sewerage services. A given community will require a certain capacity for treating its wastes. This capacity need will depend on daily and seasonal times of use. The treatment plant will be engineered to perform at a design capacity to include peak days and seasons for some predetermined time in the future. Once the design capacity is determined, its cost becomes the marginal cost for the new or added system. This cost is fixed for the life of the plant. At first glance these fixed costs should be averaged for the expected users as an "average marginal cost." Economic theory based on pricing services at marginal cost break down at this point. However, it is interesting to speculate on variations of pricing pollution control capacity. At the one extreme the nth user requiring added capacity should pay the full cost. Another variation is that of allocating the costs based on BOD, COD, or other components of the waste based on tons of material, water quantity and load or other measures related to cost of control. Perhaps the more common approach is to average the cost of base load to base load users and then price additional peak type or growth loads in waste treatment as an additional fee for those imposing costs or loads that stress the design capacity. This is a more complex pricing system but one in which user behavior can be best directed. A user requiring above average service would then be paying an estimated marginal cost that would most likely be above average fixed cost and thus contribute net funds to amortization or fund accumulation for future expansion. The user would have alternatives when working within a municipal system to install pretreatment, effluent storage (to avoid system stress), reuse or other actions to minimize his costs of waste disposal.

Some economists would ignore capacity charges, seeking to recover all costs through true marginal costs that exceed average costs. This approach to recover capacity costs works well with short life, low capital assets but not with the long life, high capital needs of utilities. This approach leads to short range planning, undersizing of facilities, underinvestment and biases that seem to favor higher volume or higher load users. The alternative is to construct a capacity price (one-time or periodic) that evens out the flows of net revenues to the utility owner/operator throughout the life of the system. These capacity charges, when modified to reflect varying and above average loads, will recover sufficient fixed costs to meet financial services (debt) loads in the early years of operation and to provide excess revenues for future expansion as the system capacity nears its upper limit. The economic ideas of using an administered capacity charge as a part of the price are to force more efficient sizing of facilities and to spread the fixed cost of capacity among classes of users and timing of users in the most equitable manner.

COMMODITY PRICES

Commodity pricing of sewerage services is more straightforward than capacity pricing. Commodity prices more nearly approximate the economic ideal of a competitive market, marginal cost price. The commodity price will approximate the variable costs of the sewerage services. In practice there will be a charge for the resource used (the water) and the direct operating, maintenance, repair (OMR) costs for the pollution control system. In cases where dilution is practiced the marginal costs will be the values of the downstream uses foregone. Commodity prices are based on the conventional downward sloping demand curve of economics. It is a willingness to pay concept wherein the first units fetch a higher price than the later or lower valued units. There are many good arguments for the inverse price-quantity relationships of normal goods. If pollution control can be made a normal good by regulations or by education then the conventional commodity pricing (as in water supply services) systems will be appropriate. Otherwise the commodity price for pollution control will be taken from the conventional supply curve in economics. As pollution increases, the supply of pollution control services must increase, raising the marginal cost successively. Thus we price up the supply curve as marginal costs increase.

This makes the commodity pricing of pollution more difficult to handle in economic theory but not necessarily in practice. Prices should be set to cover marginal costs of treatment and control but no less than total average costs. In this scenario, one could approach the problem on a market basis by establishing one price for all users at one of two points. The first point is to establish a price where the marginal cost just equals the marginal revenue for all users on a per unit basis. This point will yield the maximum net revenue (minimum losses) for the utility. This is the point, also, where water supply service (a normal type service) would be priced as a commodity. The second point would be to charge that price where the rate of change in price is equal to the rate of change in quantity of effluent supplied -- an elasticity coefficient of 1.0. This is the point where gross revenue from a single market price would be maximum -- without regard to costs.

Other variations in commodity prices are available. The most common is the increasing block rate (decreasing block rate for water supply) wherein revenue can be further increased (over the maximum net or gross explained above) by charging sequentially higher prices for successive units of waste treatment. Since the treatable wastes are "flows" the user must pay for the first unit before getting the subsequent unit(s). The utility owner/operator is then able to extract the potentially full economic value of the service with infinitely small price-quantity blocks. If this method were used alone there would be no financial need to use a capacity price or a two-part tariff system.

These suggestions are offered as introductory material to the issues of pricing pollution control services. Pricing of such services offer the best possible vehicle for integrating most of the problems of water pollution. Economics is the great integrator of engineering, hydrology, geology, policy, law, planning and other fields. However, economics, as reflected in benefit cost analyses and pricing, is a complex tool that must be used carefully, from the beginning, to serve best as an integrator of ideas and solutions for water pollution problems.

Unfortunately economists have not spent much time and effort on pricing water pollution control services. So much of these services have been provided by the governmental sector from a general revenue or special tax base that the issue has been neglected. A great deal of research as been done on pricing water services -- much of which is directly applicable to water pollution only in the use of water supply surcharges for waste water (Sellers and North, 1988). Pricing and demand studies have been available for some time: Headley (1963), Gottlieb (1963), Howe and Linaweaver (1967) Gardner and Schick (1967), and Ware and North (1968). However, these tools have not been used for purposes of developing public water pollution control pricing structures to (1) generate predetermined amounts of revenue, (2) limit subdivision development, (3) limit water use, and/or (4) provide "profits" for public vendors. Recent practices in the USA (Mercer and Morgan, 1985), suggest that public water suppliers and waste treatment vendors now engage in pricing to provide market rates of returns on capital. The precise determinants of water service price variables and how they should be measured are not agreed upon in most cases. The exact relations and the model specifications that properly express the relations are complicated by the inadequate data bases available for analysis and the problems associated with aggregate data, time-series data, cross-sectional data, and customer reactions to signals, even when sample data are generated.

Given that several relations are inexact, let us add to the confusion the problem of infrastructure-expansion needs under conditions of rising costs, along with decreasing Federal and state involvement in providing financing. New sources of revenue as well as price increases to generate more direct revenue must be considered as new management tools. Of particular interest at this time are the "advance capital" charges now being proposed and implemented by utilities (cities) on developers, which are popularly known as impact fees.

IMPACT FEES FOR WATER AND WASTE WATER SERVICES

It is not often clear as to exactly what an impact fee is. An impact fee is "any charge or exaction unrelated to a direct point of use service charge." A clear instance of an impact fee would be a charge for future expansion of system capacity. In some instances, impact fees are proposed for other external costs, such as connector roads, school expansion, police protection, anticipated inflation and other types of soft or social infrastructure services that may or may not be directly related to the project, normally a subdivision. In these instances, the developer becomes the identifiable marginal-cost passenger, in the sense of Coase (1946), wherein the nth-plus-one contributor must pay the full cost of added capacity to the system. In the former case, for expanded water system capacity, the Coasian case is adequate. But in the latter case, the impact fees levied for total social infrastructure costs incorrectly assign all social costs (fixed and variable) to the identifiable marginal contributor, even though many of the future users may be external or nonchargeable. A connecting road may serve many users who would be free riders, from whom the developer would be unable to collect a fee in the absence of a toll system. The most likely result would be that buyers (consumers) of the new subdivision would be charged a fee in excess of their true average marginal costs for these nonexclusive facilities and services. If the consumer (or developer in this case) can respond directly to a connection fee that is related to the marginal cost of service, the fee is considered not to be an impact fee (Sellers and North, 1988).

Burby et al. (1987, p. 33) suggests that the use of impact fees is much more widespread and more common than the literature suggests. Part of this understatement evolves from confusion in terms such as privatization, private financing, extension fees and land-use planning. The currently available literature on impact fees relates largely to land-use planning, where empirical evidence of the extent of use of impact fees and the effects of target charges are not known. Burby et al. (1987) found that 35 percent of the water and sewer utilities in nine southeastern states imposed an "impact fee" in 1986. The highest rate of use of impact fees was in Florida, where 71 percent of the utilities reported the use of impact fees for water and sewer services. The lowest rate was 13 percent of the utilities in Kentucky and Georgia. Burby et at. (1987) adjusted their data and that of earlier studies to reflect a weighted factor for size of state and number of cities. These adjusted data then show that 12 to 13 percent of the utilities in Florida, California, Colorado, Washington, and Oregon imposed impact fees in 1984.

Another indication of the "sudden maturity" status of impact fees is the March 1988 publication by the National Association of Homebuilders of a 90-page Impact Fee Manual (NAH, 1988). This manual urges developers and homebuilders to "...have an effective voice in the control of impact fees..." and suggests a dozen or so specific issues that should be part of a statewide impact fee bill or a local ordinance (Link, 1988). This approach, which favors a statewide impact-fee-enabling act or law that will be "fair and equitable," is reminiscent of the waterways operators in the mid 1970's who began to urge themselves to help structure a navigation user fee when such fees were inevitable by 1978.

We don't have 40 years of studies and data on water-services impact fees to help determine: (1) the alternative structure for impact fees; (2) the proper applications, estimations and uses of impact fees; (3) the mechanisms for insuring that impact fees are earmarked and accounted for; (4) the potential consequences of either good or bad impact fees on; (a) growth, on land use, on utility planning, on other services such as housing; (b) on the overall management, control and regulation of water and sewer services.

The response and knowledge situation with impact fees is yet an infant. We do not have a good sense of what a "correct" impact fee should be. We do not know much about the economic efficiency or equity impacts of various impact-fee structures. We do know that there is great confusion over the raison de etre' for impact fees. Are they intended to finance expansion? Are they intended to raise revenue to continue historical Federal/state/local subsidies to existing users (and to new users when costs cannot be passed through)? Are they intended to support land-use planning and growth management (Sellers and North, 1988).

We know, from economic theory and application, that impact fees are a fixed cost to the consumer. Therefore, the fee structure is not one of the decision variables affecting water pollution services directly. More likely, such fixed costs will be responded to only in the consumers choices of housing, plant location, etc. Some will argue that new consumers should pay the full cost of water services, including capacity and associated external costs. This concept introduces a significant bias of public infrastructure costs toward new users with respect to old users. Imagine a long-time user, whose water services were financed partially or wholly through subsidized bonds and taxes, enjoying subsidized rates while a new user pays, in advance, at private market interest rates, the full cost of capacity, plus levelized taxes, plus, in many cases, a water rate 1.5 to 3 times that of the long-time user. Impact-fee financing of water services will offer several years of economic and financial study in order to address the structural, efficiency and equity issues raised by this largely untried method of financing (Sellers and North, 1988).

SUMMARY

The art and science of pricing pollution control services need much more attention by engineers and economists. The pricing system integrates the planning, the operations, the financing and the larger interests of the community very well. The most difficult pricing studies and decisions are yet to come as we seek to achieve the levels of efficiency, equity and acceptability in pricing pollution control as we have in pricing water supply services. Finally, the economic theory for normal goods must be readdressed to fit the special aspects of negative goods such as waste disposal.

BIBLIOGRAPHY

BURBY et al. - "Financing Water and Sewer Extensions in Urban Growth Areas: Current Practices and Policy Alternatives." Center for Urban and Regional Studies. University of North Carolina, Chapel Hill. 1987.

COASE,R.H.(1946). - "The Marginal Cost Controversy." Economica,13,1946,pp.169-182. (Also, See HOTELLING,H. - "The General Welfare in relation to Problems of Taxation and of Railway and Utility Rates." Econometrica,6,1938,pp.242-269.)

GARDNER,B.D.;SCHICK,S.H. - "Factors Affecting Consumption of Urban Household Water in North Utah". Bulletin,449,1964,43 pp. Agricultural Station, Utah State University, Logan, Utah.

GOTTLIEB,M. - "Urban Domestic Demand for Water: A Kansas Study." Land Economics,39, 1963,pp.204-210.

HEADLEY,J.C. - "The Relation of Family Income and Use of Water for Residential and Commercial Purposes in the San Francisco-Oakland Metropolitan Area." Land Economics,39,1963,pp.441-449.

HOWE,C.W.;LINAWEAVER,JR.F.P. - "The Impact of Price on Residential Water Demand and Its Relation to System Design and Price Structure." Water Resources Research,3(1),1967,pp.13-32.

National Association of Homebuilders - Impact Fee Manual,1988,90pp. Washington D.C.

SELLERS,J.;NORTH,R.M. - "Commodity Prices, Demand Charges, Impact Fees and Water Services," in Water-Use Data for Water Resources Management, edited by M. Waterstone and J. Burt., AWRA Proceedings, Bethesda, Maryland.

WARE,J.E.;NORTH,R.M. - "Price and Consumption of Water for Residential Use in Georgia." Atlanta Economic Review,58(10),1968,pp.9-13.

Chapter 15

FINANCIAL IMPLICATIONS OF EUTROPHICATION: RECREATIONAL USE OF WATER OFFERS ECONOMIC INCENTIVES FOR POLLUTION CONTROL

R E Haynes, F C Viljoen (Rand Water Board, Johannesburg, Republic of South Africa)

ABSTRACT

The paper will deal with the financial implications of eutrophication in relation to the supply of potable water, waste water purification, recreational use, riparian ownership and industrial and agricultural use of water.

It will be shown that the economic value placed on unpolluted water for recreational use offers strong incentives to reduce pollution from waste water treatment plants.

KEY WORDS: Eutrophication, potable water purification, waste water purification, recreation, financial implications.

INTRODUCTION

The Republic of South Africa is a water poor country and optimal use must be made of the available water sources. The enrichment of water with plant nutrients such as phosphorus and nitrogen, generally known as eutrophication creates many problems. The basic justification for the evaluation of the cost of eutrophication to the community should be based on the cost of providing water and the "value" of the water to other users.

To evaluate such use it is necessary to identify those activities that may be affected by eutrophication and to determine the financial importance of these activities to the community.

It is generally accepted that the priorities accorded to water usage are the following:-a) Potable b) Agriculture c) Industry d) Recreation.

It is the object of this paper to highlight the direct and indirect financial implications of eutrophication not only to the water industry but to other water users and the incentive role that recreational and riparian use has on the reduction of eutrophication.

1. WATER PURIFICATION (POTABLE)

Problems related to the water supply industry as a result of eutrophication are well documented. Although excessive macrophyte growth in flumes and canals may cause problems, they are relatively easily solved, however, excessive algal growth as a result of enriched water poses a far greater problem and is more difficult and costly to control.

Filamentous algae attach to filter walls and launders and have to be removed. Detached algae form a sand algal agglomerate on the filters that is not removed during backwashing and has a tendency to penetrate the upper filter media. Other forms of algae cause tastes, odours and colour in water and both algae and their degradation products may give rise to unwanted organic compounds such as trihalomethanes upon chlorination.

A study carried out by (Steynberg, et al., 1985) showed that with an expected increase in water eutrophication due to natural industrial and population growth the Board would be faced with increased costs to combat algal problems.

Algal control options are therefore evaluated to ascertain the present and possible future costs.

The Rand Water Board is the largest Regional Water Supply Authority in the Republic of South Africa. The Board has purification capacity for 4170 000 m^3/day of which 32% of its raw water supply comes from the Vaal River Barrage Reservoir (an impounded river downstream of the Vaal Dam) and the remaining 68% from the Vaal Dam. As most of the algal related problems originate from the Barrage Reservoir the following costs are based on the 600 000 m^3/day that is currently being pumped from that impoundment.

1.1   Pre-oxidation

To ascertain the financial impact of pre-oxidation on purification, it was assumed that such treatment was necessary for 25%, 50% and 100% of the time and such costs are shown in Table 1.

1.1.1   Chlorine

A chlorine dosage level of 5 mg/ℓ was selected as it was found necessary to satisfy the chlorine demand of the raw water and to assure a residual chlorine level of 0,25 - 1,0 mg/ℓ which according to literature (Courchene, 1971; Holden, 1970; McKee and Wolf, 1963 and White, 1972) was necessary for effective control. Costs are based on an average chlorine price of R1,83/kg.

To facilitate future pre-chlorination requirements the Board is at present constructing pre-chlorination plants at it's three river intakes at an estimated total cost of R800 000.

1.1.2   Chlorine dioxide

During recent studies by the Board the algicidal properties of chlorine and chlorine dioxide were compared and it was found that chlorine dioxide had a dramatic affect on chlorophyll values after 1 hour contact time and that very little regrowth had occurred after incubation under ideal conditions for 7 days. With regard to chlorine only a marginal change was discernible after one hour contact time. Regrowth after 7 days was, however, notably reduced at chlorine levels above 3 mg/ℓ.

Taking cognisance of the fact that the tolerance limits of various algal species differ, it was assumed that a chlorine dioxide dosage equal to 20% of the chlorine dosage would be required for the same measure of control. Therefore a chlorine dioxide dosage of 1 mg/ℓ was taken for cost purposes. The production costs of chlorine dioxide were calculated to be R18,16/kg of ClO_2.

2.   MICROSCREENING

From literature and preliminary work done by Cravens and Van Den Broeck, (1983), in South Africa it seems that the maximum mesh size that can be used in microscreens for the effective removal of algae such as Tetraselmus Chlamydomonas and Euglena is 6 μm.

Removal efficiencies on samples monitored by the Board during the above tests were of the order of 50% to 60%. It would thus seem that the mesh size used may have to be reduced to 1 μm. If this would be required it would imply that the number of units required would increase five fold. Using data obtained from a company supplying microscreens in South Africa it was calculated that the Board would require sixty 3 x 4,9 m units. These calculations were based on microscreens using a 6 μm mesh at a loading rate of 15 m^3/h.m^2. Estimated unit costs are 2,3 cents/m^3 of water treated.

3.   DISSOLVED AIR FLOTATION

Because of the buoyant nature of algae, flotation is considered one of the most suitable methods for the removal of algae and at present there are quite a number of units in operation throughout the country (Van Vuuren, et al, 1981; Bernstein, et al, 1985; and Vosloo, et al, 1985). Experimental work done by the Board using an on filter flotation pilot plant designed and built by the National Institute for Water Research indicated that algal removal efficiencies of up to 95% were possible.

One of the advantages of dissolved air flotation is that it can be combined with rapid gravity sand filters into a single process unit.

Using data obtained by the Board during its pilot plant studies, data obtained from the Mzingazi Works (Rademan and Vosloo - personal communication) and information based on work done by Van Vuuren, et al, (1983), a cost estimate was prepared and it was estimated that the unit costs would be 3,15 cents/m³ of water treated.

4.   ACTIVATED CARBON

The presence of algae in water may give rise to colour, tastes, odours and other undesirable organic compounds such as trihalomethane precursors (Palmer, 1970; Viljoen, 1984; Hoehn, et al, 1980; Oliver, 1980). With regard to the Board's water supply increased chlorine demand and regrowth in the distribution system, as a result of the presence of organic substances in the water, would be the major reasons for the inclusion of activated carbon as a unit process.

As colour, taste and odour problems are experienced in the production of water from both the Barrage and Vaal Dam, costs of activated carbon treatment are based on the total amount of potable water presently being supplied by the Board (2200,000 m³/day). The capital outlay of a GAC plant to meet the Board's needs would probably exceed R200 000 000 and the running costs inclusive of capital redemption is estimated to be 9 cents/m³. Due to the large capital requirement for a granular activated carbon plant it may be possible, because of the intermittent occurrence of taste and odour problems, to use powdered activated carbon.

Burlingame, et al., (1986) reported on the removal of the odourous metabolites associated with geosmin and found that 43% removal required a dosage rate of 12 mg PAC/ℓ and 73% a dosage rate of 42 mg PAC/ℓ. The Umgeni Water Board who are experiencing geosmin related problems has found a PAC dosage rate of 17 mg/ℓ sufficient to solve their problem.

The capital outlay for a PAC dosage plant capable of the required dosage rate is estimated at R10 000 000 and depending on the concentration levels of

TABLE 1:   A SUMMARY OF INCREASES IN WATER PURIFICATION COSTS AS A RESULT OF EUTROPHICATION APPLYING VARIOUS ALGAL CONTROL MEASURES.

MEASURE ADOPTED	PERIOD OF TIME IN USE % OF YEAR	TOTOAL COSTS PER ANNUM	UNIT COST c/m³
Chlorination+	25%	R 504 000	—
	50%	R 1 007 000	—
	100%	R 2 014 000	0,92
Chlorine dioxide+	25%	R 0 985 000	—
	50%	R 1 971 000	—
	100%	R 3 942 000	1,8
Microscreening+	100%	R 5 040 000	2,3
Flotation+	100%	R 6 895 000	3,15
Activated Carbon PAC*	25%	R 26 097 000	—
Based on Dosage of 42mg/ℓ	50%	R 52 195 000	—
	100%	R104 390 000	13
Activated Carbon GAC*	100%	R 72 270 000	9

+ Costs are for a 600 000 m³/day plant.
* Costs are for the total amount of water supplied by the Board. (2200,000 m³/day)

the offensive odours and other organics present in the water the unit treatment costs could vary from 3,6 cents/m³ to 13 cents/m³ with total costs of between R72 000 and R260 000 per day.

Using the preceding information table 1 was compiled summarising the impact that eutrophication may have on the cost of water purification. It should, however, be pointed out that increased disinfection costs (of the water in supply) as a result of the presence of algae and costs due to additional man-power requirements were not included in all cases.

From table 1 it is evident that the increases in water purification costs as a result of eutrophication can vary drastically due to the extent of the measures that need to be taken to assure a wholesome water.

5.   WATER PURIFICATION (SEWAGE)

The catchment of the Vaal Barrage Reservoir is densely populated and heavily industralised, this has resulted in a point source production of phosphorus that exceeds the diffuse source (Grobler and Silberbauer, 1984). A 1 mg/ℓ phosphorus discharge standard is mandatory and for this reason sewage purification works have to add chemicals; periodically in the case of activated sludge units and continuously in the case of old biological filter plants.

It is estimated that due to seasonal and operational difficulties it may be necessary to chemically dose activated sludge plants at least 25% of the time. Costs are based on a return flow to the Barrage of 302 529 m³/day from bio-filter plants and 327 743 m³/day from activated sludge units.

Although it is difficult to arrive at a very accurate figure it is roughly estimated that the cost to chemically dose for phosphorus control is approximately R4 500 000 per annum. (Chemical cost 3,2 cents/m³).

6.   INDUSTRY

Eutrophication related problems experienced by industry includes blockages of intake pumps, canals and pipes and general unacceptability of the water quality for specific industrial applications. It is necessary to mention that the water quality required for many industrial applications is far more stringent than the requirements for potable water. Unfortunately, data applicable to this study are unavailable.

7.   AGRICULTURE

Many algal species are toxic to stock and the presence of these species may render water unsafe for animal watering. Over-abundant growths of filamentous algae and macrophytes may block pump intakes and irrigation systems. Bruwer, (1979) studying the effect of abundant growth of algae and macrophytes in irrigation canals has shown that the cleaning of these canals may amount to several thousands of rands. As for industry very little applicable data for the purposes of this study are available.

8.   RECREATION

During discussion with various sociologists, officials of the Department of Sport and Development and the Human Sciences Research Council, the importance of recreation was stressed. Water-based activities such as boating, boating related sport, riverbank recreation and angling, constitutes an important component of a communities recreational requirements. Unfortunately constraints, mainly as a result of eutrophication related problems, significantly affect the quality of water-based recreation both on or near any waterbody. The costs related to recreation are determined for the Rand Water Board Barrage Reservoir which has an area of 1 683 hectares, is 62 km long, and has a shoreline of 140 km.

8.1   Boating and related activity

Boating activity and associated sport such as water skiing is one of the most significant recreational demands made on an aquatic environment. For a waterbody to be acceptable for boating and related sports certain requirements are considered necessary namely, a water surface free of plant growth, bank accessibility, and to a lesser extent clean water, except in the case of water contact sport such as water skiing where algal rich waters would be less acceptable.

In a boating situation certain weed growth could be tolerated for example limited emergent bank vegetation, however, excessive submerged growth (eg. Myriophyllum) or floating growth (eg. Eichhornia) would preclude any form of boating due to the snagging of propellers or obstruction of water ways.

Algal rich waters although navigable would be aesthetically unacceptable and could stain the boats etc.   Such conditions would be less acceptable to water contact sportsmen for various reasons such as staining of clothes possible dermal reactions, etc.

The financial implications of boating and related sport is very difficult to define and the amount spent on boating tabulated below only reflects the available information on power boating and does not include the sums of money spent on yachting, board sailing, purchase of equipment or the casual boat that may be launched at a public launching ramp.   All the costs and petrol consumption reflected below are average figures compiled after discussions with boat users, boat dealers, etc.

TABLE 2:   A SUMMARY OF THE ESTIMATED EXPENDITURE BY BOATERS

	Unit cost	Total cost	Remarks
Boats*	R30 000	R102 000 000	Both clubs and riparian
Fuel usage (average consumption 14 kg/h)*	118c/ℓ	R 941 640	Total fuel costs based on boating hours
Running costs including repayment of boat	R8290	R 28 186 000	
		R 29 127 640	

* Estimated number of power boats using the Barrage per annum 22 800
   1 Litre = 0,8 kg.

The above sum of money represents a yearly expenditure of R17 307 per hectare based on this surface area of the Barrage or R208 054 per kilometre based on the length of the Barrage (Both banks).

8.2   Fixed property

The Vaal Barrage is considered to be unique, and is often called the Venice of South Africa with its 62 kilometres of navigable river kept at a constant level by the Barrage structure. The riparian plots have free access to the river and as such command high prices. The value placed on proximity to the river is reflected in the elevated prices paid for houses built near the river in comparison with the prices paid for houses situated a distance from the river. The most compelling reason why people buy these riparian properties is that they wish to be near water and a close second is the availability of the river to practice water sport.

From similar surveys carried out in America (Harris, et al., 1979) it would appear that riparian property owners place a very high premium on the aesthetic aspect of living near a waterbody; it therefore stands to reason that any diminution in the quality of the riverine surroundings would have a drastic affect on the value of properties riparian to a river or lake.

The average value of a house riparian to the river is R550 000 and land values R80 000/ha. Chalets, etc. leased by clubs are in excess of R80 000 and villas currently being built will be sold in excess of R320 000 per unit.   The total value of all the fixed property excluding farms is approximately R345 106 000 and if the redemption of capital is taken over twenty years at say

18% interest then the annual input into the economy from this source would be R62 000 000; an investment per kilometre river frontage in the Barrage of approximately R443 000. The monetary value of fixed property riparian to the Barrage, works out approximately R2 500 000 per kilometre of water frontage. If, therefore, the aesthetic quality of the Barrage would reduce the property value by only 5%, this would amount to R125 000/kilometre river frontage or R17 500 000 for the Barrage.

8.3 Angling

The principal effect of eutrophication on fish is one of oxygen depletion that may result in fish mortality. Welch E.B., (1980) is of the opinion that the long-term effect of eutrophication on fish would be one of changed species composition that would tend to favour the less desirable fish species such as the detritus/bottom-feeding fish. On both accounts the angler would be adversely affected as both the number of fish and the desirable species would be affected. Excessive algal and macrophyte growth may render a particular site aesthetically unacceptable to anglers or impede angling because of snagging of line resulting in a reduction in the available angling sites.

Cadieux J.J. (1980) compiled a report outlining the economic impact of angling in the Transvaal for the 1977/78 season (Table 3). Using the values shown in table 3 and the 140 km river frontage of the Barrage the estimated financial impact of anglers per 1 km Barrage river frontage is calculated to be R51 171 per annum.

TABLE 3:   EXPENSES INCURRED BY ANGLERS DURING THE 1977-78 SEASON AND ESTIMATES OF EXPENSES FOR THE 1987-88 SEASON BASED ON AN ANNUAL ESCALATION RATE OF 16% PER ANNUM.

	TRANSVAAL			BARRAGE ESTIMATE FOR SEASON
	1977/78 SEASON		1987/1988	
	Rands/Angler	Total spent*		
Entrance fees	R22,80	R2 587 572	R 6 373 584	R 583 200
Boat expenses	R16,90	R1 917 981	R 4 666 880	R 427 032
Refreshments	R68,30	R7 751 367	R18 864 038	R1 726 110
Transport	R82,80	R9 396 972	R22 868 773	R3 092 550
Accommodation	R19,50	R2 213 055	R 5 385 678	R 492 804
Fishing equipment	R72,90	R8 273 421	R20 133 444	R1 842 264
Total	R283,20	R32 140 368	R78 292 397	R7 163 960

* Figures obtained from Cadieux J.J. (1980)

8.4 River bank recreation

It is very difficult to arrive at a reliable figure as to the number of people utilising the areas upstream of the Barrage for recreational purposes. Figures obtained from various authorities indicated that for the year ending December approximately 510 000 vehicles entered the resorts under their control, this does not include private clubs, private resorts or people using private riparian property for recreational purposes. If we assume an average of 3 persons per vehicle then at least 1 530 000 persons per annum visited the resorts, to this can be added at least another 100 000 from private clubs, etc. and those attending international and local boating events. If a family of 3 spent an average of R32 on entrance fees, refreshments, etc, then the 1 600 000 people spent approximately R 52 200 000 per annum, this represents an expenditure of approximately R373 000 per kilometre of water frontage or R31 000 per hectare of water surface.

When considering the acceptability of various recreational facilities it must be borne in mind that there are several factors which can affect the personal preferences of people with regard to aesthetic value. Because of the difficulty in categorising the individual response to tangible degradation of the physical environment three arbitrary conditions have been chosen to represent a 25, 50 and 100 percent reduction in usage due to any of the many manifestations of eutrophication. If the reduction in usage is taken into account then the financial loss to the economy relative to the Barrage only is reflected in table 4.

TABLE 4   THE FINANCIAL IMPACT OF A REDUCTION IN RECREATIONAL USAGE OF THE BARRAGE DUE TO EUTROPHICATION. (ESTIMATES ARE FOR THE 1987/88 SEASON)

TYPE OF ACTIVITY	25% REDUCTION	50% REDUCTION	100% REDUCTION
Boating	R 7 281 910	R14 563 820	R29 127 640
River bank recreaction	R13 050 000	R26 100 000	R52 200 000
Fishing	R 1 790 000	R 3 581 982	R 7 163 964
Total	R22 122 900	R44 245 800	R88 491 600

9.   CONCLUSIONS

The Rand Water Board's Barrage Reservoir is probably the most exploited waterbody in the Republic of South Africa. As a result of this factor it may not be possible to relate the financial assessments made in this paper to other waterbodies, however, the authors are of the opinion that should the same intensity of exploitation take place then the cost pattern would be of the same order.

One of the most significant aspects that has come out of this survey is the tremendous monetary value that can be placed on the exploitation of water for recreational purposes. The money spent on recreational activities in the Barrage area far exceeds the expenditure of local authorities (in the Barrage catchment) on reduction of phosphorus for nutrient control; this factor on its own is a strong enough incentive to reduce eutrophication without taking into account other requirements.

REFERENCES

BERNSTEIN, M., VAN VUUREN, L.R.J. and BOTES, V. 1985   Pilot and Full-scale Applications of DAF Technology to treat Eutrophic Waters from Lake Nsese. Paper presented at IWPC Biennial symposium.   27-30 May, Durban RSA.

BRUWER, C.A. 1979   The economic impact of eutrophication in South Africa. Technical Report No. TR 94, Department of Water Affairs.

BURLINGAME, A, ROGERS, A.D. and GEOFFREY, C.E. 1986 A Case Study of Geosmin in Philadelphia's Water. Jour. AWWA (March 1986). pp. 56-51.

CADIEUX, J.J. 1980   Freshwater Angling in the Transvaal.   Project TN 6/4/7/3. Nature Conservation Division. Transvaal Provincial Administration.

COURCHENE, J.C. and CHAPMAN, J.D. 1975   Algae Control in Northwest Reservoirs. Water Technology/Quality. Journal AWWA. March 1975.

CRAVENS, J. and VAN DEN BROECK, J. 1983   Pilot Evaluation of Microscreens for algae in South Africa. Biennial Conference and exhibition of the Institute of Water Pollution Control (S.A. Branch).   East London. 16-19 May 1983.

GROBLER, D.C. and SILBERBAUER, M.J. 1984   Impact of Eutrophication Control Measures on the Trophic Status of South African Impoundments. Report No. 130/1/84 of the Water Research Commission., South Africa.

HARRIS, G., GROVER, A., HALE, B and HEDIN, R. 1979   Environmental Management. Vol 3. No. 3.

HOEHN, R.C. et al. 1980 Algae as Sources of Trihalomethane Precursors. Journal AWWA. June 1980. pp. 344-350.

HOLDEN, W.S. 1970 Water treatment and Examination. J.A. Churchill, London.

McKEE, J.C. and WOLF, H.W. 1963 Water Quality Criteria. 2nd Ed. State Water Quality Control Board. Publication No. 3A.

OLIVER, B.G. and SHINDLER, D.B. 1980 Trihalomethanes from the Chlorination of Aquatic Algae. Environ. Sci. Technoly. Vol 14, No. 12. December 1980.

PALMER, C.M. 1980 Algae and water pollution. Castle House publications.

STEYNBERG, M.C., VILJOEN, F.C. and PIETERSE, A.J.H. 1985 Eutrophication in the Vaal Barrage: Past, Present and Future. Paper presented at IWPC biennial symposium 27-30 May, Durban. RSA.

VAN VUUREN, L.R.J. and DE WET, F.J. 1981 The application of the flotation/filtration process to water and effluent treatment. Paper presented Symposium on Industrial Effluent: Control and Treatment, 23 November, Pretoria.

VAN VUUREN, L.R.J., DE WET, F.J. and CILLIE, G.G. 1983 Treatment of Water from Eutrophied Impoundments. Water Supply. Vol. 1 No. 1 pp. 145-156.

VILJOEN, F.C. 1984 The necessity of phosphate restrictions within the Vaal River Barrage Catchment. Imiesa 9. (9) 11-29.

VOSLOO, P.B.B., WILLIAMS, P.G. and RADEMAN, R.G. 1985 Pilot and full-scale investigations on the use of combined dissolved air flotation and filtration (DAFF) for water treatment. Paper presented at IWPC biennial symposium 27-30 May. Durban. RSA.

WELCH, E.B. 1980 Ecological effects of waste water. Cambridge University Press, Cambridge.

WHITE, G.C. 1972 Handbook of Chlorination. Van Nostrand Reinhold Company.

Chapter 16
WATER POLLUTION AND HEALTH

Z J Brzezinski (Institute of Mother and Child, Poland)

1 – INTRODUCTION

Water pollution is defined as an alteration in its composition or condition so that it becomes less suitable for any or all of the functions and purposes for which it would be suitable in its natural state [1]. This alteration includes changes in the physical, chemical and biological properties of water resulting from discharge of liquid, gaseous or solid substances into water which create nuisances or render such waters harmful to public health.

Water pollution is mainly caused by the disposal of sewage and other liquid wastes originating from domestic use of water industrial wastes, agricultural effluents from animal husbandry and drainage of irrigation water, and urban run-off. Another cause of pollution is spreading of chemicals on the land to increase crop yields or the addition of chemicals to water control undesirable organisms.

Human health may be affected by ingesting water directly or in food, by using it in personal hygiene or for agriculture, industry or recreation.

2 – AN OVERVIEW OF HEALTH HAZARDS OF WATER POLLUTION

Health hazards originating from water pollution can be divided into two main categories: biological agents that may affect man following ingestion of water or other forms of water contact, or through insect vectors; and chemical and radioactive pollutants resulting from discharges of industrial wastes.

2.1 – Biological agents

The most common health risk associated with drinking-water is contamination by sewage, by other wastes, or by human or animal excrement. Faecal pollution may introduce a variety of intestinal pathogens, bacterial, viral and parasitic.

2.1.1 – Bacterial pathogens

Intestinal bacterial pathogens include, strains of Salmonella, Shigella, enterotoxigenic Escherichia coli, Vibrio cholerae, Yersinia enterocolitica and Campylobacter fetus. They may cause diseases that vary in severity from mild gastroenteritis to severe dysentery, cholera and typhoid.

Other organisms, naturally present in the environment and not regarded as pathogens, may also cause occasional disease. Such organisms in drinking-water may be responsible for infection among people whose local or general natural defence mechanisms are impaired.

Potable water, if it contains excessive numbers of organisms such as Pseudomonas, Flavobacterium, Acinetobacter, Klebsiella and Serratia is capable of producing infections involving the skin, and mucous membranes of the eye, ear, nose, and throat [2].

2.1.2 – Virological pathogens

Viruses of main concern regarding waterborne transmission of infectious disease are that multiply in the intestine and are excreted in large numbers in the faeces of infected individuals. Enteric viruses may cause various conditions including gastroenteritis, myocarditis, meningitis, respiratory disease and hepatitis. Asymptomatic infections are common and serious manifestations rare. If drinking-water is contaminated with sewage gastroenteritis and infections hepatitis may occur in epidemic from [3].

2.1.3 – Pathogenic protozoa

Three intestinal protozoa pathogenic for man may be transmitted by drinking-water: Entamoeba histolytica, Giardia spp., and Balantidium coli.

Infections with the above protozoa may be asymptomatic or may result in clinical manifestations of amoebiasis, giardiasis and balantidiasis, respectively.

Free-living amoeba can also be waterborne agents of disease. Infection with these organism is associated with recreational contact.

2.1.4 – Helminths

A variety of helminth eggs larvae infective to man may be present in drinking-water. They are divided three groups. Group I includes helminths acquired by man drinking-water containing the intermediate host crustacea. The most important member of the group is Dracunculus medinensis which causes an infection (dracontiasis) leading to a major disabling disease.

Spirometra, another member of group I, is responsible for less common disease sparganosis.

Group II comprises flukes and roundworms whose infective larvae are able to penetrate the human skin and mucous membranes. The main genus concerned is Schistosoma responsible for human schistosoma infections.

Group III includes intestinal helminths. In the case of all the species of this group, other methods of transmission (food and direct faecal-oral routes) are more common than drinking-water route.

2.1.5 – Free-living organisms

The free living organisms important in water supplies include plankton and macroinvertebrates. There is evidence that some toxic substances produced by certain algae in water supplies may have a significant adverse effect on public health [2].

2.2 – Chemical agents

The health risk due to toxic chemicals in drinking-water differs from that caused by biological contaminants. It is very unlikely that any one substance could result in an acute health problem except under exceptional circumstances, such as massive contamination of the supply.

The problem associated with chemical constituents arise primarily from their ability to cause possible adverse effects after prolonged periods of exposure [4].

2.2.1 – Inorganic constituents

In a recent WHO publication on drinking-water quality quoted above [2.4] the health related information on 37 inorganic constituents were examined and for only 9 of them guideline values could be recommended (arsenic, cadmium, chromium, cyanide, fluoride, lead, mercury, nitrate and selenium). Comments were also given on other 7 constituents that are the subject of current debate (asbestos, barium, beryllium hardness, nickel, silver and sodium).

Most of the evidence concerning possible adverse effect of inorganic constituents on human health comes from toxicological studies, occupational exposure or episodes of acute poisoning. Less often it has bean demonstrated by population based epidemiological studies. In this connection asbestos can be mentioned because of its possible causal relationship with cancer of gastrointestinal tract [5].

Water hardness attracted an interest of many investigators because of its inverse correlation with mortality from cardiovascular disease. However, as some more recent studies show the size of the effect of hardness appears to be rather small and it has not been possible to identify the specific water constituent that is responsible [6].

It has been shown that excessive amounts of nitrate in drinking-water may cause infantile methaemoglobinaemia [7].

Nitrate has been also implied in causation of cancer, although a study failed to find an evidence of a positive association between nitrate levels in the treated drinking-water and mortality from all cancers or stomach cancer in particular, in the urban areas of the UK [8].

Raised sodium intake has been linked with hypertension and it has been found that school children living in areas with higher levels of sodium in the drinking-water in comparison with those living in areas of lower levels had higher blood pressure [9].

It has been though that orally ingested aluminium compounds do not have deleterious health effects on normal individuals [4]. It has been associated, however, with dialysis encephalopathy [10]. Recently, a relationship has been found between Alzheimer's disease and aluminium in drinking water [11].

2.2.2 – Organic contaminants

More than 3000 organic chemicals have been identified as water pollutants and of these over 600 were found in drinking-water. Many of them are pharmacologically active and hazardous to health: 20 are recognized carcinogens, 23 suspected carcinogens, 18 carcinogens promoters and 56 are mutagenic.

Due to the progress in analytical technology many of organic pollutants can be identified, characterized and quantified at very low level. Nevertheless, the identified organic contaminants represent some 10% of the total. The problem of organic contamination has grown in importance since the discovery that disinfection of public water supplies has led to the proliferation of by-products resulting from interaction between natural organic material in the water and the disinfectant [12].

The experts preparing new WHO standards found the task of making recommendations increasingly complicated because of the uncharacterized organic fraction consisting of non-volatile substances the determination of which still presents a considerable analytical challenge [4]. They selected for detailed evaluation the organic substance that met following criteria:
   – well-founded evidence that the substance can cause acute or chronic illness,
   – evidence that the substance is known to occur at significant concentrations,
   – evidence that the substance has been detected relatively frequently in drinking-water,
   – the availability of analytical methods for monitoring and control purposes,
   – evidence that the concentrations of the substance in water can be controlled.

Guideline values were established for 15 compounds (aldrin and dieldrin, benzene, benzo(a)pyrene, chlordane, chloroform, 2,4 D, DDT, 1,2-dichloroethane, 1,1-dichloroethane, heptachlor and heptachlor epoxide, hexachlorobenzene, lindane, metoxychlor, pentachlorophenol and 2,4,6-trichlorophenol) and tentative values were set for 3 others (carbon tetrachloride, tetrachloroethene and trichloroethene).

Similarly, as in case of inorganic contaminants, evidence that organic compounds can be harmful to health is based on the results of toxicological studies, occupational exposure and acute poisoning. Epidemiological evidence deriving from population based studies is rather limited.

2.3 – Radioactive materials

The effects of radiation exposure may be somatic if they become manifest in the exposed individual or hereditary if they affect his descendants. The most important delayed somatic effects of radiation exposure are malignant neoplasms. The WHO recommended guidelines values for gross alpha activity and for gross beta activity are 0.1 Bq/litre and 1 Bq/litre respectively [4].

3 – HEALTH IMPACTS OF COMMUNITY WATER SUPPLIES

Following the sanitary revolution there was a decline in the occurrence of diseases associated with poor sanitation and crowded environments [13]. Much of this decline was due to the improvement of water supply and sewage disposal. For example, introduction of water filtration in 20 American cities resulted in 65% fall in typhoid mortality. Similar fall was observed in 14 Indian towns following introduction of water purification half century later than in the U. S. cities [14].

There were numerous similar observations showing that the experience of developed countries with regard to the reduction of gastroenteric infections was shared by less developed countries with proper time adjustment [15].

However, although available literature shows beneficial health impacts following improvements in water supply, the exact quantification of the health effect of this factor alone presents serious difficulties.

First, it is linked very closely with the overall sanitation. Secondly, parallelly to the improvement in water supply, other important changes in living conditions, nutrition or health care took place.

Therefore, the actual health benefit represents the outcome of many influence and depends on variety of factors. To clarify this issue a general theory on the relationship between water supply and sanitation and health, the threshold-saturation theory has been proposed [15]. It takes into consideration three variables: health status, socioeconomic status, and sanitation level, and attempts to encompass, for the first time in one general theoretical framework, numerous conflicting empirical findings. The two-tiered S-shaped logistic form of the relationship that is proposed assumes that at the lower end of the socioeconomic spectrum there is a threshold below which investments in community water supplies and/or excreta disposal facilities alone result in little detectable improvement in health status. Similarly, at the higher end of the socioeconomic scale, it is suggested that a point of saturation is reached beyond which further significant health benefits cannot be obtained by investments in conventional community sanitation facilities.

In a recent review Cvjetanovic [16] considering the same problem has pointed out that the direct effects of water supply and sanitation on water related and sanitation-related diseases are only one of the components of health benefits measured by a decrease in incidence of these diseases. Another frequently more important effects is the indirect impact of water supply and sanitation through socioeconomic, educational and other improvements on the health status (including nutrition) of the population.

The author has underlined that the impact of water supply depends on the economic status and the epidemiological situation of the community. On the other hand socioeconomic is and important factor for the success of water supply and sanitation programmes and for achieving health benefits.

4 – METHODOLOGICAL PROBLEMS

Most health effects result from multiple influences. Similarly, environmental hazards are frequently interrelated. This creates a need for strict methodological discipline in any environmental health research including studies in water and health [17]. Many investigations have provided inconclusive, confusing or even contradictory results a large proportion of which are due to methodological deficiencies [18]. A review of the published literature on the impact of water supply and/or excreta disposal facilities on diarrhoeal diseases, or on infections related to diarrhoea revealed several methodological problems such as lack of adequate control, the one to one comparison, confounding variables, health indicator definition, failure to analyse by age, or failure to record usage and the seasonality of impact variables [19].

In many studies aggregated data were used. Some more recent examples of such investigations include studies in cancer mortality and fluoridation of water supply [20], or chemical toxic waste disposal sites [21] or type of water source [22].

The interpretation of the results of studies based on aggregated data presents difficulties because of possible confounding variables. An interesting example is a study on the incidence of cancer and reuse of drinking-water [23]. Areas supplied with water of different degrees of reuse were correlated with cancer incidence. Positive associations were found between the average percentage of domestic sewage in the water supplied to an area and the incidence of stomach and urinary cancers in females. However, there was also an association between two variables and socioeconomic characteristics. The associations between water reuse and cancer were reduced when social factors and variations in area size were taken into account.

Due to methodological difficulties there are few investigation using more refined epidemiological designs. In this connection two recent examples both based on case-control design, are of interest. One was a study of colon cancer and drinking-water trihalomethanes [24]. Second, investigated the association between chemicals in material drinking-water consumed during pregnancy and congenital heart disease in the offsprings [25]. In both studies efforts were made to quantify individual exposure. This problem appears to be primary importance as it has been demonstrated by a study in which the assumption that people using the same water supply have similar intakes of minerals from drinking water was examined. Duplicate samples of all water drunk during a 24 hour period, including that boiled for beverage preparation, were collected. The results showed that there may be more than a tenfold variation in the amount of water people drink daily. In addition it was found that the mineral concentrations in the 24 hour samples can very markedly from those in water collected from the source of supply or from household taps which are the usual sampling points for epidemiological studies [26].

The advantage of the improved design of an epidemiological study was well demonstrated by a recent investigation on water supply, sanitation and the risk of infant mortality from diarrhoea [27]. This population-based case-control study revealed an association between the availability of piped water and diarrhoea mortality of a greater magnitude than had been suggested by most earlier studies of overall infant mortality. After allowing for the piping of water, both source of the water supply or type of sanitation were no longer associated with mortality.

5 – CONCLUDING REMARKS

The increased availability of controlled water supply and sanitation along with other public health and social measures have reduced the incidence and effects of water related infectious diseases both in developed and less developed countries. However, in spite of the progress made in the safe water availability there are deficiencies resulting in outbreaks of waterborne diseases and the risk related to the consumption of drinking water which do not meet the bacteriological standards still exists even in the developed parts of the world [28,29].

The new challenge is created by the technological development leading to increasing environmental pollution by chemical compounds and in particular by synthetic organic chemicals.

The volume of knowledge on the relationship between health water is impressive. However, in spite of much evidence that water pollution affects human health, there are still gaps and uncertainties concerning the nature of the effect of many pollutants or the identify of factors involved in causation of health damage.

More and better studies are needed in the health impact of water supply and sanitation. These may best be undertaken by the combined efforts of engineers, social scientists and epidemiologists [19]. The increasing pollution of surface waters used for recreational purposes also requires more attention and research [30,31].

REFERENCES

[1] Health hazards of the human environment. Geneva, World Health Organization, 1972.

[2] Guidelines for drinking-water quality. Vol. 2. Health criteria and other supporting information. Geneva, World Health Organization, 1984.

[3] Lund, E. – Waterborne virus diseases. Ecology of Disease, i: 25-35, 1982.

[4] Guidelines for drinking-water quality. Vol. 1. Recommendations. Geneva, World Health Organization, 1984.

[5] McCabe, L. J. and Millette, J. R. – Health effects and prevalence of asbestos fibres in drinking water. Proceedings of the American Water Works Association Annual Conference, San Francisco, 24-29 June 1979, Denver, CO, AWWA.

[6] Lacey, R. F. and Sharper, A. G. – Changes in water hardness and cardiovascular death rates. *International Journal of Epidemiology*, 13: 18-24, 1984.

[7] Nitrates, nitrites and N-nitroso compounds. Environmental Health Criteria 5. Geneva, World Health Organization, 1978.

[8] Beresford, S. A. A. – Is nitrate in drinking water associated with the risk of cancer in the urban UK? *International Journal of Epidemiology*, 14: 57-63, 1985.

[9] Tuthill, R. W. and Calabrese, E. J. – Drinking water sodium and blood pressure in children: a second look. *American Journal of Public Health*, 71: 722-729, 1981.

[10] Schreeder, M. L. *et al.* – Dialysis encephalopathy and aluminium exposure an epidemiologic analysis. *Journal of Chronic Diseases*, 36: 581-593, 1983.

[11] Martyn, C. C. *et al.* – Geographical relation between Alzheimer's disease and aluminium in drinking water. *The Lancet*, 1: 59-62, 1989.

[12] Drinking-water quality and health related risk. Copenhagen, World Health Organization, 1987.

[13] McKean, T. and Record, R. G. – Reasons for the decline of mortality in England and Wales during the nineteenth century. *Population Studies*, 9: 119-141, 1955.

[14] McJunkin, F. E. – Water supply and health: an overview. The impact of interventions in water supply. Proc. Sem., PAHO, Washington, U. S. Agency for International Development, 1981.

[15] Shuval, H. I. *et al.* – Effect of investments in water supply and sanitation on health status: a threshold-saturation theory. *Bulletin of the World Health Organization*, 59: 243--248, 1981.

[16] Cvjetanovic, B. – Health effects and impact of water supply and sanitation. *World Health Statistics Quarterly*, 39, 105-117, 1986.

[17] Guidelines on studies in environmental epidemiology. Environmental Health Criteria 27. Geneva, World Health Organization, 1983.

[18] Esrey, S. A. and Habicht, J-P. – Epidemiologic evidence for health benefits from improved water and sanitation in developing countries. *Epidemiologic Reviews*, 8: 117-128, 1986.

[19] Blum, D. and Feachem, R. G. – Measuring the impact of water supply and sanitation investments on diarrhoeal diseases: problems of methodology. *International Journal of Epidemiology*, 12: 357-365, 1983.

[20] Chilvers, C. – Cancer mortality and fluoridation of water supplies in 35 US cities. *International Journal of Epidemiology*, 12: 397-404, 1983.

[21] Najem, G. R. *et al.* – Clusters of cancer mortality in New Jersey municipalities; with special reference to chemical toxic waste disposal sites and per capita income. *International Journal of Epidemiology*, 14: 528-537, 1985.

[22] Carpenter, L. M. and Beresford. S. A. A. – Cancer mortality and type of water source: findings from a study in the UK. *International Journal of Epidemiology*, 15: 312-320, 1986.

[23] Beresford, S. A. A. – Cancer incidence and reuse of drinking water. *American Journal of Epidemiology*, 117: 258-268, 1983.

[24] Young, T.B. *et al.* – Case-control study of colon cancer and drinking water trihalomethanes in Wisconsin. *International Journal of Epidemiology*, 16: 190-197, 1987.

[25] Zierler, S. et al. – Chemical quality of maternal drinking water and congenital heart disease. *International Journal of Epidemiology*, 17: 589-594, 1988.

[26] Gillies, M. E. and Paulin, H. V. – Variability of mineral intakes from drinking water: a possible explanation for the controversy over the relationship of water quality to cardiovascular disease. *International Journal of Epidemiology*, 12: 45-50, 1983.

[27] Victora, C. G. *et al.* – Water supply, sanitation and housing in relation to the risk of infant mortality from diarrhoea. *International Journal of Epidemiology*, 17: 651-654, 1988.

[28] Hartemann, Ph. *et al.* – Epidemiology of infectious diseases transmitted by drinking water in developed countries. *Rev. Epidem. et Santé Publ.* 34: 59-68, 1986.

[29] Ferley, J. P. *et al.* – Étude longitudinale des risques lies a la consommation d'eaux non conformes aux normes bacteriologiques. *Revue Epidemiologique et Santé Publique*, 34: 89-99, 1986.

[30] Calderon, R. and Mood, E. W. An epidemiological assessment of water quality and "swimmer's ear." *Archives of Environmental Health*, 37: 300-305 (1982).

[31] Philipp, R. *et al.* – Health risks of snorkel swimming in untreated water. *International Journal of Epidemiology*, 14: 624-627, 1985.

PART III

Groundwater and river contamination

Chapter 17

INTEGRATED APPROACHES TO SOIL AND GROUNDWATER POLLUTION ABATEMENT AND REMEDIATION

P Korfiatis (Stevens Institute of Technology, New Jersey, USA)

ABSTRACT

Soil and groundwater contamination has become a problem of multifaceted complexity in the United States and in many other parts of the industrialized world. This paper presents and analyzes integrated approaches to the solution of the two fundamental problems, namely the reduction of the amount of contamination entering the subsurface environment and the clean-up of existing contaminatd sites. Several components of integrated approaches such as education and training, research and technology development, public awareness and education, legal, regulatory and compliance issues, international cooperation and social-economic impacts are discussed and their importance and interrelationships are analyzed.

Key Words: Groundwater, contamination, remediation, environment.

1. - THE PROBLEM IN PERSPECTIVE

Groundwater is a natural resource that is essential for the survival and well-being of humanity. Although both quantity and quality are important considerations in the management of the groundwater supplies, the issue of quality will be examined in this paper. Over the past half of this century groundwater quality has been degraded, in some cases very severely, in many parts of the globe. The deterioration of this valuable resource has been accelerated by several factors including but not limited to urbanization, industrialization, exploration of mineral resources, groundwater pumpage, agriculture, land disposal of chemical wastes, and atmospheric deposition.

The groundwater pollution problem is complex with no quick fix solutions. Its complexity stems from its multidimensional nature. In order to understand the magnitude of the problem, the technical, institutional and socio-economic components are presented and analyzed. Although these three components are treated separately they are very much inter-dependent and they influence each other dramatically.

1.1 - Technical

The technical complexity of the groundwater pollution problem can be better understood by examining the fundamental processes governing contaminant transport and fate in the subsurface environment. The problem very often has three distinct components, namely: (a) identification of the pollution source, (b) evaluation of the extend and magnitude of the contamination and (c) remediation and restoration. The magnitude and degree of difficulty of each phase varies significantly. Finding sound solutions for all three components hinges upon the understanding and quantification of the physical, chemical and biological processes that control the soil-water (vapor) - contaminant interactions. The major factors hindering the mathematical description and quantification of these processes include:

. The highly heterogeneous, and anisotropic nature of soil and rock geologic formations

. The enormous numbers of natural and synthetic chemicals with diverse adsorption, desorption and biodegradation properties

. The complex electrochemical forces controlling chemical - clay mineral interactions

. The inherent complexities of multi-phase transport and associated competitive effects

. The highly non-linear equations that govern multi-phase transport in porous media

Although all the phases of the groundwater pollution problem are complex the most difficult phase is the remediation and restoration. This difficulty stems from the absence of well developed, appropriate technology for aquifer restoration and the high cost and long time required for its development. Although the technology for groundwater treatment exists, based on developments in wastewater technology, the removal of dissolved or insoluble components of contaminants from the soil matrix is an enormous task. The following factors are responsible:

. The very low flow velocities characteristic of groundwater systems

. The high capillary, and other surface, forces that keep dissolved and undissolved contaminants tightly bound in the soil matrix

. The very often enormous soil surface areas being contaminated

1.2 - Institutional

Aquifers have been polluted throughout the post-industrial revolution period all over the world. The realization however of the severity of the groundwater pollution problem is very recent. The institutional infrastructure therefore, has not developed adequately to provide sound groundwater protection strategies.

For example, it was not until 1972 that significant federal legislation affecting groundwater was passed in the United States. Under the Federal Water Pollution Control Act of 1972, the quality of all U.S. waters became a federal responsibility to be achieved by a joint federal/state/local mechanism. This was followed by the Safe Drinking Water Act of 1974 which created national standards for public water quality and the Resource Concentration and Recovery Act of 1976 which regulates the management of hazardous and toxic wastes. Statutes within these three acts deal either directly or indirectly with the management of the water quality. It was not until 1980 that the Comprehensive Environmental Resource, Compensation and Liability Act (CERCLA) was enacted to discover, investigate and remediate hazardous waste sites. In addition to federal legislation which is enforced by the Environmental Protection Agency, individual states have passed numerous statutes which are enforced by state regulatory agencies. It is interesting to note that water pollution prevention legislation was developed first, followed by legislation targeted toward the clean-up of hazardous waste sites threatening groundwater supplies. Enforcement of mitigation legislation involves recovery of clean-up costs from parties responsible for the contamination. This has proven to be a very difficult task, hindering the clean-up efforts by entering into lengthy and very costly litigations.

The legislators in the United States have underestimated the number of sites requiring clean-up, the cost and difficulty of remediation and recovery of money from responsible parties.

1.3 - Socio-economic

Polluted groundwater and other environmental problems are having an extraordinary impact on the United States society. This impact is realized in two areas: (a) concerns for health implications and (b) cost for remediating and restoring the aquifers.

Concerns for health implications are sometimes based on facts and other times are perceived fears. More and more environmental policies and decisions are made based on risk assessment. The public has been very reluctant to accept added risk, no matter how small, that may result from past or proposed future waste disposal practices. The public lacks the appropriate education and does not trust decisions of the regulatory agencies, in part because they have failed in the past to come up with satisfactory solutions. There is therefore, a large gap in the public's understanding of real and perceived risks which hinders the solution of environmental problems. The problem of waste disposal becomes more complex when public interest groups on the one side and waste producing/handling organizations on the other, influence political powers to shape environmental policy.

In the economic arena, the cost for remediating and restoring the subsurface environment is staggering. The important issue however is that this cost is passed either directly or indirectly to the public. Since 1980 more than four billion dollars have been spent under CERCLA to clean-up only 38 of the 1,175 hazardous waste sites in the national priority list. Several more millions have been spent under state sponsored clean-up programs. It is estimated that it will cost over $500 billion to remediate the more than 30,000 toxic waste sites that have been discovered to date. The list of hazardous waste sites is expanding and the clean-up costs are rising.

2. - INTEGRATED APPROACHES

The interdisciplinary nature of the groundwater pollution problem requires integrated approaches for its solution. Figure 1 identifies six entities which have a major role in the restoration and maintenance of groundwater quality. These entities very often represent opposing interests and are driven by different motivations. The development of integrated solutions begins by indentifying their common interests and needs and assessing their responsibilities. Groundwater management policies must be formulated often taking into account the major concerns of each of these entities. These policies are also linked to the broader environmental picture involving overall water resource management. The composition and role of each entity will be examined in the context of groundwater pollution mitigation, prevention and resource management.

2.1 - Environmental Regulatory Agencies

Environmental regulatory agencies are vested with the authority to formulate, promulgate and enforce regulations and develop policies and management strategies for the protection and restoration of groundwater quality and quantity. The overriding concern in this process is protection of public health. The federal Environmental Protection Agency and Departments of Environmental Protection in each state are the major regulatory organizations in the United States. In addition there are several other agencies on the federal, state, county and municipal levels that are either directly or peripherally concerned with the environment. Regulatory agencies as policy makers must interact effectively with all other entities which influence the process and keep them in a delicate balance. There are several fundamental practices that regulatory agencies must adhere to in order to be efficient and effective, including:

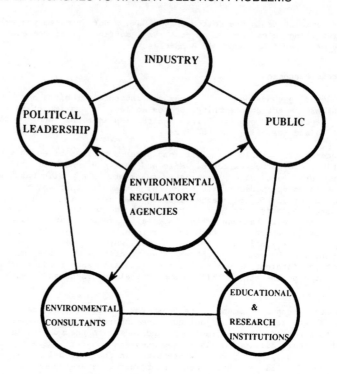

Fig. 1 - Components of an Integrated Approach to Groundwater and Soil
Remediation and Restoration

. Employ people with vision and deep understanding of the technical,
  regulatory and management aspects of groundwater pollution.

. Develop policy and make decisions within a clear and concise
  institutional context.

. Formulate groundwater quality standards and regulations which are
  distinct and are based on sound scientific and engineering analyses.

. Effectively integrate the public in the decision making process and
  gain public acceptance for policies and regulations.

. Provide funding and direction for basic and applied research aimed
  toward the solution of short and long term groundwater quality
  problems.

. Assume leadership in educating all other responsible entities on the
  needs and their participation in integrated approaches for
  groundwater protection.

Very recently, the State of New Jersey released for public comment a
groundwater strategy, elements which are used as an example, NJDEP (1989).
The foremost principles of this strategy are:

a) The groundwater resources are public assets held in trust for the
   citizens and are essential to the health, safety, economic welfare
   or recreational and aesthetic enjoyment of the people.

b) The chemical, physical and biological integrity of groundwaters
   shall be maintained and restored as necessary to protect public

health and the environment to safeground and enhance the beneficial uses of groundwater.

The programmatic objectives for groundwater protection and remedial action have been stated in the New Jersey strategy as follows:

1 - Prevent induced contamination of aquifers due to the effects of pumping wells.

2 - Control discharges to groundwater from facilities or activities with intended discharges that are necessary to the function of the facilities or activities.

3 - Minimize the risk of discharges to groundwater from facilities or activities which are not intended to discharge (e.g., storage tanks).

4 - Control discharges to groundwater from highly diffuse sources (also known as nonpoint pollution sources, such as urban runoff and the land application of chemicals).

5 - Protect public and private water supply wells from pollution.

6 - Minimize the potential for groundwater pollution by improved handling, reduced production and recycling of hazardous substances and solid wastes.

7 - Control the areal extent and concentration of groundwater pollution so as to minimize remedial costs and the impairment of water uses.

8 - Restore polluted groundwater to quality which supports designated existing or potential uses.

9 - Refine understanding of the transport, fate and mitigation of pollution in groundwater and soils.

Although these objectives have been formulated based on the needs and priorities of a specific region with unique problems, KORFIATIS (1987), they can serve as guidelines for targeting groundwater protection efforts in other parts of the world.

2.2 - The Public

The public plays a very vital role in all aspects of groundwater protection. The active, organized participation of the public can have a positive impact on waste reduction, allocation of public funding for groundwater protection, policy making and legislation development. Irrational and irresponsible public involvement however may be detrimental. The following recommendations are believed to make the public participate more effectively in the integrated approach for groundwater restoration.

. The public should be educated as to the severity of the groundwater pollution problem, its role and obligations in the solution of the problem, and the consequences of negative public participation.

. The public should understand environmental risks, in the context of the dynamic nature of today's societal demands and the steps that must be taken to minimize it.

. The public should aggressively participate in recycling, reuse and other waste minimization programs.

. The public should utilize its collective power in a positive fashion to incorporate its concerns in the legislative and regulatory process.

2.3 - Educational and Research Institutions

Educational and research institutions contribute significantly to solutions of groundwater quality problems by providing training, developing new technology and facilitating technology transfer. Training and soil/groundwater remedial technology development are in the forefront of the present environmental field needs.

The multidisciplinary nature of the groundwater quality problem requires scientists and engineers with diversified backgrounds to work cooperatively to develop sound solutions. The following recommendations are made on the basis of present and anticipated future educational needs:

. Undergraduate and graduate curricula in environmental science and engineering must be designed such that students gain hands-on experience and are exposed to the problem in its totality rather than only individual aspects of it.

. Programs must be developed to train the labor force required to carry out the hazardous waste clean-up and aquifer restoration work.

. Continuing professional education programs must be designed to provide systematic re-training of existing professionals to meet short term demand.

. Public environmental awareness programs, especially those targeted to the next generation, must be developed to help facilitate implementation of waste minimization strategies, and teach environmental conservation.

The needs for research and development of new technology for soil and groundwater remediation are immense. A large volume of basic and applied research is presently conducted in the United States and other countries. Because of the urgency of the problem however, the majority of the efforts are aimed to answering short term questions to the detriment of longer term needs. In this context, the following research needs are highlighted:

. Development of appropriate waste disposal technology and practices that will eliminate the introduction of wastes in the groundwater environment.

. Development of in-situ soil and groundwater remediation technology such as bioremediation on the basis of long term benefits.

. Establishment of soil and groundwater clean-up criteria and standards on the basis of minimizing exposure.

. Enhancing our understanding of the processes governing the transport and ultimate fate of various contaminants in the subsurface environment.

. Development of better in-situ monitoring techniques, early warning systems and emergency response technology to minimize the risk for large scale soil and groundwater contamination from leaking underground or above ground storage facilities.

2.4 - Industry

Industries in the United States that produce, store or handle products that can potentially become soil or groundwater pollutants are becoming aware of the high cost of groundwater pollution. This cost is realized frequently as direct expenditures for clean-up, fines to federal or state governments, litigation and court settlements or awards, and loss of competitiveness. This has forced industry to frequently re-evaluate and alter waste treatment and disposal practices in order to comply with strict environmental regulations. The effectiveness of industry as a key player in the development of integrated approaches for soil and groundwater quality protections can be enhanced by:

. Taking steps for reducing the waste stream quantity and strength

. Developing long range plans for environmentally sound waste disposal

2.5 - Political Leadership

Environmental protection has become a highly socio-political issue in the United States. The environment is one of the top issues in most election campaigns across all levels of government as a result of the increased public awareness, sensitivity and concern about the future.

The political machine affects the environmental remediation and protection process by writing environmental legislation and influencing public perception. The sensitivity of elected officials to public opinion forces them to respond to public pressure by making very short term solutions their highest priority. With very few exceptions, legislators have underestimated the complexity of groundwater pollution and have repeatedly promised to the public much more than state-of-the-art technology and the existing infrastructure is able to deliver. The political leadership therefore can become more effective by addressing long term problems within the context of realistic expectations, technical soundness and optimum benefits. Disbursement of funds and political appointments must also be made within the same context.

2.6 - Environmental Consultants

The realization of the magnitude and complexity of the groundwater pollution problem has resulted in a realignment in the priorities of many existing scientific and engineering consulting organizations. The large amount of government and private funding made available for hazardous waste site and groundwater remediation has benefited the environmental profession. The following difficulties however have been encountered:

. Lack of expertise in the areas of groundwater pollution investigations and remediation.

. Excessively high liability risks and strict regulations.

. Lack of well developed technology.

. Lack of previous experiences and case studies upon which to draw conclusions.

Environmental consultants play a vital role in the development and transfer of new technology from experimentation to field application. They therefore must:

. Stay at the cutting edge of the latest developments and state-of-the-art technology.

. Continuously train their employees on new technologies.

. Interact effectively with educational and research institutions and government regulatory agencies.

3. - CONCLUSIONS

Groundwater is a hidden resource and as such is very easily abused. Its quality has deteriorated in many parts of the world. The challenges that lie ahead entail prevention of any further groundwater pollution and discovery and remediation of existing contamination. Both issues are complex and require integrated strategies.

The role of the major entities which influence the solution of the groundwater quality problems has been examined in this paper. The contribution of each of these entities to an integrated approach is unique and necessary in order to maximize the benefits. Although examples are

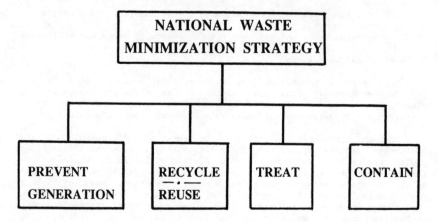

Fig. 2 - Components of a Typical National Waste Minimization Strategy

given on the basis of experiences in the United States, other countries can benefit from these experiences.

Solutions to soil and groundwater contamination problems must be developed in the context of a long term waste minimization strategy that each nation must formulate. The components of a such typical national strategy which are shown in Figure 2 include:

. Prevention of generation of hazardous and toxic wastes that persist in the environment.

. Recycling and reusing at maximum possible rates.

. Treating wastes before discharging into the environment.

. Containing wastes in order to minimize exposure and risk.

4.- REFERENCES

KORFIATIS, G.P., "Groundwater Environmental Concerns in New Jersey U.S.A.", Water for the Future, W.O. Wunderlich and J.E. Prins, Eds., A.A. Balkema Publishing Co., 1987, pp. 533-540.

NJDEP (New Jersey Department of Environmental Protection), "A Groundwater Strategy for New Jersey" Draft for Public Comment, February 1989.

Chapter 18

CONTRIBUTION TO STUDY OF MINERAL WATERS OF GALICIA (SPAIN)

C Baluja-Santos, A Gonzalez-Portal (University of Santiago de Compostela, Spain)

ABSTRACT

This paper study the mineral waters of Galicia (N.W.Spain) flow through granitic and dibasic rocks. In thirty-six mineral waters, co llected directly the springs and classified using physico-chemical and atomic absorption criteria. Most had medium mineral content.

Key words : Mineral waters,classification, physico-chemical and atomic spectrometry analysis.

1 - INTRODUCTION

Determination of the chemical components of natural waters throws light on their origen. The mineral content of waters is proportional to the concentration and distribution of soluble salts in the rocks through which they flow, and also depends on physical factors such as temperature, pressure, flow rate, contact time, diffusion parameters and the solubility of the materials of the rock face, CATALAN-LAFUENTE (1981).
Various classifications of natural waters have been made SHCHUKAREV (1934), SLAVYANOV (1948) on the basis of the predominant ions or salt or dissolved gas content, SCHOELLER (1962). In the first geological study of the medicinal-mineral waters of Galicia (N.W. Spain) IGLESIAS-IGLESIAS (1924) grouped them in four categories on the basis of the salt content data of CASARES-RODRIGUEZ (1837 to 1864) in MAIZ-ELEIZEGUI (1952) TABLE I.
The present communication they are classified in accordance with NOI-SSOTE'S (1961) system on the basis of conductivity, dissolved salts and hardness (in French degrees), moreover iron and manganese, GONZALEZ-PORTAL et al. (1988).

2 - EXPERIMENTAL

2.1 - Sampling and screening

Thirty-six Galician mineral springs were chosen for sampling on the basis of their historical, therapeutical or possible commercial interest.Samples were collected and stored following standard procedures, APHA, et al. (1976) (fig. 1).
The oxidation state of iron and manganese in natural waters depends on the redox character of medium GONZALEZ-PORTAL, et al. (1988), in the study of samples were screened in situ for iron (III) using a plate test with 1 % (w/v) ammonium thiocyanate as reagent. Only five of the samples gave a positive result to test, and it was concluded that all the iron in most of the samples was present as iron(II).

TABLE I. The geological classification of Galicia medicinal
mineral waters, IGLESIAS-IGLESIAS (1924)

Major salts	Location	Nº of springs	Altitude (m)	Rock
SODIUM CHLORIDE and SULPHUR SALTS	Caldelas de Tuy	1	25	Granites and recent alluvia.
	Caldas de Reyes	6	40	Granites.
SODIUM SULPHIDE	Caldas de Cuntis	Several	164	Granites with diluvial outcrops.
	Carballino	1	500	Granites.
	Cortegada *	6	200	Granites.
	Carballo **	2	100	Granites with gneiss and siluric outcrops
	Lugo	4	370	Siluric rock.
SODIUM CHLORIDE	Loujo or Toja	7	0	Granites with fissures and infiltration from the sea
	Arteijo	3	12	Granites.
SODIUM BICARBONATE	Mondariz	2	121	Granites
	Verín	3	380	Granites.
	Molgas	3	580	Granites
	Incio	1	700	Iron-rich primary rock.

* Now under a reservoir.
** Abandoned and on the verge of drying up.

2.2 - Conductivity

The conductivity is a numerical expression of the ability of a water
sample to carry an electric current. This number depends on the total concentration of the ionized substances dissolved in the water * total dissolved solids* and the temperature at which the measurement is made. For the
determination "in situ" using as conductivity cell HACH CHEMICAL Co.(USA)
in ranges 0-200, 200-2000 and 2000-20000 μS/cm and temperature sensor.

2.3 - Dissolved salt content a 103-105 ºC

A well mixed sample is evapored in a weighed capsule and dried to cons
tant weight in an oven at 103-105 ºC, EPHA(1976).

2.4 - Hardness

For determination of hardness using the EDTA (ethylenediamine tetraace-
tic acid) titration method, which measure the Ca and Mg ions EPHA(1976).

2.5 - Calcium, Magnesium,Iron and Manganese

The atomic absorption spectrometry (AAS) is ideal for the determination
of metals in mineral water due to simplicity, rapidity, specificity, and was
consequently rapidly adopte as a standard method by the American Public

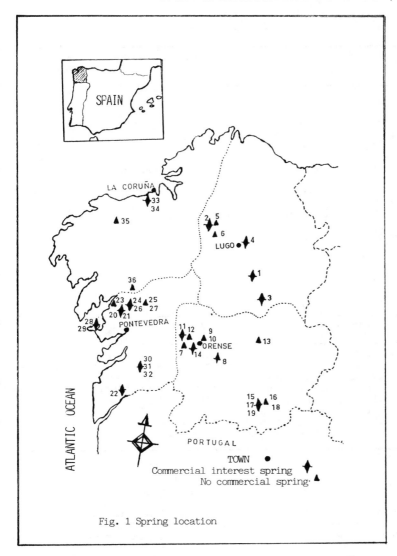

Fig. 1 Spring location

Health Association (1976) and American Society for Testing and Materials (1979). The values obtained for hardness are compared favorably with those found using AAS with an flame nitrous oxide-acetylene for calcium and magne sium GONZALEZ-PORTAL et al. (1985). The AAS with an air-acetylene flame, to determine iron and manganese of all samples is used GONZALEZ, et al(1988). In almost al samples the Mn/Fe ratio lies between 0.05 and 0.20 regarded by PLOCHNIEWSKI (1972).

The values of conductivity, total residue at 103–105 ºC, and hardness found in the present study were as follows in TABLE II.

3 - CONCLUSION

The waters studied seldom fitted well into the six category cla-ssification scheme developed by NOISSOTE (1961) for natural waters in ge-neral. Instead, a ten category scheme was found preferable. TABLE II and III lists the waters in accordance with this scheme, together with the the geo-

TABLE II Conductivity, mineral content, hardness and Mn/Fe ratio
of the mineral waters studied

Class	μS / cm	Mineral content (mg/l)	Hardness (º French)	Ratio Mn/Fe
I	140–260	136.4–275.6	0.8–1.7	––
II	340–660	324.0–520.4	1.0–3.0	0.01–0.22
III	228	118.0	3.5	0.05–0.27
IV	890–1020	576.0–780.0	2.5–3.0	––
V	950–1100	624.2–694.7	3.2–4.0	0.08–0.40
VI	1150–2010	1205.0–2160.6	3.7–7.4	0.06–0.24
VII	1500–2400	1370.2–2089.0	14.0–27.3	0.02–0.15
VIII	2150	2220.5	40.1	0.04–0.05
IX	1500	1178.4	93.0	––
X	$(29.6–30.4)\times10^3$	28.75–29.80[*]	301.8–317.6	––

(*) Mineral content expresed in g/l

I : Very soft water with negligible content.
II: Very soft with slight content.   III: Soft with very little mine-
IV: Hard little with asses content.   ral content
V : Hard with asses mineral content. VI: Hard with high mineral content
VII: Medium hard with high mineral content.
VIII:Very hard with high mineral content.
IX: Very hard with high mineral content.
X : Salt with very high mineral content.

logy of the sites at which they emerge. Two of the waters are highly saline
due to their proximity to the sea in an area where fissures in the rock
allow seawater to reach deep subterranean levels and form extensive sodium
chloride rich layers.
        The manganese/ iron ratio is in the range 0.05 to 0.27 as typical
of Jurassic and Cretaceous waters.

4 - REFERENCES

APHA, AWWA, WPCF " Standard Methods for the Examination of Water and
Wastewater" 14th Edition 1976, Washington (USA),1976, pp 71, 91,202,
and 301.

ASTM " 1979 Annual Book ASTM Standard " American Society for Testing
and Materials. Philadelphia, Pa.(USA), 1979.

CASARES-RODRIGUEZ, A. " Análisis de las aguas minerales de Caldas de
Reyes y Caldas de Cuntis ". Imp. V. é H. de Compañel. Santiago de
Compostela(España) 1837, pp. 59

Ibiz. " Análisis de las aguas ferruginosas del Incio " Imp. de Jacobo
Souto é Hijo. Santiago de Compostela(España) 1864, pp 21.

CATALAN-LAFUENTE, J. " Química del Agua " 2ª Ed. Editorial Blume,Madrid
1981, pp. 389.

TABLE III. Classification of the mineral waters of Galicia

Class	Location	Rock
I	Guitiriz-Bath-(Lugo)[a] Incio (Lugo)[a] Guitiriz Pardiñas(Lugo)[a] Berán(Orense)[b] Carballino(Orense)[b] Caldas de Partovia(Orense)[c] Fuente Piñeira(Orense)[a] Catoira (La Coruña)[a]	A + B with areas of calco-alkaline granite, sands and siliceous alluvia.
II	Céltigos(Lugo)[a] Lugo- Bath- [c] Canedo(Orense)[c] Laias(Orense)[c] Cuntis Burgas(Pontevedra)[d] " fountain(Pontevedra)[d] " Castro 1(Pontevedra)[c] " Castro 2(Pontevedra)[c] Carballo(La Coruña)[c] Puentevea(La Coruña)[a]	A + B with areas of Creta-ceous and Tertiary calca-reous rocks.
III	Parga (Lugo)[a]	Feldspathic or calcareous granites, basalts,silico-calcareous sand and alluvia.
IV	Burgas (Orense)[d] Verín.Spring of toad(Orense)[a] Caldas de Reyes- Acuña-(Pontevedra)[c]	A + B with calcareous Tertiary areas.
V	Baños de Molgas(Orense)[c] Caldas de Reyes-Davila-(Pontevedra)[c] Caldelas de Tui(Pontevedra)[c]	B + calco-alkaline grani tes,basalts,with calcare ous Tertiary areas,alluvia
VI	Verín Caldeliñas(Orense)[a] " Fontenova(Orense)[a] " Sousas(Orense)[a]	B+calco-alkaline granites,basalts, and silico-calcareous sand with Ter tiary lacustrine beds.
VII	Verín Cabreiroá(Orense)[a] Mondariz-Bath-(Pontevedra)[a] Arteijo fountain(La Coruña)[a] Arteijo-Bath-(La Coruña)[c]	Calcareous and granites Tertiary and Cretaceous rock with lacus-trine beds.
VIII	Mondariz Troncoso (Pontevedra)[a]	Calcareous Tertiary rock lacustrine deposits.
IX	Mondariz.F del Val (Pontevedra)[a]	Deep waters from calca-reous grounds with Ter-tiary lacustrine deposits
X	La Toja exterior(Pontevedra)[c] La Toja interior(Pontevedra)[c]	Deep waters from strata rich in Na clays with carboniferous cal-careous rock

A: Acidic granites volcanic rocks, gneiss acid schists
B: Mica schists.   (a) Temperature < 20ºC (b) Temperature 20-30 ºC
(c) Temperature 30 to 50ºC   (d) Temperature > 50ºC

GONZALEZ-PORTAL, A.;BALUJA-SANTOS,C.;BERMEJO-MARTINEZ,F. " Calcium and Magnesium analysis in mineral waters by AAS" XXIV C.S.I. Garmisch-Partenkirchen (Germany) 1985, pp.532.

GONZALEZ-PORTAL, A.;BALUJA-SANTOS, C.;BOTANA-LOPEZ, A.; BERMEJO-MARTI-NEZ, F. "Determination of iron and manganese in mineral waters of Galicia (Spain) by atomic absorption Spectrophotometry" Water S A., 1988, 14, 13.

IGLESIAS-IGLESIAS, L. " Estudio Geológico de las fuentes minero-medi-
cinales de Galicia" Editado by Eco Franciscano. 1924 , Santiago de
Compostela, pp 26.

MAIZ-ELEIZEGUI, L. "Estudio Bibliográfico del Dr. A. Casares-Rodriguez"
An. R. Acad. Farm., 1952, (1), 29.

NOISSETTE, G. " Agresivité des eaux et protection interieur. Clasifi-
cation des eaux " 5^{e}Congrès Ass. Int. Distribution d'eau, 1961, 2,
20 mai-2 juin. Berlin.

PLOCHNIEWSKI, Z."Iron and manganese in the subsurface water in Plio-
cene, Miocene, Oligocene, Cretaceous and Jurassic sediments of Polish
lowland" Binl. Inst. Geol., Warsaw, 1972, 256, 3-57

SCHOELLER, H. " Les eaux souterraines " Ed. Masson, 1962, París.

SHCHUKAREV, S. " Intento de una revisión general de las aguas de Georgia desde el punto de vista Geoquímico" Tr. Gos. Tsentr. in-ta
Kurortologii., 1934, 5.

SLAVYANOV, N. " Eiskii mineral Spring, its encasing and history "
Trudy Lab. Gidrogeol. Savarenskogo, 1948, 1.

Chapter 19

A GROUNDWATER POLLUTION CASE-STUDY: RIO MAIOR, PORTUGAL

J P L Ferreira (National Laboratory for Civil Engineering, Lisbon, Portugal)

ABSTRACT

The object of this paper is to present the groundwater pollution case-study of Rio Maior, Portugal. It is not an attempt to solve any real case of pollution but rather to apply the mathematical and experimental methods developed in FERREIRA 1986 and 1987 to a real hydrogeological situation. Groundwater pollution scenarios with real characteristics will be defined. The results obtained will be those expected if the decisions to be taken by the decision-makers in the Rio Maior project are similar to those considered in the selected scenarios. The potential pollution problem of the white-sand aquifer is described as well as the possible contamination of well B5, which supplies most of the water for the city of Rio Maior. The analysis of the results obtained, in the way presented in the paper, will supply an approximate view of the expected (noxious) environmental impact effects, which will simplify decisions and allow a more realistic future management of the case-study region.

KEY WORDS: groundwater pollution; environmental studies; mathematical modelling; Rio Maior; Portugal

1   INTRODUCTION

"The Rio Maior Project", to be implemented by Electricidade de Portugal (EDP), consists mainly in open-pit mining of the lignite ore of the Rio Maior region, for burning the extracted coal in a thermal power plant also designed for Rio Maior. The project, which calls for a high investment, favours the use of national mining resources during 25 or 50 years, depending on the mining extraction rates.

The area under interest for this study includes the zones of the confined Rio Maior aquifer, that corresponds to the geologic formations below the diatomitic-carboniferous aquitard, which contains the lignite ore and also the phreatic aquifer of the white (kaoliniferous) sands. The part of this aquifer that lies under the aquitard has a little-variable mean thickness of about 100 m.

The white-sand aquifer, whose average grain size is 0.5 to 1 mm, has, according to RODRIGUES 1981, an hydraulic conductivity of 4.76 m/d in the NW part of the basin, 5.4 m/d in the SE part and 5.7 near the river Maior (B5 pumping well zone). The transmissivity in this area is equal to 400 m^2/d. The storage coefficient of that zone is $S = 0.83 \times 10^{-3}$ and that of the zone confined by the carboniferous-diatomitic complex is $S = 1.20 \times 10^{-3}$.

The total porosity of the white-sand aquifer is 31%, the specific retention is 20% and the specific yield is 11%.

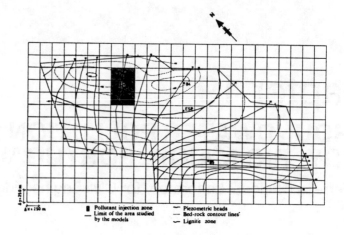

Figure 1: Mean annual piezometric surface of the Rio Maior aquifer

Fig. 1 shows a chart of the aquifer containing the equipotential lines of the mean annual piezometric surface, and the contour lines of the base of the aquifer. The difference between the elevations of the two surfaces made it possible to calculate the saturated thickness of the phreatic aquifer. Under the diatomitic-carboniferous complex the saturated thickness was considered to be equal to 100 m.

The calculation of the hydraulic gradient i was based on the equipotential lines drawn in Fig. 1. Darcy's velocity V_D was based on those calculations: $V_D = K \times i$, where K is the hydraulic conductivity of the aquifer. To evaluate the interticial velocity (V) a kinematic porosity value equal to the mentioned regional effective porosity value (S_y) of 11% was considered:

$$V = V_D/S_y$$

2   GENERAL CHARACTERIZATION OF THE RIO MAIOR AQUIFER GROUNDWATER POLLUTION PROBLEM

The potential pollution problem of the white-sand aquifer is described subsequently as well as the possible contamination of the well B5, which supplies most of the water for the city of Rio Maior.

The carboniferous-diatomitic complex will be exploited in an open-pit mine. During the mining, the piezometric levels of the excavation zones and their surroundings will be lowered. Part of the water pumped will be used to cool the equipment of the thermal power plant designed for Rio Maior, which will burn the extracted lignites. The pumped water, including that rejected by the power plant, will be discharged into the river Maior.

The ashes produced by the thermal power plant and part of the mine tailings will be dumped in the exhausted areas of the mine. This hypothesis will constitute the basis of the scenarios studied hereinafter.

It is stressed that the scenarios presented are working hypotheses, which may later on be put in practice if the entities responsible for the project decide to do so. In that case, the main long-term factor of groundwater pollution would be the leaching of the ashes due to rainwater infiltration followed by deep recharge of the aquifer. The quantification of these values is presented in FERREIRA (1982).

Note that the lowering of the piezometric levels during the mining period will prevent the flow of the leached products through the aquifer since these, even percolating through the confined aquifer (subjacent to the mine), are collected subsequently. After exhausting the carboniferous stocks the situation will be reversed and simultaneously with the leaching of the ashes there will be flowing of pollutants by the aquifer, because then the pumping of groundwater to lower the piezometric levels will have ceased.

Another phenomenon which is a matter of concern to experts working on similar projects is acid water infiltration due to mine drainage.

However, to exemplify the application of mass transport models it will be enough to consider any of the numerous conservative pollutants or pollutants with decay that exist in ashes: As, Be, Cd, Co, Cr, Cu, F, Hg, Li, Mn, Mo, Ni, Pb, S, Sb, U, V and Zn.

2.1 Management Scenarios

2.1.1 Ash Dumping Sites

The hypothesis considered for this case-study was that the ashes produced in the thermal power plant will be dumped in exhausted areas of the mine, constituting part of the landfill needed to fill out excavations (some over 100 m deep) due to the open-pit mining. Even if the excavation is not totally filled out by the landfill, for instance to create an artificial lake, the hypothesis assumed will still be valid if use, though only partial, is made of ashes or any other material containing leachable polluting elements.

Fig. 2 shows the area marked down for dumping the ashes. In Scenario 1 an area of about 0.166×0.75 km^2 was considered. In Scenario 2 this area is approximately triplicated. In Scenario 3 the area is five times larger than the initial one. The areas considered are thus:

Scenario 1 : 0.125 km^2

Scenario 2 : 0.375 km^2

Scenario 3 : 0.625 km^2

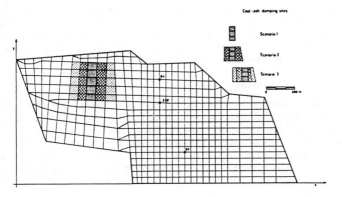

Figure 2: Disposal areas of coal ashes and DISP4.FOR model grid

2.1.2 Waterproofing of the Ash Dumping Sites

Each one of the three scenarios just presented was completed by three hypotheses regarding the waterproofing of the bottom and lateral walls of the ash containing landfill.

If care is taken to execute efficient waterproofing of the landfill areas, for instance and according to FERNANDEZ-RUBIO (1981), by using clay, concrete, asphalt, latex or plastic barriers, infiltration and recharge of the aquifer with strongly polluted water will be much reduced. To take into account this type of situation the three scenarios presented were each completed by three situations with different infiltration and groundwater recharge situations.

Leaching of the pollutants is caused by the rainwater infiltration parcel that penetrates the superficial soil layer, subjected to evapotranspiration, and that after recharges the subjacent aquifer.

In FERREIRA (1982), the rainfall recharge of the Rio Maior aquifer is evaluated by means of an original sequential dayly water balance model (BALSEQ.FOR). The value calculated for the recharge was compared with the (implicit) value of the recharge which originates the piezometric surface surveyed in the field. The agreement between the values justified the use of the model in Rio Maior. In the calculations developed in this chapter, a mean annual groundwater recharge of 287.1 mm was chosen, in accordance with that report. This value was then increased by about 25%. A value of approximately 1 mm, that is 1000 m^3/km^2, was obtained for the dayly mean infiltration rate and recharge of the white-sand aquifer. The magnification of approximately 1/4 of the initial value makes it possible to consider unfavourable situations due to the removal and digging of the natural ground and its replacement by fills, or to changes in the plant cover. To this infiltration rate a concentration value of 100 mg/l was assigned. Two infiltration rates corresponding to percentages of 50% and 10% of the maximum value were also considered.

In short, the three infiltration rates considered were:

- a) Rate A or 100%: daily mean infiltration of 1000 m^3/km^2.

- b) Rate B or 50%: daily mean infiltration of 500 m^3/km^2.

- c) Rate C or 10%: daily mean infiltration of 100 m^3/km^2.

Rates B and C correspond to the favourable effect of possible waterproofing barriers constructed on and around the landfill of the ash dumping sites or to significant changes in the local groundwater flow.

2.2 Description of the Selected Mathematical Model

2.2.1 General comments

Among the models described in FERREIRA (1987) and based on the conclusion of the comparative analysis developed in that study the finite element model DISP4.FOR was selected for application to this case-study. This model is the most versatile and accurate model tested in that dissertation. It has the possibility of the simultaneous consideration of triangular and quadrilateral quadratic elements, i.e., with six or eigh nodes per element, and was programmed to solve well mixed problems, like the one considered in this case-study. The time integration is based on the Crank-Nicholson scheme wich ensures the stability of the solution.

2.2.2 Boundary Conditions

The boundary conditions of the Rio Maior aquifer were object of CORREIA (1981). That study evaluates the groundwater flow of the Rio Maior aquifer, applying a quadrilateral quadratic finite element model. Fig. 3 presents the natural and essential boundary conditions of the Rio Maior aquifer used in this case-study. The boundaries comprised between nodes $B - C - D - E - F - G - H$ and between $I - J - K - A$ are considered impervious, corresponding to geological faults. The components of the groundwater flow velocity normal to those boundaries have a zero value. These boundary conditions are of the second type. They are also called Newmann boundary conditions.

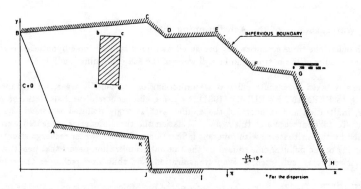

Figure 3: Boundary conditions

The boundary condition considered for boundary $A - B$ is of the first type. It is also called Dirichlet boundary condition. In this case-study it has the following reference value $C = 0$ mg/l, i.e. the recharge concentration along this boundary has set equal to zero.

Finally boundary $H - I$ is a downstream discharge boundary, no explicitly boundary condition having been imposed. Model DISP4.FOR considers for this natural boundary, as no flux across the boundary was specified, that the flux across the boundary due to dispersion is zero, i.e. the total flux across this downstream boundary is entirely due to convection. Boundary $H - I$ is considered sufficiently far away from our sections of interest and also far away from any concentration source, so that the concentration gradients are small. In the close vicinity of this boundary DISP4.FOR model results cannot be considered precise.

2.2.3 Surface Source

A recharge of mass was considered for the area located inside rectangle $abcd$ of Fig. 3. This corresponds to the leaching of ashes deposited in the project dumping area, as referred to in item 2.1.2. The recharge concentration was assumed to be equal to 100 mg/l. Recharge values were calculated for each node, corresponding to Scenarios 1, 2 and 3 (Fig. 2) and infiltration rates A, B and C, according to the size of its area of influence.

2.2.4 Initial Conditions

The initial conditions ascribed to each node of the grid were considered equal to zero, with exception of the nodes situated in the area corresponding to the ash dumping sites of Scenarios 1, 2 and 3 (Fig. 2). For these nodes an initial concentration of 100 mg/l was considered:

$$C_0 = 100 \text{ mg/l} \quad \text{for the area } abcd \text{ (Fig. 3)}$$

This initial condition corresponds to the pollution caused into the aquifer during the mining activities, before the begining of the model simulation period, that is before $t = 0$. Time $t = 0$ will correspond to the initial reference time when all mining activities will cease and the natural groundwater flow of the aquifer will start the development of a plume of polluted water percolating throughout the aquifer.

2.2.5 Finite Element Grid and Time-step

The finite element grid used in this case-study is presented in Fig. 2. This grid has 480 quadratic quadrilateral and triangular elements (461 of which are quadrilateral elements) and has 1509 nodes. The grid was drawn in such a way that the Peclet number of all elements is less than 5. The Peclet number P_e is a dimensionless number relating the space increment of the model's grid Δx with the dispersion of the porous media D and the interstitial groundwater flow velocity V, i.e. $P_e = V \Delta x / D$. This procedure was done according to the requirements needed for an acurate application of this kind of mathematical models.

The time-step was calculated according to the requirements needed for this kind of models, so that the Courant number C_r is less than 1.0. The Courant number is a dimensionless number relating the time-step Δt used in the model with the interstitial groundwater flow velocity V and the space increment Δx, i.e. $C_r = V \Delta t / \Delta x$.

2.2.6 Longitudinal and Transverse Dispersivity of the Aquifer

The dispersivity α is a tensor parameter characteristic of the aquifer that can be defined in relation to two constants in the case of isotropic aquifers, like what was considered to be the case of Rio Maior. These two contants are the longitudinal dispersivity α_L, evaluated along the groundwater flow lines of the aquifer, and the transverse dispersivity α_T that is perpendicular to the longitudinal dispersivity.

The dispersivity depends on the characteristics of the aquifer, being equal the the ratio of the aquifer's dispersion by the groundwater flow interstitial velocity.

According to the results of the tracer experiments described in FERREIRA (1987) the value of the longitudinal macrodispersivity of the Rio Maior aquifer, was evaluated as follows:

$$\alpha_L = 20 \text{ m}$$

The macrodispersivity is the regional dispersivity of the aquifer. In our case-study the regional scale was considered to be equal to 6 km.

The value considered for the transverse macrodispersivity of the Rio Maior aquifer has been set equal to 1/10 of the longitudinal dispersivity:

$$\alpha_T = 2.0 \text{ m}$$

This procedure was done in accordance with reports on field experiments, published by several researchers.

2.2.7   Calibration of the Two-dimensional Finite Element Model

The analysis of the accuracy of the results using the finite element grid referred to previously (Fig. 2), was done in two steps. First by comparing the results obtained with this model with those obtained with an analytical solution (model SA2D1.FOR). Secondly by comparing the results obtained with this model with those obtained with a one-dimensional finite element model developed along the flow lines of the aquifer. Both comparisons were done considering a boundary condition equal to 100 mg/l in the dumping area of Scenario 2 (Fig. 2).

The results obtained with model DISP4.FOR are almost coincident with the analytical values. It should be noted however that in this case the analytical solution is not an exact one because the groundwater flow velocity has not a constant value, as the analytical model considers, but has some small variations.

To verify the accuracy of the two-dimensional DISP4.FOR model in downstream areas of the aquifer were the groundwater velocity field increases from 0.15 m/d to 0.80 m/d a one-dimensional numerical model was implemented along the flow lines of the aquifer, Fig. 4. The model used was also the finite element DISP4.FOR model. The longitudinal x axe was considered along the flow lines of the aquifer starting at point B (Fig. 3). The model then crosses the dumping sites, the well B5 areas, and ends at boundary $H - I$, near point I.

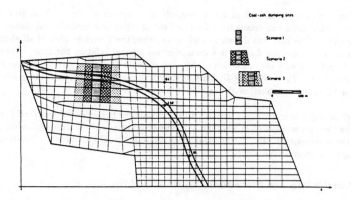

Figure 4: One and two-dimensional model grids

As may be seen in Fig. 5 the results of the two models are similar, validating the two-dimensional finite element grid under study. The precise fitting of the breakthrough curves obtained with the one-dimensional model and the two-dimensional model was not expected due to secondary effects included in the two-dimensional model, like the transverse dispersion and the influence of transverse variable flow fields.

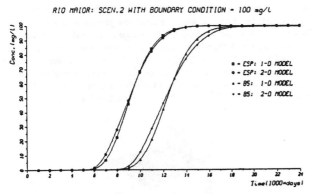

Figure 5: Breakthrough curves obtained with the one and the two-dimensional models

3 RESULTS OBTAINED WITH THE DISP4.FOR FINITE ELEMENT MODEL

Based on the characteristics defined in the previous items nine simulation runs of the two-dimensional DISP4.FOR model were executed on a VAX 8700 computer. The nine simulation runs correspond to the three Scenarios 1, 2 and 3 each one of them for the three considered infiltration rates A, B and C. The outputs of the model were analyzed in two different ways.

First comparing the results of all scenarios and infiltration rates on well B5, which supplies the water demand for the city of Rio Maior.

Secondly by observing the progress in time of the pollution plume throughout the aquifer.

In order to facilitate the first type of analysis, i.e. the variation of the pollutant concentration in time, the results were drawn simultaneously in Fig. 6, for well B5. It is worth seeing the paramount importance that the area of the dumping sites has regarding the values of the expected pollution concentration peaks. It is also relevant to observe that the waterproofing of the dumping sites is of the greatest importance for the expected long-term pollution concentration value, mainly when the dumping area increases in size. A decrease in the infiltration rate brings a significant decrease in the concentration peak values and in the long-term concentration values.

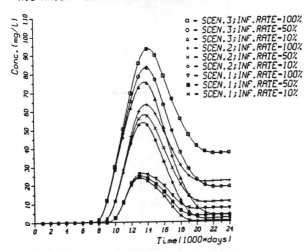

Figure 6: Output concentration curves at well B5

The second type of analysis, i.e. the variation of the pollutant concentration in space, allows the visualization of the environmental impact of the project in the groundwater quality of the region. This type of analysis is examplified for Scenario 2 with infiltration rate A. Two-dimensional contour maps and three-dimensional perspectives were drawn for three different simulation times: 4000 d, 12000 d and 24000 d, Figs. 7, 8 and 9. These three times may represent respectively the short-term, when the initial pollution plume begins to move throughout the aquifer, the medium-term, when the initial pollution plume is passing between Espadanal and well B5, and the long-term, after the initial pollution has been expelled from the aquifer.

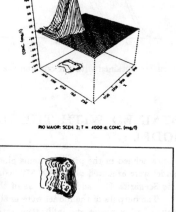

RIO MAIOR: SCEN. 2; T = 4000 d; CONC. (mg/l)

Figure 7: Two-dimensional contours and a three-dimensional perspective of Scenario 2 for a simulation period of 4000 d

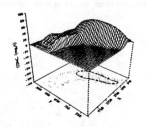

RIO MAIOR: SCEN. 2; T = 12000 d; CONC. (mg/l)

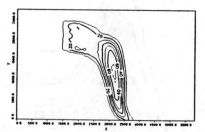

Figure 8: Two-dimensional contours and a three-dimensional perspective of Scenario 2 for a simulation period of 12000 d

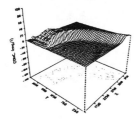

RIO MAIOR: SCEN. 2; T = 24000 d; CONC. (mg/l)

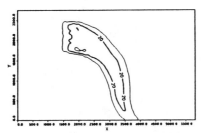

Figure 9: Two-dimensional contours and a three-dimensional perspective of Scenario 2 for a simulation period of 24000 d

Based on the mentioned values and on the European Economic Community environmental groundwater standards an adequate economic analysis would allow the selection of the best cost/effectiveness solution for the location of the Rio Maior ash dumping sites and the required type of waterproof devices, in order to reduce the short-term and medium-term pollution concentration peaks and the long-term pollution of the groundwaters of the region.

REFERENCES

Correia, R. M. (1981) *Empreendimento de Rio Maior. Modelos Numéricos de Rebaixamento dos Níveis Piezométricos.* Laboratório Nacional de Engenharia Civil, Lisboa.

Fernandez-Rubio, R. (1981) *Efecto sobre las Águas Subterráneas de las Actividades Mineras. Medidas de Prevencion,* in Analisis y Evolution de la Contaminacion de Aguas Subterráneas en España, Curso Internacional de Hidrologia Subterránea, Barcelona.

Ferreira, J. P. Lobo (1982) *Actualização do Estudo Hidrológico da Bacia Hidrográfica do Rio Maior.* Laboratório Nacional de Engenharia Civil, Lisboa.

Ferreira, J. P. Lobo (1986) *A Dispersão de Poluentes em Águas Subterrâneas.* Tese de Especialista, Laboratório Nacional de Engenharia Civil, Lisboa.

Ferreira, J. P. Lobo (1987) *A Comparative Analysis of Mathematical Mass Transport Models and Tracer Experiments for Groundwater Pollution Studies,* Doktor-Ingenieur Dissertation, Technische Universität Berlin, Berlin. (Memória N. 724, Laboratório Nacional de Engenharia Civil, Lisboa.)

Rodrigues, J. Delgado (1981) *Empreendimento de Rio Maior. Síntese dos Estudos Hidrogeológicos.* Laboratório Nacional de Engenharia Civil, Lisboa.

Chapter 20

MICROBIAL ECOLOGY OF IKOGOSI WARM SPRING, NIGERIA

O Odeyemi (Obafemi Awolowo University, Ile-Ife, Nigeria)

ABSTRACT

Ikogosi warm spring is situated in Ikogosi town in Ondo State of Nigeria. The spring has a temperature of 40°C and a pH of 7.0. Repeated sampling and analysis of the thermal water were done four times over a 6-month period. Prominent bacteria isolated from the spring were Chromobacterium violaceum, Achromobacter parvulus, Micrococcus roseus, Aerobacter cloacae and Flavobacterium sp. Fungi isolates were Geotrichum sp., Aspergillus sp., Penicillium sp. and Candida sp. No actinomycete or protozoa was isolated from the spring. Since there are plans by the Nigerian Tourist Board to bottle and market this thermal mineral water, it is pertinent to characterize its microbial profile whose innocuousness should also be ascertained.

Key words: Ikogosi warm spring, bacteria, fungi, actinomycetes, protozoa.

1 - INTRODUCTION

The Ikogosi warm spring has its three sources located about one kilometer west of the bucolic valley town of Ikogosi which is situated just north of the 7° 35'N latitude and slightly west of the 5°00'E longitude in Ondo State area of Nigeria (Fig. 1).

Detailed physico-chemical studies have been conducted on the Ikogosi warm spring (ROGERS et. al., 1969), but since the major use to which the spring is subjected are mainly recreational, close to its source, and domestic along its course, the importance of microbial analysis cannot be overemphasized. The population of Ikogosi town itself is only a little above three thousand but the ever flowing stream of tourists subject the area to an indefinite microbial territoriality.

It is therefore important as a background, basic measure, to establish data concerning the autochthonous population of micro-organisms at the source of the spring. Most of the stream and river waters of Nigeria have temperatures in the range of 20 to 30°C and Ikogosi spring is one of the very few water sources in the country with an unusually high temperature of 40°C. The identity of the micro-organisms in this milieu should therefore be a spontaneous scientific curiosity to the microbial ecologists.

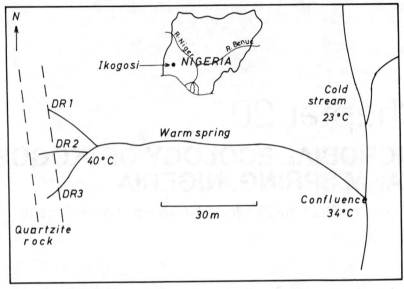

Fig·1: Diagrammatic sketch of Ikogosi spring area showing general sample locations. DR1, DR2, and DR3 are sampling points

2 - MATERIALS AND METHODS

2.1 - Sampling

Fresh water samples were collected in sterile containers from the three point sources of the warm spring (temp. 40°C) designated DRI, DR2, and DR3 (Fig. 1). Samples were also collected from the body of a nearby cold stream (temp. 23°C) with a different source, and also from the confluence of the warm spring and the cold stream which is about 100 meters from the warm spring sources. The temperature of the confluent stream was 34°C. The latter samples were taken in order to compare their microbial densities, pH values and temperature with those of the warm spring. The pH values of the samples were measured on site with a portable Coleman pH meter. Temperature values were also recorded on site with a mercury thermometer.

2.2 - Microbial analyses

All the water samples were analysed within three hours of collection The total bacteria population of the samples were enumerated using plate count agar and nutrient agar (APHA, 1981). It should be noted though that these media might not necessarily support the growth of all the autochthonous bacteria in the spring water. Presumptive test for coliform organisms was performed using lactose broth. Eosin methylene blue agar was used to confirm and differentiate the coliform organisms (BBL, 1973). Citrate utilizing <u>Aerobacter</u> species were differentiated from <u>Escherichia</u> species by the Koser citrate medium (APHA, 1981). All Petri plates for enumeration and quantification of bacteria were incubated at 37°C and 40°C. Pure colonies were isolated and subcultured repeatedly in nutrient broth for subsequent biochemical tests and identification.

Morphological characteristics of isolates were examined microscopically after Gram staining. Isolated colonies were subjected to standard biochemical tests, namely, gelatin liquefaction, action on litmus milk, production of ammonia from arginine, indole production, hydrogen sulphide production, urease activity, reduction of nitrate, fermentation of sugars, methylred and Voges-Proskauer test, citrate utilization, and catalase tests (APHA, 1981).

Fungal population of the samples was enumerated on malt extract agar using the serial dilution technique. Predominant colonies from the pour plates were subcultured on freshly prepared malt extract agar. Lactophenol stained colonies were observed microscopically to ascertain their morphological characteristics. Isolated fungal colonies were incubated at $25^{\circ}C$, $30^{\circ}C$, $35^{\circ}C$ and $40^{\circ}C$ to determine optimal growth temperatures in view of confirming autochthonous species.

The water samples were screened for the presence of protozoa by plating dilutions on mannitol soil extract agar, while soil extract agar was used to determine the presence of actinomycetes (ALEXANDER, 1977). Repeated sampling and analysis of the warm spring were done four times covering a period of 6 months.

3 - RESULTS

The bacterial and fungal populations, pH values and temperatures of the warm spring, its adjacent cold stream and the confluent water are shown on Table 1. There was little or no difference in the total bacterial densities of the warm spring, its adjacent cold stream and the confluent water as they all harbour an average population of 10^6/ml each. However, the fungal population of the adjacent stream was higher than that of the spring. It is noteworthy that the fungal numbers of about 5/ml of the spring water were extremely low when compared with those of the bacteria. The pH values of all the samples were similar, but the $40^{\circ}C$ temperature of the spring water was considerably higher than that of the cold stream ($23^{\circ}C$).

TABLE 1

Microbial population, pH and temperature of samples of Ikogosi warm spring, its adjacent cold stream and the confluent water

Sample No.	Bacterial No./ml	Fungal No./ml	pH	Temp.
DR 1	2.1 x 10^6	6	7.05	$40^{\circ}C$
DR 2	2.2 x 10^6	4	7.15	$39.5^{\circ}C$
DR 3	2.6 x 10^6	4	6.95	$39.5^{\circ}C$
Cold stream	8.5 x 10^6	12	7.05	$23^{\circ}C$
Confluent water	4.3 x 10^6	10	7.01	$34^{\circ}C$

DR 1 = Direct rock sample one i.e. water sample taken directly from the rock where the spring oozes out.

DR 2 = Direct rock sample two

DR 3 = Direct rock sample three.

Since we are primarily interested in the biota of the warm spring, only the microbial flora of the spring were isolated and characterized. Five distinct bacteria types were repeatedly isolated from the samples of the warm water issuing out of the rocks. Four were Gram negative bacilli while the remaining one was a Gram positive coccus.

Altogether, four distinct fungi isolates were prominent in all the samples. Highly recurrent was the isolation of a black-pigmented filamentous fungus with a raised colony on malt extract. It was designated as isolate 2. Fungi isolates 1 and 2 showed optimum growth at $35^{\circ}C$ while isolates 3 and 4 had optimum growth at $30^{\circ}C$.

With the selective media used, no protozoan or actinomycete was detected in the samples of the thermal water.

4 - DISCUSSION

A thorough check of the morphological characteristics of the bacteria isolates, in conjuction with HOLT's Bergeys Manual of Determinative Bacteriology (1977) strongly suggest the following identities for the bacterial isolates. The Gram negative rods with rounded ends, designated as isolate A has violet pigments (violacein). It utilizes citrate and ammonia as sole carbon and nitrogen sources for growth and is identified as Chromobacterium violaceum. The Gram negative slender rods (isolate B) that ferments few carbohydrates including glucose and is catalase positive is identified as a Flavobacterium species and resembles F. lutescens more than any other specie. Isolate C, a non-carbohydrate fermenter, non motile, Gram negative rod which does not liquefy gelatin and has a creamy-white non-chromogenic colony is identified as Achromobacter parvulus. Isolate D, forming rose-red coloured, raised colonies of small, Gram positive coccus cell that produces nitrites from nitrates and is catalase negative is identified as Micrococcus roseus. Isolate E, Gram negative rods with fecal odour and catalase positive colonies which do not liquefy gelatin is identified as Aerobacter cloaceae.

All these organisms have been repeatedly isolated from bodies of freshwaters (HOLT, 1977) and besides Chromobacterium violaceum which causes food spoilage and some mammalian infections, they are relatively harmless bacteria. From literature, only Aerobacter cloaceae grows uninhibited at $37^{\circ}C$ and $40^{\circ}C$ but strains of Flavobacterium, Achromobacter and Micrococcus species that flourish at $37^{\circ}C$ and $40^{\circ}C$ were isolated from the warm spring. This adaptation for growth at the body temperature might have conferred other properties on these strains; pathogenicity not precluded. Even though the bacterial density of the spring water is high at 2.3×10^{6} cells/ml incidences of disease outbreak from using the water are rare on record.

From the descriptions of TALBOT (1978) the fungal isolates 1, 2, 3 and 4 are recognised as species of Geotrichum, Aspergillus, Penicillium and Candida respectively. All of these fungi are fairly ubiquitous in the tropical regions and have also been implicated in mycotic infections (ALEXOPOULOS, 1962). According to ROGERS et. al. (1969) predominant algae in the warm spring are mainly diatoms which include Cymbella sp., Melosira sp., Navicula sp., Nitzschia sp. and Synedra sp.

The absence of actinomycetes and protozoa from the sources of the warm spring could be due to a somewhat narrow survival niche of these organisms. Their habitat is mainly soil (ALEXANDER, 1977) and considering a previous study (ROGERS et. al. 1969), there is nothing particularly unusual regarding the chemistry or source of the warm spring to warrant a major shift in ecological adaptation.

Since there are plans by the Nigerian Tourist Board, the Highland Waters Company of Scotland, and the Ondo State Government to bottle and market this thermal mineral water which is believed to be superior to the marketed mineral water in France, it is necessary to ascertain the innocuousness of the biotic components of the spring water especially the isolated microorganisms described above.

5 - BIBLIOGRAPHY

ALEXANDER, M. - Introduction to Soil Microbiology. 2nd edn. New York (USA), John Wiley and Sons, 1977.

ALEXOPOULOS, C. J. - Introductory Mycology. 2nd edn. New York (USA)
John Wiley and Sons, 1962.

APHA. - Standard Methods for the Examination of Water and Wastewater.
15th edn. Washington, D.C. (USA), American Publich Health Association
Inc., 1981.

BBL. - BBL Manual of Products and Laboratory Procedures. 5th edn.
Maryland (USA), BBL Division of Becton, Dickinson and Co., 1973.

HOLT, J. G. - The Shorter Bergey's Manual of Determinative Bacteriology.
8th edn. Baltimore (USA), Williams and Wilkins Co., 1977.

ROGERS, A. S.; IMEVBORE, A. M. A. ; ADEGOKE, O. S. - "Physical and Chemical
Properties of the Ikogosi Warm Spring, Western Nigeria". Journal
Mining and Geology, 4, 1969, pp. 69-81.

TALBOT, P. H. B. - Principles of Fungal Taxonomy. London (England),
Macmillan Press Ltd., 1978.

Chapter 21

LOAD ASSESSMENT OF TRACE METALS IN RIVERS: A COMPARISON OF FLOW-PROPORTIONAL AND GRAB SAMPLING

R M Harrison, G Thorogood (University of Essex, Colchester, UK), T F Zabel (Water Research Centre, Marlow, UK)

ABSTRACT

Estimates of contaminant loads transported in a UK river have been made by means of both conventional grab sampling and a novel continuous flow-proportional sampler. The data are used to evaluate the accuracy and confidence limits of load estimates based upon grab sampling of river water at different temporal intervals. Examples are also presented of the relationship of river flow to the concentration for two substances.

Key words: transport load, trace metals, nutrients, river water

INTRODUCTION

Rivers represent a major route for potentially polluting substances of land-based origin reaching coastal waters and the oceans. Information on input loads transported by rivers is important for the management of discharges and monitoring the effectiveness of control measures. River loads are usually determined from the product of water discharge rate (usually measured continuously) and concentration of specific substances in a limited number of grab samples. However, load estimation from a limited number of grab samples is associated with great uncertainties because of the variability of the concentrations depending on inputs and flow regime.

The use of a flow-proportional water sampler is described to estimate total loads of trace elements and ions at a site on the freshwater part of a UK river. Whilst continuously flushed with river water, the sampler collects a weekly composite sample derived from up to 2000 aliquots (5 ml each) of water at a frequency directly proportional to the river discharge as measured by an ultrasonic flow gauge. The composite sample and spot samples collected weekly are analysed for total suspended solids, dissolved and particulate metals, nitrogen compounds and phosphate.

The relationships between composite and grab sample composition are discussed in terms of pollutant transport processes and accuracy of load estimation.

WATER SAMPLING AND ANALYSIS

Sample Collection

Composite river water samples were collected over an interval of one week by an automated flow-proportional sampler. The sampler, designed by W.R.c., discharged 5ml aliquots of river water into a 10 litre polypropylene container at a rate directly proportional to the volumetric discharge of the river. The river discharge was sensed by a Sarasota multi-path ultrasonic river gauge installed on the river bank and connected directly to the sampler controller.

On the day of changing the composite sample container, a grab sample of equal volume to the composite sample was taken from the outlet of the sampling pump. Samples were acidified on site to 0.01M H_2SO_4 by addition of 5M Analar sulphuric acid.

Analytical Procedures

Sampling bottles, the filtration equipment and other glass and polyethylene ware were leached in a nitric acid bath and rinsed thoroughly with double-distilled deionised water prior to use.

Upon return to the laboratory, all samples were kept at 4°C and were analysed for Pb, Cd, Cr, Cu, Ni and Zn within 2-4 days. Sample preparation involved filtration through a 0.45μm membrane to provide a filtrate which was acidified and analysed directly. The suspended solids were digested with 1:1 concentrated nitric acid at 90°C for 2h in Teflon digestion bombs, filtered and the filtrates diluted prior to analysis. The walls of the sampling bottles were washed with 1:1 concentrated nitric acid which was diluted and analysed. Analyses of metals were routinely carried out by GFAAS using a Perkin-Elmer Model 280 Atomic Absorption Spectrometer with HGA 400 Graphite Furnace and AS 40 Autosampler attachment with a standard additions method employed for all samples. Procedural blanks were carried out in all determinations. Total concentrations were obtained by adding the results for filtrate, solid fraction and wall extracts.

Suspended solids were determined on a further aliquot of sample by filtering 1 litre of the river sample through a preweighed glass fibre paper (Whatman GF/C grade), drying the residue at 105°C and determining its weight by difference.

Nitrate and nitrite were analysed in the dissolved (GF/C filtered) fraction by ion chromatography using a Dionex 2000i/SP chromatograph and comparison with aqueous standards. Total phosphorus (TP) and soluble reactive phosphate (SRP) were determined following the procedures of BLAIR AND SMITH (1984).

Method Verification

Verification experiments were carried out as follows:

(a)   Multiple recirculation experiments demonstrated that the sampling pump did not significantly alter concentrations of the determinands. (effect <1%)

(b)   Testing of various types of tubing material showed that a specific type of reinforced PVC tube did not significantly enhance or deplete metal concentrations in river water samples

(c)   Detection limits for metals (defined as 3 times standard deviation of the blank) were low compared with the metal concentrations encountered in the river

(d)   Samples could be stored acidified in polyethylene bottles for a week without appreciable effects in metal recovery.

(e)   When sampling for nitrate and phosphate was commenced, samples were acidified upon collection to 0.01M with sulphuric acid. Detailed experiments showed that there was no adverse effect upon trace metal recovery, even for lead.

(f)   Concentrations of nitrate, total phosphorus and soluble reactive phosphate in a river water were well in excess of detection limits; analytical precision was determined.

(g)   Systematic measurements of cross-river transects of the concentrations of the determinands showed that, within the limits of analytical precision, the sample collected at the autosampler was representative of the cross-river transect as a whole.

LOAD ESTIMATION

For estimating transport loads within rivers, concentration data are normally only available for periodic instantaneous grab samples although flow measurements are made continuously, at least at important locations such as the freshwater limits. Until now, it has not been clear to what extent loads estimated from grab samples are representative of true loads. In this work, a flow-proportional continuous sampler has been used together with continuous flow rate measurements in order to provide a more accurate estimate of transport loads against which loads estimated from grab samples may be compared.

WALLING AND WEBB (1985) have highlighted the problems of estimating river loads of pollutants from non-continuous sampling data. They indicate five possible methods of calculation. Three of these (their methods 1,3 and 4) may be shown from the outset to introduce bias and were not considered worthy of further examination. The other two methods have been examined further using our own data set; Walling and Webb's methods 5 and 2 are respectively our methods 2 and 3.

Grab sample data were used to estimate loads and associated 95% confidence intervals using these two methods. These estimates may then be compared with loads calculated from the composite sample/averaged flow data, taken to give the "right" answer assuming that the errors associated with the analysis are small compared to the variability of concentrations and flow. This approach allows the influence of different frequencies of grab sampling to be evaluated.

The data used for the exercise comprised:
48 weekly samples for trace metals
42 weekly samples for suspended solids
30 weekly samples for total phosphorus (TP),
                 soluble reactive phosphate (SRP),
                 and nitrate ($NO_3 - N$)

Notation

The period of time over which the exercise took place comprised n intervals of equal 1 week length. Let t_i be the date of sample collection at the end of the ith interval (i=1, ..., n) and t_o be the date of the start of the study. The ith interval is thus the time from t_{i-1} to t_i

C_i the concentration of the flow proportional composite over the ith weekly interval

Q_i the average flow over the ith weekly interval

$\bar{Q}$ the average flow over the whole period = $\dfrac{1}{n}\displaystyle\sum_{i=1}^{n} Q_i$

c_i grab concentration at time t_i

q_i instantaneous flow at time t_i

$l_i = c_i q_i$ instantaneous load at time t_i

$$\bar{q} = \frac{1}{n}\sum_{i=1}^{n} q_i$$

Flow, concentration and load are expressed in consistent units.

Method 1

Using the concentrations measured in the flow proportional composite samples, together with average flows derived from continuous measurement, it is possible to calculate the average weekly load.

$$M_1 = \frac{1}{n}\sum_{i=1}^{n} C_i Q_i$$

There should be no (or negligible) error in this estimate arising from temporal variability of concentration or of flow. It provides the best estimate of the true load. There will, of course, be uncertainties arising from analytical errors and, perhaps, from spatial variations across the river but these are outside the scope of the present discussion. We shall therefore regard the standard error (SE) of this estimate as zero.

$$SE(M_1) = 0$$

The value of M_1 thus provides a base point for comparison of other methods which use grab sample data.

Method 2

A popular method of estimation of load from grab sample concentrations and instantaneous measurements of flow, but adjusting for mean flow over the period of the record, is to estimate load by

$$M_2 = \left\{ \frac{1}{n} \sum_{i=1}^{n} c_i q_i \right\} \frac{\bar{Q}}{\frac{1}{n} \sum_{i=1}^{n} q_i}$$

$$= \frac{\sum_{i=1}^{n} c_i q_i}{\sum_{i=1}^{n} q_i} \cdot \bar{Q}$$

$$= \bar{c} \ \bar{Q}$$

where $\bar{c}$ is the so-called 'flow weighted' average concentration in the grab samples.

The adjustment in this formula for the difference between $\bar{q}$ and $\bar{Q}$ may have intuitive appeal but the theoretical evaluation of the operating characteristics of resulting formulae is a severe statistical problem.

This method is believed to have negligible bias, although the proof that its bias is zero has not been demonstrated. For the standard error of the method, an approximate formula has been suggested by JOLLY and ZABEL (1986).

This formula gives an estimate for the variance of $\bar{c}$,

$$var(\bar{c}) = \left[\sum_{i=1}^{n} (c_i - \bar{c})^2 q_i / n\bar{q} \right] \sum_{i=1}^{n} q_i^2 / (n\bar{q})^2$$

and $SE(M_2) = \bar{Q} \ \sqrt{(var(\bar{c}))}$

Method 3

Using only the instantaneous measurements there is the simpler formula for estimating load

$$M_3 = \frac{1}{n} \sum_{i=1}^{n} c_i q_i$$

$$= \frac{1}{n} \sum_{i=1}^{n} l_i$$

This estimates the average load over the period as the straightforward average of the instantaneous loads, and these can be viewed as drawn from a statistical distribution of instantaneous loads. From this viewpoint the standard error would be

$$SE(M_3) = \sigma / \sqrt{n}$$

where the standard deviation σ can be estimated by the sample standard deviation s of the l_i.

Figures 1 and 2   Effect of sampling frequency on load estimates (with 95%ile confidence limits) using different methods for load calculations.

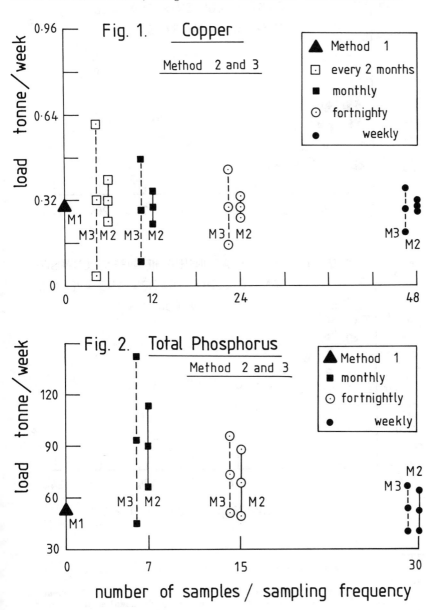

It is possible that $SE(M_3)$ may be smaller or larger than $SE(M_2)$ depending on circumstances, in particular the form of relationships between concentrations and flow.

$SE(M_3)$ may give an unduly pessimistic estimate of the accuracy of M_3 if c and q are changing at a rate that is slow compared with sampling frequency.   In order to examine the effect of frequency of sampling for each substance, load estimates with 95% confidence limits were plotted against frequency of sampling for the three methods for each substance, Figure 1 and 2.   The value of average load (M_1) estimated by Method 1 provides a base point for comparison of Methods 2 and 3.

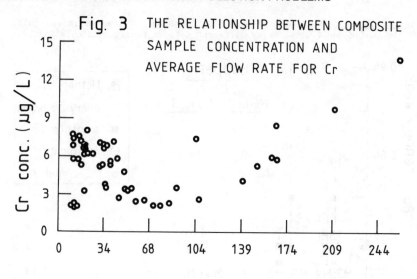

Fig. 3   THE RELATIONSHIP BETWEEN COMPOSITE SAMPLE CONCENTRATION AND AVERAGE FLOW RATE FOR Cr

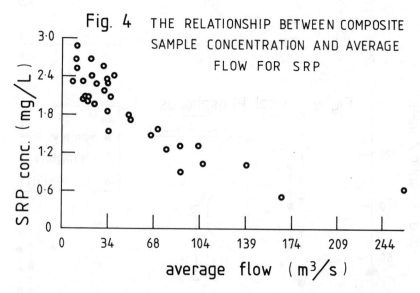

Fig. 4   THE RELATIONSHIP BETWEEN COMPOSITE SAMPLE CONCENTRATION AND AVERAGE FLOW FOR SRP

average flow (m³/s)

RESULTS

Sample results for copper and TP are shown in Figures 1 and 2 respectively.

In the figures, the load estimate and confidence interval appear on the y-axis. The x-axis shows the number of grab samples used in the calculation which for the trace metals is 48, 24, 12 and 6 (ignore the apparent offsets between the M_2 and M_3 data), representing samples taken weekly, fortnightly, monthly and bimonthly. For suspended solids, the number of grab samples considered is 40, 20 and 10, or weekly, fortnightly and monthly samples. For TP, SRP and NO_3^- - N, there were 30, 15 or 7 samples, corresponding to weekly, fortnightly or monthly grab samples.

As might be expected, the closeness of the load estimate by Methods 2 and 3 to the Method 1 load improves with the number of grab samples taken. Additionally, the confidence interval narrows for the larger number of samples. Almost invariably the confidence interval of Method 3 is wider than that of the Method 2. This appears to justify the "flow weighting" of concentration used in Method 2.

Relationship between analyte concentration and flow rate

Examples of the relationship between composite sample concentrations and average flow rate is shown in Figures 3 and 4. Figure 3, for chromium, is typical of trace metals which show a U-shaped profile. This appears to result from reduction in concentration with increasing flow rate at lower concentrations due to dilution, with a subsequent increase at high flow rates as suspended sediment loads increase. In contrast, soluble reactive phosphate (Figure 4) shows a steady decrease in concentration with flow rate. This primarily dissolved component has a relatively constant source strength and thus higher flows serve only to provide dilution.

LIST OF SYMBOLS

c_i grab concentration at time t_i

C_i the concentration of the flow proportional

composite over the ith weekly interval

$l_i = c_i q_i$ instantaneous load at t_i

q_i instantaneous flow at time t_i

$$\bar{q} = \sum_{i-1}^{n} q_i/n$$

$\bar{Q}$ the average flow over the whole period $= \sum_{i=1}^{n} Q_i/n$

Q_i the average flow over the ith weekly interval

ACKNOWLEDGEMENTS

The study was partly funded by the UK Department of the Environment whose permission to publish has been obtained. The authors gratefully acknowledge the cooperation of Thames Water in providing access to the test site and assistance with setting up the sampling equipment. They also acknowledge the advice of Mr R F Lacey concerning the statistical procedures used in analysing the data.

BIBLIOGRAPHY

BLAIR A.J. and SMITH P.R. "The Determination of Low-Level Soluble Reactive Phosphate in Surface Waters using an Autoanalyser II and Total Phosphorus by a Pre-Digestion in an Autoclave". Water Research Centre Report, ER-799-M, 1984

JOLLY P.K. and ZABEL T.F. "Estimates of the Loads of Certain List I and List II Substances Discharged to the North Sea". Water Research Centre Report, (DoE 1350-M/1), 1986

WALLING D.E. and WEBB, B.W. "Estimating the Discharge of Contaminants to Coastal Waters by Rivers: Some Cautionary Comments". Marine Pollution Bulletin, 16, pp 488-492, 1985

Chapter 22

SEDIMENTS AS INDICATORS OF THE RIVER AVE CONTAMINATION BY HEAVY METALS

E P R Gonçalves (Centre of Chemical Engineering, Porto, Portugal) R A R Boaventura (Faculty of Engineering, Porto, Portugal)

ABSTRACT

The river Ave contamination by trace elements such as cadmium, chromium, copper, lead and zinc was evaluated using sediments as indicators in surveys carried out in 1985 and 1986 Summers.

Two grain fractions were tested for metal analysis (<63 μm and 63 - - 180 μm) in 1985 survey. In 1986 survey the < 63 μm fraction ("informative one") was selected.

The preferential heavy metals bonding to sediment organic matter was estimated and it was verified that the organic matter , expressed as volatile matter, is a reference parameter against which the metal concentration may be expressed.

From background levels - the average metal concentrations in sediments from non-polluted zones - the contamination factors were determined in twenty sampling stations and a metal pollution index was calculated.

A high contamination of river Este by Cd, Cr, Cu, Pb and Zn, river Vizela by Cd, Cu and Pb and river Selho by Cr was detected.

The contamination indices of Ave river were not so high. However at Santo Tirso the river was polluted and downstream (Trofa, Ponte d'Ave and Vila do Conde) the metal pollution was still considerable.

Key words: heavy metals, river sediments, micropollutants.

1. INTRODUCTION

The use of sediments as environmental indicators is widely recognized. They can accumulate or integrate organic or mineral pollutants allowing the determination of these substances when they occur in water at ppb level and may escape detection by water analysis.

Over the past few years trace metal contamination of sediments has been studied in Portugal by different research teams, including the evaluation of the mercury level in the coastal lagoon of Aveiro (A. Hall, 1987), the characterization of metal pollution in sediments from the river Tagus estuary (Carrondo, 1984) and a preliminary study in a reservoir of the river Douro - dam of Miranda do Douro (Gonçalves, 1987).

Sediment analyses in the river Ave basin are particularly useful to detect pollution sources and to select critical sites for routine water sampling. Aquatic mosses will be also used as metal pollution indicators in this basin, so a comparison between results is intended.

Four intensive surveys were carried out, the first one in September 1985 and the others between June and October 1986.

Trace metals such as cadmium, chromium, copper, lead and zinc were analysed in the fraction of the sediment with grain size less than 63 µm. In the first survey the analyses were also performed in the fraction with grain size between 63 and 180 µm.

The preferential bonding of metals to organic matter was also investigated.

2. INTEREST OF THE RIVER AVE BASIN AS A STUDY REGION

The river Ave drainage basin is situated in the Northern Region of Portugal. It covers an area of about 1388 Km2 of which 245 Km^2 correspond to the hydrographic basin of the river Este, the most important tributary on the right bank, and 340 Km^2 to the hydrographic basin of the river Vizela (left bank). The lenght of the Ave is 98 Km.

The basin presents stringent water management problems, mainly in the aspects related to water quality. Due to the high industrial densities in the middle and lower part of the river basin, problems related to the lack of water are also quite serious. Textiles absorb over 70% of all the people employed in industry. Most of the industrial wastewaters are discharged directly and without any treatment into the receiving streams. In what concern sewage discharges the situation is quite similar although only a small fraction of the total population benefits from a sewerage network.

3. FIELD AND LABORATORY WORK

3.1 Sampling stations

Twenty sampling stations were selected in the river Ave and in the most contaminated tributaries (Fig.1). In 1985 only eight stations were considered but in 1986 the sampling and analytical programme was extended to eighteen stations.

3.2 Sampling procedure

Samples were taken by an Ekman-Lenz dredge, which is specially indicated for studying the superficial layer of bottom deposits. The collected material was then transfered to plastic bags and refrigerated until the start of the analytical procedure.

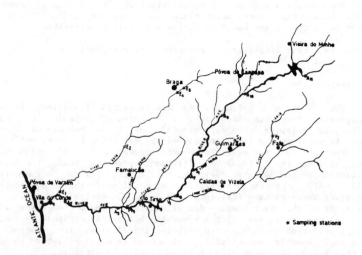

Fig. 1 - Hydrographic basin of the River Ave: location of the sampling stations

3.3 Sample processing

Samples were spread on Petri dishes, dried at room temperature and sieved. Fractions less than 63 µm and between 63 and 180 µm were separated and analysed in the first intensive survey. In the other surveys only the fraction less than 63 um was analysed.

3.4 Analytical methods

a) Organic matter

Organic matter was calculated as the volatile matter (VM) at 550°C, taking as reference the weight of the dried sediment at 105°C.

b) Trace metals

The dried and weighed samples were digested with a mixture of $HClO_4/HNO_3$ in the proportion of 2:3 (LNEC, 1986). Digestion was performed during 4 hours in mini-autoclaves of teflon, at 140°C.

With this procedure the sample solubilization is almost complete, resulting in a significative economy of time and reagents when compared with digestion in open vessel.

Atomic absorption spectrophotometry was used for metal analysis. Chromium, copper, lead and zinc were determined with air-acetylene flame; cadmium was analysed by the graphite furnace method.

4. RESULTS AND DISCUSSION

4.1 First field work session (September 1985)

Two grain size fractions (F1< 63 µm and 63>F2<180 µm) were analysed in the sediments from eight sampling stations. Five stations were located in the main stream and the remaining three stations in the mouth of the most important tributaries: river Este, river Pele and river Vizela. The results of the metal contents expressed as the ratio of the metal concentration to the volatile matter concentration are given in Table 1. Volatile matter results are also presented.

Figures 2 to 6 show the distribution of metal concentrations in the fraction F1 along the river.

4.2 Field work sessions in 1986

Sediment samples were collected in June, September and October, covering the main stream and several tributaries. Only the fraction F1 (<63 µm) was analysed. The results reported to volatile matter are presented in Table 2.

Table 1 - Results of the sediments analysis (first survey)

Sample station	VM %		Cd mg/kgVM	Cr g/kgVM		Cu g/kgVM		Pb g/kgVM		Zn g/kgVM	
	F1	F2	F1	F1	F2	F1	F2	F1	F2	F1	F2
E2	18.9	13.7	12.2	.88	.80	7.87	6.94	1.47	1.07	7.29	6.50
A2	11.9	8.3	3.4	.50	.45	4.20	1.37	.48	.30	2.95	2.35
P11	13.7	9.0	3.9	.50	.56	1.45	1.20	.39	.36	1.93	2.29
A5	3.6	1.4	5.0	.39	.42	1.61	1.71	1.64	2.00	2.86	4.43
V1	16.9	9.7	2.4	.12	.15	1.04	.98	.41	.39	1.07	1.31
A7	12.2	9.0	3.9	2.43	2.26	1.43	1.14	.66	.28	2.52	2.72
A8	11.8	3.9	4.6	1.46	1.31	6.89	2.33	2.43	.77	6.61	2.62
A9	10.0	2.3	6.2	.24	.65	1.86	1.78	.64	1.09	1.38	2.65

F1 - fraction with grain size less than 63 µm
F2 -   "   "   "   "   between 63 and 180 µm

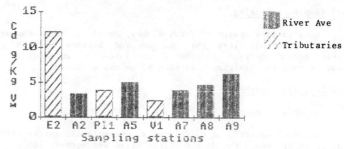

Fig.2 – Cadmium concentration in the sediments (fraction Fl<63 μm) Sept 1985

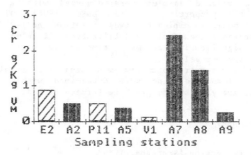

Fig.3 – Chromium concentration in the sediments (fraction Fl<63 μm) Sept 1985

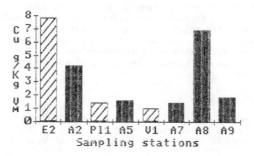

Fig.4 – Copper concentration in the sediments (fraction Fl<63 μm) Sept 1985

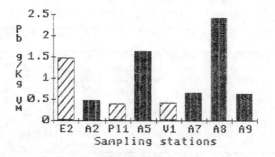

Fig.5 – Lead concentration in the sediments (fraction Fl<63 μm) Sept 1985

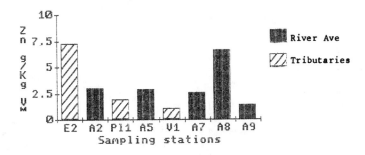

Fig.6 - Zinc concentration in the sediments (fraction F1<63 μm)
       Sept 1985

Table 2 - Results of the sediments analysis (1986 surveys)

Sample station	VM %	Cd mg/kgVM	Cr g/kgVM	Cu g/kgVM	Pb g/kgVM	Zn g/kgVM
A1	10.5	3.6	.68	1.50	.71	2.49
E1	11.7	1.8	.56	15.86	1.21	17.90
E2	21.5	27.0	1.14	8.86	1.94	12.57
E4	10.6	2.0	.31	2.95	.89	2.96
E5	13.6	1.2	.12	1.86	.53	1.25
A2	10.0	4.4	1.32	1.12	.63	1.72
A3	12.4	2.3	1.09	1.23	.62	1.51
Ph1	12.8	3.9	.30	1.45	1.04	2.12
P11	18.3	2.7	.44	1.68	.34	1.00
A5	12.2	2.5	1.47	.99	1.07	1.76
V1	15.6	4.4	.21	3.19	3.64	1.28
V2	7.2	2.5	.14	.72	.54	1.26
A6	7.5	1.9	2.68	.85	.60	1.59
S1	10.0	3.6	11.87	1.65	1.24	2.88
S2	11.7	2.8	.14	.72	.54	1.26
A9	3.6	5.3	.75	2.97	1.17	2.78
A10	13.5	1.6	.12	.99	.51	.92
A11	5.2	3.7	.27	1.33	.79	3.60

Figures 7 to 11 show the evolution of metal concentrations in the
river basin. As it was done with the previous results, a different
graphic pattern was chosen for the main river and the mouth of the
tributaries. The values occuring in the river Este reveal its importance
in what concerns metal pollution.

4.3 - Metal accumulation in the organic material

The measured concentrations of metals in fraction F1 and F2 are,
with a few exceptions, well correlated with the volatile matter content
in each fraction (Fig.12), suggesting that organic matter may be mainly
responsible for metal accumulation in sediments.

4.4 - Background levels

An important problem when discussing the analytical results is the
evaluation of the natural background level of the metals being
determined. River sediments from non-polluted areas (sources of main
stream and tributaries) were used for the assessment of the background
values. Sampling stations presenting the lower concentrations were
selected for this purpose. The average metal concentration in those
stations was assumed as the natural background level or reference
level.In Table 3 are presented the calculated values for the background
levels.

River Ave

Tributaries

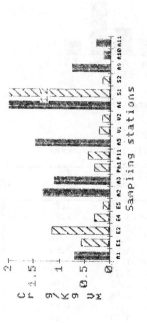

Fig.8 - Chromium concentration in the sediments
(fraction Fl<63 μm) 1986 surveys

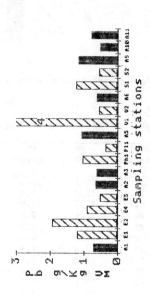

Fig.10 - Lead concentration in the sediments
(fraction Fl<63 μm) 1986 surveys

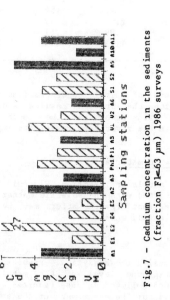

Fig.7 - Cadmium concentration in the sediments
(fraction Fl<63 μm) 1986 surveys

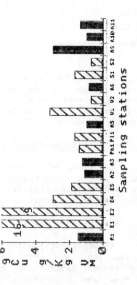

Fig.9 - Copper concentration in the sediments
(fraction Fl<63 μm) 1986 surveys

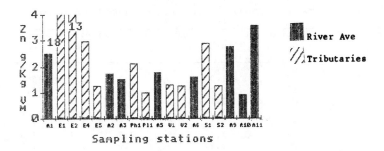

Fig.11 - Concentrations of zinc in the sediments (fraction F1≤63 μm)
1986 surveys

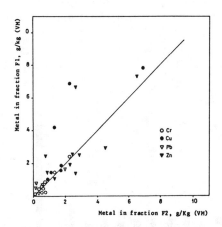

Fig.12 - Correlation between the metal concentrations in the fractions
F1 and F2

Table 3 - Background levels in the river Ave basin (mg/KgVM)

Cd	1.9
Cr	160
Cu	920
Pb	520
Zn	1130

4.5 - Contamination factors and metal pollution index

It is possible to calculate for each metal a contamination factor
FC defined as

$$FC = C_M/C_B$$

where C_M is the metal concentration and C_B the background value. If
$C_M \ll C_B$, FC is supposed to be equal to 1.
The arithmetic mean of the five contamination factors for a given
sampling station represents a metal pollution index (Robbe, 1984).
In Table 4 are presented the individual contamination factors for
each station in 1986 and in Fig. 13 can be observed the variation of the
metal pollution index along the river Ave and at the mouth of the most
important tributaries.

Table 4 - Contamination factors - 1986 field sessions

Station	Contamination factors				
	Cd	Cr	Cu	Pb	Zn
A1	1.9	4.3	1.6	1.4	2.2
E1	1.0	3.5	17.2	2.3	15.8
E2	14.2	7.1	9.6	3.7	11.1
E4	1.1	1.9	3.2	1.7	2.6
E5	1.0	1.0	2.0	1.0	1.1
A2	2.3	8.3	1.2	1.2	1.5
A3	1.2	6.8	1.3	1.2	1.3
Ph1	2.1	1.9	1.6	2.0	1.9
P11	1.4	2.8	1.8	1.0	1.0
A5	1.3	9.2	1.1	2.1	1.6
V1	2.3	1.3	3.5	7.0	1.1
V2	1.3	1.2	1.1	1.2	1.1
A6	1.0	16.8	1.0	1.2	1.4
S1	1.9	74.2	1.8	2.4	2.5
S2	1.5	1.0	1.0	1.0	1.1
A9	2.8	4.7	3.2	2.3	2.5
A10	1.0	1.0	1.1	1.0	1.0
A11	1.9	1.7	1.4	1.5	3.2

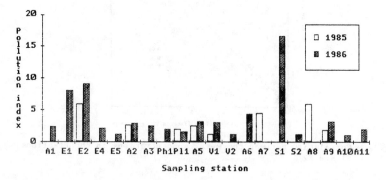

Fig.13 - Variation of the metal pollution index along the river Ave
and at the mouth of the most important tributaries.

5. CONCLUSIONS

The use of sediments to evaluate the metal pollution in the river
Ave basin constitutes a valuable approach with advantages over the
direct measurement in the water.

It was verified that the metal levels in the fine grain fraction
are well correlated with the volatile matter.

The contamination factors for Cd, Cu, Pb and Zn, expressed as the
ratio between metal concentrations and background values, are below 4 in
almost all the sampling stations. In the station E1 this value was
exceeded for Cu and Zn and in the station E2, also in the river Este,
the contamination factors, unless that for Pb, are all quite over 4. The
contamination factor for Pb in the station V1 is also higher than 4. In
what concerns Cr, the FC values in the mainstream are all over 4
downstream the station A9, which is explained by the wastewater
discharges from the textile and tanning industries.

Comparing the metal pollution indices in 1985 and 1986, it can be
observed that with the exception of the station P11, the values are
slightly higher in 1986. Along the mainstream chromium is the main
responsible for the increase; in the river Este, it is the copper; in
the river Vizela all the metal concentrations present a small increase.

REFERENCES

CARRONDO, M. J. T., REBOREDO, F., GANHO, R. M. B. and OLIVEIRA, J. F. S. (1984) - "Analysis of sediments for heavy metals by a rapid electrothermal atomic absorption procedure", Talanta, 31, 7, pp 561-564.

GONÇALVES, E., BOAVENTURA, R. e GUEDES DE CARVALHO, R. A. (1987) -"Estudo de sedimentos da albufeira de Miranda do Douro", Recursos Hídricos, 8, 1, pp 51-57.

HALL, A., DUARTE, A. C., CALDEIRA, M. T. M., LUCAS, M. F. B. (1987), "Sources and sinks of mercury in the coastal lagoon of Aveiro, Portugal", Sci. Total Environ., 64, pp 75-87.

LNEC (1986), "Estudo de sedimentos da bacia hidrográfica do rio Ave", Relatório 92/86-NHHF, Lisboa.

ROBBE, D. (1984), "Interprétation des teneurs en élements métalliques associés aux sediments", Rapport des laboratoires, Série: Environnement et genie urbain EG-1, 149 p., LPCP, Paris.

Chapter 23

EFFECTS OF LAND USE ON NITROGEN, PHOSPHORUS AND SUSPENDED SOLIDS CONCENTRATIONS IN TROPICAL RIVERS

H Orth, L Sukhanenya (Asian Institute of Technology, Bangkok, Thailand)

ABSTRACT

Effects of land use on nitrogen, phosphorus, and suspended solids concentrations were analyzed. The three main land use classes were forests, deforested land, and agricultural land. The results show the expected effect of land use, but this effect is subject to a very distinct annual cycle of concentrations which can be explained by the ecology of tropical areas.

Key Words: land use, water quality, nitrogen, phosphorus, suspended solids

1. BACKGROUND AND OBJECTIVE OF THE STUDY

The importance of land use on water pollution originating from non-point sources is recognized worldwide. A comprehensive overview of the effect of land use is provided by the study on non-point source stream nutrient level relationships by Omernik (1977). In this study, which covers the entire USA, the average total nitrogen (TN) and total phosphorus (TP) concentrations were about nine times higher in predominantly agricultural watersheds than in predominantly forested watersheds (0.598 and 5.354 mg/l respectively for TN; 0.018 and 0.161 mg/l respectively for TP). The effect of deforestation has been demonstrated by Likens et al. (1969) who reported an average nitrate concentration of 53 mg/l in a experimentally deforested watershed as compared to 0.9 mg/l in the undisturbed forest. Besides the general impact of land use, several influencing factors and mechanisms were identified which account for differences between various types of land use and even between similar types of land use under varying conditions. Rainfall pattern, seasonal variations and stream flow rates are examples. Different results were reported with respect to the influence of the geology and slopes (e.g. Dillon and Kirchner, 1975; Omernik, 1977).

Research on non-point pollution sources, however, is very much concentrated in industrialized countries, partly due to the notion that pollution is more severe in these countries and that the management of point sources is still the primary goal in developing countries. In many situations in developing countries, however, pollution from non-point sources is at a comparable or even higher level. Agricultural and forestry practices have changed rapidly in many countries, resulting in intensive use of fertilizers and machinery.

Since most developing countries are located in tropical areas, the intensified anthropogenic impact on land resources coincides with very sensitive ecological conditions and effects may be even more immediate and more severe than in a temperate climate.However, the actual impact of non-point sources and land use practices on water bodies in tropical areas is largely unrecorded. It is this lack of information and assumed detrimental developments that has created concern rather than many demonstrated cases. Limited knowledge on the behaviour of tropical water bodies, furthermore, requires control measures to be based largely on experience from temperate climates.

It is with this background that a study was undertaken to investigate the effect of different types of land use on nitrogen, phosphorus and suspended solids concentrations in tropical rivers. Priority was given at this stage to recording effects of different types of land use rather than to the analysis of specific mechanisms which determine the input from non-point sources. This objective requires a rather large study area to obtain sufficiently variant and representative conditions. The main steps of the study were;
- the analysis of nitrogen, phosphorus, and suspended solids concentrations, and of several supporting parameters at various streams;
- the analysis of land use in the respective catchment areas;
- the evaluation of effects of land use, specifically of forest, deforested and agricultural land on the nitrogen, phosphorus, and suspended solids concentrations in streams.

2. DATA COLLECTION AND EVALUATION

The Study Area

The study area is located in northern Thailand. The area covers the upper part of the river Ping catchment area and tributaries with a total area of about 19,055 km^2. The climate is tropical with the rainy season lasting from May to October. The average monthly temperature is between about 20 °C in January and 29 °C in April. The water temperature in the rivers is mostly in about the same range and only a few values above or below were recorded.

Forests occupy about 78% of the river basin. About 82% of these forests are deciduous forests with limited and low undergrowth. Uncontrolled deforestation is frequent, but deforested areas are usually covered by fast growing pioneer plants. Agriculture is the predominant land use in the flat river valleys with rice being the main crop. First crop rice is planted in the first half of the rainy season during June and July. Second crop, mainly upland crops and vegetables, is planted in November and December. Nitrogen, phosphorus and potassium fertilizer is used by most farmers.

Land Use Analysis

Land use maps are available only of a small percentage of the area, mainly urban or otherwise intensively used areas. Thus, satellite photography was used as the only immediately available data source for land use analysis. Land use was at first grouped into nine classes. For the final evaluation, the four classes describing forest areas were combined into the two classes forests and deforested areas. Similarly, the two classes of agricultural land were combined into one. Settlements and waterbodies, covering 0.7 % of the area, were according to the objective of the study not included in the further analysis and 5.2% of the area remained unclassified.

Water Quality Analysis

Twenty-two sampling points were at first located on the river Ping and twelve tributaries. Based on experience from the first sampling period, the number of sampling points was,then, increased to thirty six. The sampling points were generally placed upstream of settlements in order to minimize the impact of point sources as far as possible. Water samples were taken at the sampling points six times over the period of one year. For each measurement, the average of three replicates was taken. The seventeen parameters of the sampling program are given in Table 1.

1.	Ammonia nitrogen	10.	Total phosphorus
2.	Nitrite	11.	Suspended solids
3.	Nitrate	12.	Conductivity
4.	Inorganic nitrogen	13.	Turbidity
5.	Organic nitrogen	14.	Settleable solids
6.	Total nitrogen	15.	Dissolved oxygen
7.	Orthophosphate (SRP)	16.	Temperature
8.	Inorganic phosphorus	17.	pH
9.	Organic phosphorus		

Table 1: Parameters of the Water Quality Sampling Program

Supporting Data

Some additional data were collected as supporting information; average ground slopes for the catchment area in the vicinity of individual sampling points were estimated from topographic maps in the scale of 1 : 50,000 or 1 : 250,000, as available. A classification of available phosphorus concentrations, pH-value and surface runoff could be obtained from soil maps available for a few sampling points.

Preliminary data on rainfall, humidity, and evaporation during the sampling program could be obtained from the Meteorological Center for Northern Thailand. The data do not show any extreme situation. However, rainfall was less than in an average year, particularly in May and August. The annual rainfall in the Chiang Mai province was 1005 mm as compared to an average of 1226 mm for the years 1972 to 1981.

Evaluation

Regression analysis was employed besides the general evaluation for illustrating the dependence of nutrient and SS concentrations on different types of land use. Agricultural and deforested land, combinend in one class, represents the types of land use with higher nutrient and SS losses. This class is primarily used in this report for summarizing area characteristics when illustrating land use effects.

Linear regression, multiple linear regression and two forms of logarithmic regression were compared. Best results were achieved by an equation of the following form:

$$\ln(y-y_o) = a + b \ln x \qquad\qquad (1)$$

where y is the nutrients or SS concentration in mg/l, y_o is a constant basic concentration, and x is the type of land use in %. However, the limitations of the statistical analysis must be noticed. It is an illustrative regression analysis in support of the data interpretation. The study covers one annual cycle only and was performed in a specific area. Any generalization would, therefore, require confirmation by further studies.

Five sampling points were located in the vicinity of flow recording stations of the Thai Royal Irrigation Department. The flow rates of these stations and the concentrations of the respective sampling points were used for a rough estimate of area related export rates of TN, TP, and SS.

3. RESULTS AND DISCUSSION

Seasonal Variations of Water Quality

Significant seasonal variations of water quality parameters require this aspect to be discussed first. The seasonal variations exceeded the variations due to land use effects. Land use effects can, therefore, not be discussed on the basis of annual figures but only seperately for different periods within the year.

Table 2 gives an overview of the average nitrogen, phosphorus and SS concentrations for the six sampling periods. Minimum and maximum values were 0.25 and 4.10 mg/l for TN, 0.03 and 0.46 mg/l for TP, and 2 and 711 mg/l for SS respectively. With the exception of inorganic nitrogen fractions, the concentrations reached peak values during the May or June sampling periods, which mark the early rainy season. They, then, fall rapidly during the rainy season, reaching the minimum in the November sampling period at the end of the rainy season. Only the minimum concentration of SS appears one sampling period later in January.

It can be assumed that dilution contributed to the observed annual cycle of nutrients and SS concentrations. However, the estimated nutrient export rates suggest that dilution was not the main factor contributing to the parameter variations. If dilution were the main factor, nutrient and SS export rates should be largely constant. However, they were high during the first half of the rainy season and fell rapidly during the latter part of the rainy season. They remained on a low level during the dry season until the beginning of the next rainy season.

Sampling Period						
	June	Aug	Nov	Jan	Mar	May
Ammonia-N	0.02	0.11	0.12	0.09	0.08	0.09
Nitrate-N	0.10	0.10	0.08	0.09	0.06	0.07
Inorganic-N	0.12	0.21	0.21	0.19	0.16	0.17
Organic-N	1.93	0.55	0.27	0.39	0.36	0.60
Total N	2.04	0.77	0.47	0.58	0.52	0.76
Ortho P	0.05	0.02	0.02	0.03	0.04	0.07
Inorganic P	0.08	0.04	0.02	0.04	0.05	0.08
Organic P	0.10	0.04	0.04	0.05	0.06	0.07
Total P	0.18	0.08	0.06	0.08	0.11	0.15
Susp. Solids	127	63	40	19	25	71

Table 2: Summary of Nitrogen, Phosphorus and Suspended Solids
Concentrations in mg/l (Average of all Sampling Points).

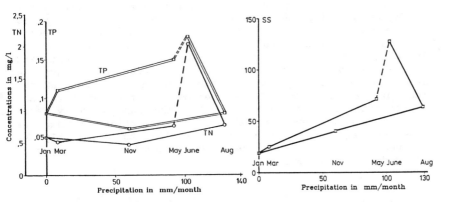

Fig.1: TN, TP and SS concentrations versus the average monthly
precipitation

The presentation of TN, TP and SS concentrations versus
precipitation in Fig. 1 indicates that also precipitation
exerts some, but again, only limited influence on the observed
concentration variations. SS concentrations correspond largely
to precipitation except in the first half of the rainy season.
In the case of TN and TP, the increase of concentrations after
the first rainfalls may be influenced by precipitation.
However, the concentration variations which do not correspond
to precipitation are not irregular but follow a similar annual
cycle. They are highest during the June sampling period but,
then, fall rapidly in the August sampling period in spite of
increasing precipitation. In relation to precipitation, the
concentrations are lower during the second half of the rainy
season and higher during the dry season.

As a basic factor contributing to the observed annual cycle
of concentrations, the ecological background of tropical areas
and the interaction between soil and climate is to be
considered. The concentration of nutrient minerals in tropical
soil is generally low. The main minerals required for
bioproduction are mainly incorporated in organic material and

subject to a process of immediate recirculation, if sufficient
moisture is available. During the dry season, organic material
only undergoes partial decomposition and accumulates on the
surface. At the beginning of the rainy season, heavy downpours
wash-off the accumulated material from the soil surface and
some of the top soil minerals.In the later part of the rainy
season, recycling of nutrient material is in full activity and
run-off and sub-surface effluent carries less materials. The
observed annual cycle of nutrient and SS concentrations may,
thus, be largely explained as a combined result of the ecology
of tropical areas, dilution by varying flow rates, and wash-
off from the surface due to precipitation run-off, with the
first factor exerting a dominant influence.

Effect of Land Use on Nutrient and Suspended Solids Levels

As mentioned earlier, the strong influence of the annual
cycle requires the separate evaluation of land use effects for
different sampling periods. With the June sampling period as
an example, the expected effect of land use on TN concentrations is demonstrated in Fig. 2a; the TN concentrations
increase with increasing percentage of agricultural and
deforested land. However, the TN concentrations during the
June sampling period represent the case of the strongest effect of land use. During other periods and for phosphorus and
SS concentrations, the effect of land use was generally less.
At the opposite extreme, the TP concentrations during the
November sampling period are shown in Fig. 2b. Apparently, no
influence of land use on the TP concentrations can be recognized during this period.

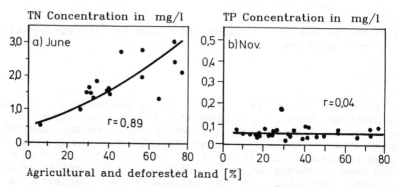

Fig.2:   Correlation   of   TN   and   TP   concentrations   with
        agricultural and deforested land

To give an overview of the effect of land use, regression
lines according to Eq. (1) are presented in Fig. 3 seperately
for the different sampling periods. The excepted influence of
land use is clearly visible. Depending on the percentage of
agricultural and deforested land, TN concentrations e.g. vary
in the June sampling period in a ratio of more than 1 : 4. In
most cases, however, the parameter variations are influenced
more by the sampling period than by land use. The same result
was observed for SS concentrations.

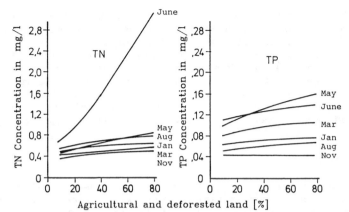

Fig. 3: TN and TP concentrations versus the percentage of agricultural and deforested land

A stronger influence of land use might have been expected but the conditions in the area provide an explanation for the above results. In areas denoted as deforested land, the primary forest has been destroyed but secondary vegetation, mainly grassland, usually develops quickly, protecting the soil. Agriculture is concentrated in the flat river valleys and, since rice is the main crop, the fields are additionally levelled. Only limited wash-out is expected under this condition and the result confirms the experience of Alberts et al. (1978) who mentioned that a level terracing system reduces sediment and nutrient losses.

Nitrogen, Phosphorus, and SS Export Rates

Flow records were available for five sampling points only. This limited amount of data allows only a rough estimate of nitrogen, phosphorus, and SS export rates. However, some trends will be reported here: The average TN export rates for the six sampling periods were estimated to be between 0.16 and 5.87 kg/ha/yr. The values for TP and SS were between 0.04 and 0.78 kg/ha/yr and 13 and 856 kg/ha/yr respectively. These variations of export rates of the different sampling periods exceed the variations observed for concentrations. The reason is that the high concentrations during the first half of the rainy season coincide with high flow rates. As a result, more than half of TP and SS and more than two thirds of TN was exported during this period. The influence of land use was also most pronounced during this period.

Subsoil, Slopes, and Fertilizer Application

Although subsoil, area slopes, and fertilizer application are generally considered as having a significant effect on nutrient and SS export, no effect could be recognized in this study. Among other factors, the following conditions of the study area may have contributed to this seemingly unexpected result: The soil type may be less important in tropical areas where most of the main minerals are incorporated in organic material. Regarding the influence of area slopes, it is to be noted that forests cover mainly the mountaineous areas, whereas agriculture is concentrated in the flat river valleys. Thus, high slopes coincide with a type of land use which is associated with low loss rates, whereas the main contributing type of land use is concentrated in flat areas. These conditions may tarnish the impact of slopes. Fertilizers are used intensively. However, they are applied during the growth season only. Furthermore, since rice is the main crop, most fertilized land is levelled land where fertilizers applied are less prone to wash-off.

4. CONCLUSIONS

The study confirmed the influence of land use on nitrogen, phosphorus, and SS concentrations and export rates. However, these effects were closely linked to an annual cycle of parameter variations. This cycle is attributed to the tropical ecology and climate. Concentrations and export rates were particularly high at the beginning and during the first half of the rainy season. It was also during this period that land use exerted the strongest effect. If this result can be confirmed by further studies, the identified period of high nutrient and SS concentrations and export rates may require specific attention in water quality management.

The observed values are rather moderate in comparison to other studies. This and the fact that no influence of fertilizers could be recognized indicate that the agricultural methods applied in the area are favorable with respect to the influx of nutrients and suspended solids into rivers resulting from non- point sources.

ACKNOWLEDGEMENT

This paper originated from a study entitled "Effect of Deforestation and Agricultural Land Use on the Nutrient Level and Suspended Solids Load of Tropical Streams" which was supported by the Commission of the European Communities within the framework of the program on Science and Technology for Development.

REFERENCES

ALBERTS, E.E., SCHUMAN, G.E. and BURWELL, R.E. (1978). "Seasonal Runoff Losses of Nitrogen and Phosphorus from Missouri Valley Loess Watersheds". J. Environ. Qual., vol. 7 (2), pp. 203-208.

DILLON, P.J. and KIRCHNER, W.B. (1975). "The Effects of Geology and Land Use on the Export of Phosphorus from Watersheds". Water Research, vol.9, pp.135-148.

LIKENS, G.E., BORMANN, F.H. and JOHNSON, N.M. (1969). "Nitrification: Importance to Nutrient Losses from a Cutover Forested Ecosystem." Science. vol. 163, pp. 1205-1206.

OMERNIK, J. M. (1977). Nonpoint Source - Stream Nutrient Level Relationships: A Nationwide Study. Corvallis Environmental Research Laboratory, Office of Research and Development, USEPA, Corvallis, Oregon.

PART IV

Industrial and urban
pollution

Chapter 24

SOLID WASTE MANAGEMENT IN A COKERY: ENVIRONMENTAL IMPACT ON SURFACE AND GROUNDWATER

J Dobosz, M Sebastian (Technical University, Wroclaw, Poland)

ABSTRACT

The source of origin, as well as the physicochemical composition of the solid wastes produced by a cokery are characterized. Based on data of the physicochemical composition water-extracts, most of the solid wastes have been classified as highly hazardous to the natural waters.

Key words: industrial solid wastes, disposal methods, physicochemical composition of water-extracts.

1. Introduction

The Cokery Entprise "Wałbrzych" develops blast—furnace coke, coke—oven gas, crude benzol, ammonium sulphate, as well as tar products of various kinds. The separate technical processes are accompanied by wastes formation having unfavourable effects on the ecosystem because of coal and dust along with other gases emission, formation of polluted waste waters, also pollution of surface and ground waters with leakages out of solid and semi—solid wastes. The cokery enterprise is located on the formed Upper Carbon stratum with hard coal beds, most of them visible on the surface.

2. Experimental

The physicochemical composition of the wastes has been determined according to Polish Standards , performing at the same time a water extract from the wastes [Kempa(1976)] that being analized with full particulars enables the determination of the waste hazard range for ground and surface waters.

3. Results and Discusion

The group of solid cokery wastes consists of: sewage sludge, tar wastes, waste tar from the production of ammonium sulfate, as well as ceramic and brick rubbles. Table 1. shows the physicochemical composition of wastes with the arith — thmetic average for separate indeces from several to tens values.

TABLE 1

Physicochemical Composition of Wastes.

WASTE	Weight loss at 378 K	Weight loss at 823 K	Water cont. xylene method	SiO$_2$	Solids insol. in toluene	Ether extract	Sulfur	Nitrogen NH$_4$	Nitrogen Ogranic	Volat. phenols	Naphta- lene
	%	%	%	%	%	%	%	%	%	%	%
Sewage sludge	32.78	59.97	20.75	4.65	53.62	4.51	0.54	0.01	0.34	0.34	12.41
Tar waste	35.12	96.39	13.60	0.88	24.87	7.27	0.54	n.d.	0.55	0.61	17.65
Waste tar from the production of (NH$_4$)$_2$SO$_4$	39.29	98.58	35.88	0.43	44.12	8.86	12.32	5.43	0.25	0.36	11.24
Ceramic and brick rubbles 1	0.08	0.20	n.d.	95.54	n.d.	n.d.	n.d.	n.d.	n.d.	n.d.	n.d.
2	5.18	2.96	1.72	81.28	n.d.	n.d.	n.d.	n.d.	n.d.	n.d.	n.d.
3	0.63	0.75	n.d.	96.37	n.d.	n.d.	n.d.	n.d.	n.d.	n.d.	n.d.

TABLE 1

Physicochemical Composition of Wastes (contd)

WASTE	Metals (% dry solids) Na	K	Ca	Mg	Fe	Pb	Zn	Ni	Cu	Cr	Co	Mn	Cd
Sewage sludge	33.00	3.00	130.1	57.50	2.50	0.02	0.40	0	0	0-0.32	0	0.18	0
Tar wastes	0.06	0.01	0.15	0.02	0.17	0.05	0.08	0	0	0.003	0	0.04	0
Waste tar from the production of (NH$_4$)$_2$SO$_4$	0.31	0.10	0.04	0.01	0.21	0.05	0.02	0.003	0.014	0	0	0.001	0
Ceramic and brick rubbles 1	0.10	0.05	0.21	0.03	0.46	<0.001	0.005	0.002	0.002	0.005	0.001	0.004	0
2	0.09	0.06	0.85	0.50	1.19	0.001	0.004	0	trace	0.038	0	0.014	0
3	0.06	0.05	0.80	0.05	0.22	0.001	0.145	0	0.001	0.013	0	0.008	0

Sewage sludge

Sewage sludge that had been stored for tens of years in the sewage form was transported to a dumping ground. The samples were taken in various spots and depths, and also in different seasons of the year. Six sewage samples were determined, each being the equolibricum mixture of five individual samples.

The determination results of physicochemical composition for water extracts are presented in Table 2.

Here the great differenciation of pollution coefficient values for separate samples can be observed. And so, the content of dissolved substances equals from 45 – 2496 g/m^3, sulfates : 0 – 1358.4 g SO$_4$ /m^3, cyanides: 0 – 4.4 gCN/m^3,phenols: 10.6–112.1 g/m^3, COD: 86–330 g O$_2$ /m^3.The sodium content equals :1– 145 g/m^3, calcium :8–334 g/m^3, magnesium: 1.06 – 115.00 g/m^3, ferrum:0.6 –6.4 g/m^3, lead:0 – 0.1 g/m^3,chromium:0 – 0.32 g/m^3,zinc: 0.20 –0.50 g/m^3. The only values not so fluctuant are those of pH and they keep in the range 7.12 – 8.13.

Thus, taking into account the presented above pollution indeces values, leakages from sewage sludges proved to be severe danger for natural surface and ground waters. This is futher confirmed by the fact that the area of dumping ground, as big as 0.5 ha, is exposed to the impact of rainfall waters. The only method of these wastes neutralization, most effective from the environment protection viewpoint, is their combustion, or storage on some special dumping ground after achieving the appropiate consistence.

TABLE 2

Physicochemical Composition of Water-extracts from sewage sludges

Sample	pH	Alka-linity	Dissolved solids			Chlo-rides	Sul-phates	Cya-nides	Total Phenols	Perman. COD	Dichrom. COD
			total	mineral	volat.						
		val/m3		g/m3		gCl/m3	gSO4/m3	gCN/m3	g/m3		g O2/m3
1	7.3	2.40	1080.0	728.0	298.0	n.d.	485.4	1.40	64.3	175.0	393.1
2	7.12	1.40	496.0	1922.0	574.0	n.d.	1358.4	2.20	112.1	330.0	667.9
3	7.40	3.20	660.0	416.0	244.0	n.d.	152.2	4.40	34.5	132.0	328.2
4	7.57	3.80	450.0	240.0	210.0	n.d	70.4	1.40	25.1	86.0	206.1
5	7.30	1.30	2005.0	1634.0	371.0	1.0	1079.5	1.40	34.9	260.0	597.8
6	8.13	0,60	45.0	16.0	29.0	3.0	0	0	36.8	175.0	255.8

TABLE 2

Physicochemical Composition of Water extracts from sewage sludge (contd)

Sample	Metals (g/m3)												
	Na	K	Ca	Mg	Fe	Pb	Zn	Ni	Cu	Cr	Co	Mn	Cd
1	20.0	2.8	144.0	44.0	3.90	0	0.37	0	0	0	0	0.17	0
2	34.5	3.6	334.0	115.0	0.60	0	0.33	0	0	0	0	0.13	0
3	7.5	3.6	74.0	33.0	6.40	0	0.50	0	0	0	0	0.57	0
4	6.2	2.2	23.0	43.0	1.10	0	0.37	0	0	0	0	<0.10	0
5	145.0	4.3	168.0	92.5	1.00	0.06	0.20	0	trace	0.32	0	0,16	0
6	1.0	1.0	8.0	1.0	1.10	0.10	0.50	0	0	0	0	0	0

Tar wastes

They come from the department of coal derivatives recovery.
They consists of sewages from decanters and dividers of various
reservoirs of dehydrated and non-dehydrated tar. The additional
factors are as well tar leakages and its destilants separated
from sludges.
Their physicochemical composition is presented in Table.1,and
the results of water extract examination in Table 3. These
wastes proved to be dangerous for natural waters, mainly
through the content of phenols, seriously exceeding the content
standards (up to 236.76 g/m^9) and zinc (up to 0.7 g/m^9), as
well as COD value (up to 829.5 g O /m^9).
The simplest and most commonly used method of this kind of
wastes neutralization is the addition of coal charge. They can
be also combused in the incinerating plants of the non thypical
arrangement [Wolany(1985)] what is of poor efficiency with
regard to supply management, but guarantees the whole final
neutralization of waste. Other possibilities of tar wastes
utilization are shown in [Dobosz(1988)].

Waste tar from the production of ammonium sulfate

Its main components are : free sulfuric acid, ammonium sulfate, pyridine bases and organic substances of resins characteristics. The physicochemical copmposition of such a kind of
wastes is given in Table 1.The chemical composition of water
extract (Table 3) of ammonium sulfate tars resembles highly
contaminated sludge. The strongly acid reaction (pH=1.87),the
quantity of dissolved substances (8750 g/m^9),sulfates content

TABLE 3

Physicochemical Composition of Water-extracts from tar wastes

Sample	pH	Alka-linity	Dissolved solids			Chlo-rides	Sulfa-tes	Thio-sulfa-tes	Total Phenols	Perman. COD	Dichrom. COD
			total	mineral	volat.						
		val/m3	g/m3			gCl/m3	gSO4/m3	gSO/m3	g/m3	g O2/m3	
1	6.20	0.40	174.0	8.0	166.0	7.0	144.8	0	56.8	775.0	217.0
2	6.75	0.40	13.0	1.0	12.0	1.0	0	4.2	41.5	145.0	170.5
3	7.82	1.60	129.0	57.0	72.0	4.0	38.7	11.2	141.1	460.0	720.9
4	7.35	1.20	66.0	28.0	38.0	4.0	37.0	8.4	236.8	570.0	829.5
5	7.35	1.00	65.0	24.0	41.0	3.0	21.4	4.2	160.7	355.0	589.2
6*	1.87	9.00#	8570.0	83.0	8487.0	3.0	6483.7	7.5	24.3+	176.0	440.8

* - waste tar from the production of ammonium sulfate
# - acidity
+ - volatile phenols

TABLE 3

Physicochemical Composition of Water Extracts from tar wastes (contd)

Sample	Metals (g/m3)												
	Na	K	Ca	Mg	Fe	Pb	Zn	Ni	Cu	Cr	Co	Mn	Cd
1	3.0	1.0	4.0	0.44	0.60	0.12	0.70	0	0	0	0	0	0
2	2.0	1.0	2.0	0.24	0	0.04	0.70	0	0	0	0	0	0
3	5.0	1.0	18.0	1.40	0.20	0.06	1.04	0	0	0	0	0.10	0
4	4.0	1.0	8.0	1.20	0.60	0.16	0.40	0	0	0	0	0.10	0
5	3.0	1.0	8.0	1.20	0.60	0.06	0.40	0	0	0	0	0	0
6*	12.3	3.3	46.0	2.35	10.00	0.07	0.40	0.10	0.10	0.38	0	0.10	0

* - waste tar from the production of ammonium sulfate

(6483.7 g SO/m^3), ferrum (10.0 g/m^3), zinc (0.40 g/m^3) and chromium (0.38 g/m^3) exlude the possibilities of dumping the wastes of this kind. After the neutralization they can be used for coal charge briquetting or as bonding material for other industry branches. The other methods of ammonium sulfate tars utilization are available in [Dobosz (1988)].

Ceramic and brick rubbles

It is formed as the results of demage and major repairs of coke oven batteries. The physicichemical composition of rubbles coming from three different factories is presented in Table 1. The results of analyses of water extracts physicochemical composition are set in Table 4. The water rubble extracts from the enterprises 1 to 3 are not dangerous for natural waters — except for the value pH = 4.53 in test 1, and zinc content up to 0.32 g/m^3 in some tests overcross the standards.
The great variety of determined coefficients of pollution appears in case of rubble from the enterprise 2. The pH value, from 4.8 to 12.5, the quantity of dissolved substances from 82.0 to 86400.0 g/m^3, sulfates content: 9.9-66844.4 g SO_4/m^3, ferrum:0-20 g/m^3, lead:0-20 g/m^3.
The mass increament noted during the volatile substances determination is connected with hydrooxides formation in strongly based water solution wich fullfils the condition 2F>M. Thus, taking into account the great variety of wastes of this type, they should not be applied as used object,e.g., as constructing material they can prove dangerous for the

TABLE 4

Physicochemical Composition of Water-extracts from ceramic and brick rubbles

Test	pH	Alka-linity val/m3	Dissolved solids total g/m3	mineral	Chlo-rides gCl/m3	Sul-phates gSO4/m3	Fluo-rides gF/m3	SiO_2 collo-idal gSiO/m3	Phos-phates gPO/m3	Cyani-des gCN/m3	Ferman. COD gO2/m3
1.1	4.53	0.10	236.0	226.0	6.0	137.0	0.29	3.50	trace	n.d.	2.50
1.2	6.53	0.20	70.0	65.0	1.0	34.5	trace	4.50	0.01	n.d.	2.80
2.1	7.10	1.10	271.0	230.0	2.0	181.8	n.d.	n.d.	n.d.	n.d.	7.80
2.2	12.25	24.50*	1189.0	1700.0	3.0	12.3	7.00	2.0	0.02	0.044	1.90
2.3	6.70	0.10	82.0	58.0	9.0	43.6	<0.10	1.0	0.01	0.012	1.30
2.4	12.25	23.40*	1082.0	1459.0	6.0	11.9	6.00	2.0	0.01	0.052	3.70
2.5	8.86	0.30*	2957.0	2755.0	172.0	1613.9	0.64	0	0	0.104	5.30
2.6	9.10	0.20*	1508.0	1417.0	6.0	918.7	0.56	1.0	0.03	0.034	3.00
2.7	4.80	0	88347.0	95.0	30.0	66811.4	<0.10	0	0.02	0.019	3.50
2.8	10.50	2.30*	478.0	366.0	87.0	9.9	1.50	1.0	0	0.031	2.40
2.9	7.40	0.10	268.0	220.0	4.0	160.4	<0.10	2.0	0	0.007	1.70
3.1	6.54	0.10	70.0	51.0	4.0	22.6	<0.10	3.0	0.01	n.d.	3.40
3.2	9.29	0.30	59.0	44.0	2.0	9.9	<0.10	6.0	0.26	n.d.	3.60
3.3	6.56	0.10	90.0	78.0	2.0	42.4	<0.10	4.0	0.07	n.d.	2.69
3.4	6.50	0.10	53.0	45.0	1.0	17.3	<0.10	4.0	0.03	n.d.	3.80
3.5	7.65	0.20	112.0	99.0	2.0	24.3	<0.10	7.0	0.35	n.d.	6.60
3.6	7.11	0.20	101.0	90.0	2.0	30.8	<0.10	7.0	0.36	n.d.	4.80
3.7	7.50	0.20	53.0	47.0	2.0	9.9	1.30	5.0	0.34	n.d.	2.50

* - F Alkalinity (phenolphthalein as indicator)

TABLE 4

Physicochemical Composition of Water Extracts from ceramic and brick rubbles (contd)

Test	Metals (g/m3)												
	Na	K	Ca	Mg	Fe	Pb	Zn	Ni	Cu	Cr	Co	Mn	Cd
1.1	19.0	31.2	16.0	2.1	2.40	0	0.30	0	0.50	0	0	0.10	0
1.2	3.0	2.0	12.4	0.3	0.80	0	0.20	0	0	0	0	0	0
2.1	7.0	4.0	27.0	18.8	3.20	0.08	0.80	0	0.10	0	0	0	0
2.2	72.0	33.0	390.0	0.1	trace	0.08	0.40	0	0	0	0	0	0
2.3	6.0	8.0	11.4	0.2	0	0.08	0.80	0	0	0	0	0.20	0
2.4	11.0	4.0	440.0	0.1	0.20	0.08	0.20	0	0	0	0	0	0
2.5	22.0	30.4	704.0	3.0	0.20	0.10	0.40	0	0	0	0	0	0
2.6	11.0	16.6	360.0	4.4	0.40	0	0.40	0	0	0	0	0	0
2.7	230.0	100.0	180.0	10.0	20.00	4.00	20.00	0	0	0	0	0	0
2.8	4.2	2.0	74.0	0.3	1.00	0.12	0.30	0	0	0	0	0.20	0
2.9	10.0	5.4	48.4	1.2	0.20	0.06	0.34	0	0	0	0	0	0
3.1	5.4	2.0	10.0	0.6	0.54	0	0.32	0	0	0	0	0	0
3.2	5.8	2.0	7.0	0.4	0.50	0	0.30	0	0	0	0	0	0
3.3	3.4	2.0	15.6	0.3	0.34	0	0.32	0	0	0	0	0	0
3.4	3.4	2.0	6.2	0.7	0.50	0	0.32	0	0	0	0	0	0
3.5	4.2	2.0	6.2	0.4	0.5	0	0.32	0	0	0	0	0	0
3.6	2.8	2.0	10.0	0.4	0.5	0.02	0.26	0	0	0	0	0	0
3.7	4.0	2.0	9.6	1.1	0.8	0	0.20	0	0	0	0	0	0

environment as the result of rainfalls influence. These wastes should be storage on dumping ground that is protected against inflow of surface waters.

Summing up, it should be noted that the wastes generated in cokeries, unappropiatelly managed, can be a severe meance for the environment, especially for surface and ground waters. The location of the cokery is propitious to quick infiltration of rainfall and surface waters around this spot. The high water permeability of carbon layers results from natural teutonic craks and faults, as well as from coal mining. Drifts, gangways and other excavations form draining system all over the Wałbrzych basin.

Bibliography

DOBOSZ,J.; SEBASTIAN,M.-*Basic Research and a Concept of Solid Waste Management in a Cokery "Walbrzych"*, Report of the Institute of

Environment Protection Engineering Technical University of Wrocław,
Poland, 1988,

KEMPA,E.- *Classification of Industrial Wastes*,Wrocław, 1876,
Papers of the Institute of Environment Protection Engineering,
Technical University of Wrocław , Monograph No. 12.

WOLANY,B.- *Koks, Smola, Gaz*, 6, 1988, 122.

Chapter 25

TRANSPORT OF LEACHATES FROM A LANDFILL COUPLED WITH A GEOCHEMICAL MODEL

Eckart Bütow, Ekkehard Holzbecher, Volker Koß (Technical University, Berlin, Fed. Rep. Germany)

ABSTRACT

Transport of groundwater contaminants from a landfill closed in 1960 was model-led with the geochemical code MINEQL and the transport code FAST. The model was based on groundwater analyses of eight wells downstream the landfill and the hydrological situation of the area. The two codes were linked by elemental K_d's calculated by MINEQL and used by FAST to model transport.

Available data seldom satisfy modeller's demands. Parameters necessary for the linked code were judged whether they are sufficiently known or can be derived from groundwater analysis or have to be estimated. Sensitivity of results was assessed in regard to hydrogeological parameters.

Key words: contaminant transport in groundwater. dispersion. sorption, modelling

I. Introduction

Modelling the transport of leachates from an abandoned landfill often meets the difficulty that measurements were not carried out according to modeller's demands. Detailed informaton on waste components and distribution coefficients of ground-water contaminants are seldom available whereas groundwater analyses and hydro-geological information are abundant. Linking a conventional transport code with a geochemical code provides a tool to model the transport of leachates from an abandoned landfill based on specific hydrogeological information and groundwater analyses of the test site and global information of the area (Bütow/Holzbecher/Koß 1989).

The transport code describes the behaviour of advection and dispersion for porous media neglecting density effects induced by high concentration of the aqueous solution. Retardation. decomposition and the formation of new substances are to be modelled with additional restrictions. Retardation processes are influenced by complexation. sorption. precipitation or solution of minerals and by filtration too. A simplified mathematical approach to include these different processes is the retardation factor or the distribution coefficient.

If no experimental distribution coefficients are available they can be calculated by a geochemical code. Direct coupling of both different types of codes is not possible because of computer storage needed and execution times. The use of a geochemical code to calculate element specific sorption coefficients offers the opportunity to model a field situation near to reality with commonly used transport codes. In addition a geochemical code is of great help in judging the quality of experimental data as it checks e.g. electroneutrality and solubilities.

This paper tries to analyse essential parameters of the linked geochemical transport model that either have to be determined experimentally or can be estimated. It focusses on hydrogeological parameters. Further papers focussing on geochemical parameters and waste parameters will follow.

II. Modelling approach

To model transport of leachates from a landfill, advection, dispersion and sorption processes are generally to be taken into account. Some of the necessary hydrogeological parameters are used for the transport code FAST, others for the geochemical code MINEQL. Figure 1 shows which subset of data is needed by the different codes. MINEQL uses the initial data to calculate distribution coefficients (K_d-values). FAST models on the basis of these K_d's the concentration distribution in space and time. MINEQL delivers in addition information on Eh and oversaturation Sorption and complex formation equilibria of main groundwater components and pollutants are computed simultaneously by the EIR version (Schweingruber 1982) of the solution equilibrium computer program MINEQL (Westall/Zachary/Morel 1976). The EIR database is used. Sorption is taken into account according to a simple surface complexation model (Koß 1988). The model is based on groundwater analysis, cation exchange capacity (CEC) and pore volume of the sediment-groundwater system. Element concentration in the sediment-groundwater system is known for groundwater only. Sorption is calculated without affecting groundwater concentrations.

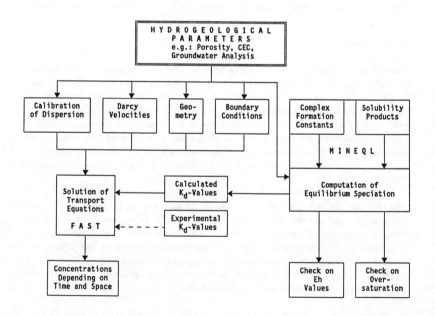

Figure 1: Overview on modelling approach

$$K_d = \frac{V}{m} \cdot \frac{\sum\limits_{\substack{i,r,v=1 \\ l=0}} r \cdot K_{r,l,v}^{app,i} \cdot [Me]^r \cdot [L_i]^l \cdot [Sorp]^v}{\sum\limits_{\substack{i,n=1 \\ p=0}} n \cdot \beta_{n,p}^i \cdot [Me]^n \cdot [L_i]^p \cdot f_{Me}^n \cdot f_{L_i}^p} \qquad (1)$$

([ ] indicates concentration in mol/dm^3, K^{app} apparent surface complex formation constant based on bulk solution concentrations, β thermodynamic overall complex formation constant, f activity coefficient in solution, v volume of solution, m mass of sediment)

The FAST-code was developed at the Technische Universität Berlin. The code is based on the differential equation for transport in porous media (Bear 1972):

$$\nabla \cdot (\rho E \cdot \nabla C) - \nabla \cdot (\rho u C) + q - \lambda R \Phi \rho C = \partial(\Phi \rho R C)/\partial t \qquad (2)$$

(ρ density, E dispersion tensor, u velocity, λ constant of decay, R retardation factor, Φ porosity, q source/sinkrates)

The dispersion tensor E is included according to Scheidegger(1961). Retardation factors are calculated from:

$$R = 1 + \rho_d \cdot K_d (1-\Phi)/\Phi \qquad (3)$$

(ρ_d rock density, K_d sorption coefficient)

The FAST-code is built according to the method of finite differences. In contrast to other transport codes FAST offers two main advantages: it performs faster especially if used with a vectorizing compiler on a CRAY (which is due to an iterative sparse matrix solver). Furtheron there is a special option included to reduce numerical dispersion. This optin is very effective (Holzbecher 1988).

III. Analysis of hydrogeological parameters

When an integrated model is used in practice, many parameters have to be known concerning the aquifer in question. Modelling may become useless when some specific parameters cannot be determined to a sufficient degree.

Parameters will be classified according to three criteria:

- Is the variable in question available? If so: how precisely?
- How sensitive is the variable for the transport model?
- How should the variable be treated in transport modeling?

It should be noted, that transport models cannot be made for all types of aquifers. This stands even when scientific progress in this field will increase tremendously.

A decision-maker should know the main criteria for not using a model:

- Inhomogenities in the geological formation. More precisely: when small scale structures are not known but crucial for the flow system.
- Great temporal changes of the flow-pattern, if they are not sufficiently monitored by measurements. Commonly these unsteady effects are caused by precipitation, leakage, wells, or from boundary conditions i.e. water-tables of surface water.

– Various sources. which cannot be located sufficiently. They may interact in
a way that distinct effects cannot be recognized.
– The time period. for which data are measured, is to small.

Flow parameters

Fluid flow has to be known to make a transport-model. Mostly it has to be
given as input-parameters. More precisely: the average linear velocities are the
flow-variables, which have a great influence on the results of a transport model.
Therefore it will be analysed, how other parameters of the aquifer effect the
average linear 'real' velocities.

– Average linear velocities (Freeze/Cherry 1979) can be determined directly by
measurements. This is done by tracer experiments. If field-measurements are
available, the data should be used for the transport-model.
– Average linear velocities can be calculated from Darcy-velocities and porosities.
In most cases knowledge of both is limited. An error of Darcy-velocity with
factor k results in an error of average linear velocity with factor k. Porosities
Φ are included with their reciprocal $1/\Phi$. Porosities in the field almost always
have values between 15% and 45% (Bowen 1986). Therefore $1/\Phi$ is found in the
intervall [2.2 . 6.7]. Even if porosities in the observed region are not known,
the choice of a medium value of 22.5% yields an error never greater than factor 2.
– Darcy-velocities can be determined from values for piezometric head and
hydraulic conductivities (K_f). It is easier to get the heights of piezometric head.
In many places local authorities take measurements by routine and store the
results in a database. More obstacles are there to get hydraulic conductivities.
These can be determined by pumping tests, which are expensive and require a
much work. Furtheron their results are doubtful. Local inhomogenities may
cause great errors in the measured data. Geological formations may differ in
their K_f's up to a magnitude of 10^7 (Verruit 1970). An error of factor k in K_f
results an error of factor k in both Darcy-velocities and average linear velocities. If the aquifer can be assumed to be homogeneous and no measurements
are taken, only the order of magnitude can be determined out of the literature.

All considerations are valid for a phreatic aquifer which in most cases is concerned first, when contaminations occur (exception: high-level waste storage in deep
geological formations). Nevertheless the considerations above are valid for a confined aquifer as well. In this case another parameter becomes important : the
transmissivity, the product out of conductivity and aquifer thickness. The later may
induce, if not known, another uncertainty for the model.

Transport-modeling in an unsteady flow field has to take the storage coefficient
S into account. Data for S vary within a range of factor 100 (Bowen 1986). Storage
coefficients can be determined by pumping tests as well – or by evaluation of
unsteady processes within the aquifer in case they are sufficiently documented.

Transport parameters

First of all it is necessary to be sure about the flow within the aquifer. If
uncertainty in the flow field is too high no transport model can be reliable.
Beside the flow parameters the following variables have to be considered for
transport modelling:

– concentrations of natural components and contaminants
– source/ sinkrates

- diffusion coefficient
- dispersion lengths
- sorption parameters
- background levels for concentration
- constant of (radioactive) decay

The parameters will be treated separately in the following section.

Concentrations

The concentration of suited groundwater components have to be determined precisely. Groundwater concentrations are usually available because often a real contamination is the reason for the modeling. There always remains the question, though, if the quality of data is sufficient. This is a crucial point because most other transport parameters can only be determined by calibration with measured concentrations. Therefore the whole model depends indirectly on the quality of the groundwater analyses.

Source- and sinkrates

Transport modelling in porous media nowadays simulates distribution of components in the saturated zone only. The contamination often occurs from the unsaturated zone (i.e. landfill sites, pollution by industrial plants or agriculture). As the processes which take place above the groundwater level are very complicated, it is usually not possible to determine the real source-rate of the contaminant. Beside the uncertainty about the dominating processes within the unsaturated zone, two further parameters may not be known: total stored amount of the substance and infiltration rates. Without any reliable information about both of these it is impossible to model the distribution within the unsaturated zone. For these reasons it is necessary to determine source-rates by calibration in most cases.

Diffusion

Coefficients of molecular diffusion are generally very small (Kinzelbach 1987). Therefore it is justified to neglect diffusion against dispersion.

Dispersion

The two parameters longitudinal and transversal dispersion-lengths can be determined by field experiments. There is a very strong correlation between lengthscale of observed phenomena and the value of dispersion-lengths (Kinzelbach 1987). If there are no other clues about the values, a proceeding as follows has been proved to be practicable and successful: the value of longitudinal dispersion should be chosen two orders of magnitude lower than the entire model and the value of transversal dispersion again one order of magnitude less. This is a good choice to start some jobs for calibration - if possible with measurements for an ideal tracer.

Sorption

Transport models mostly use the Henry isotherme, which assumes that sorbed mass and not-sorbed mass are proportional. This assumption is justified as long as free sorption sites of the sediment outweigh the occupied sites. Input parameters are retardation factor or sorption coefficients K_d which are connected by formula (3).

K_d's can either be determined in experiments or be calculated if there are sufficient informations about the geochemical environment. With calculated K_d's it can be checked easily, how a change in the geochemical environment influences sorption.

The application of K_d assumes that there is an equilibrium in every block of the model. Usually it is justified to restrict the retardation-processes to the so-called 'fast sorption' (Kinzelbach 1986). Therefore the assumption of an equilibrium is admissible.

A classification of components into three sets is reasonable:

Set 1: $\rho_d K_d (1-\Phi)/\Phi$ < 1/10
Set 2:   1/10 <   "   < 1   (4)
Set 3:   1 <   "

In set 1 sorption does not play an important role. In set 2 the effect of sorption is moderate: time delay is not greater than factor 2. In set 3 sorption has to be considered. An example in typical values: if rock-density is $2000 kg/m^3$ and porosity 20% the K_d's which mark the borderlines between the three sets are 10^{-2} and 10^{-1} g/cm^3. With data from the test-site Cl^- is found in set 1, Na^+, K^+ in set 2. Mg^{2+} and Ca^{2+} can be in set 2 or 3. Sr^{2+}, Ni^{2+}, Zn^{2+}, Pb^{2+} are classified into set 3.

Figures 2 and 3 show the influence of K_d on transport of a groundwater component. Input parameters for the test site for which the results are shown are listed in Bütow/Holzbecher/Koss (1988).

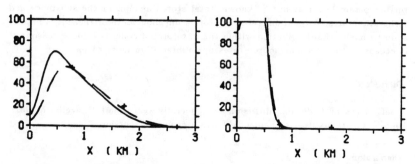

Figure 2: Na^+ concentrations (mg/l) downstream

Figure 3: Ni^{2+} concentrations (µg/l) downstream

(+ field measurements, —— calculated for 22 years after begin of leakage
-- calculated for 24 years after begin of leakage)

The classification according to (4) is not valid only for the importance of sorption but also for the sensivity due to variation of K_d. In set 1 an error in the sorption coefficient does not play any role, in set 2 it can be of minor importance. But in set 3 an incorrect value with factor k results in an error in temporal retardation almost of factor k.

It has to be noted that for the same component, sorption can be important in one part of the model and be negligible in the other part and that the borderline between these parts can move with time, depending on the geochemical environment.

Background levels for concentration

Background levels of groundwater components are given in the literature. In most cases these values are too general to be used for a model of a specific aquifer. They can only give a hint on the order of the background-level. It is important for the modeller, that there are measurements of the background-level within the aquifer in question. For example it is convenient to take measurements where flow has not yet entered the contaminated area.

Figure 4 shows agreement of experimental and modelled results when background concentration of Cl$^-$ is neglected. The dispersion length α_L determined by calibration is 300m. Taking background levels into account results in a more realistic dispersion length of 50m (figure 5).

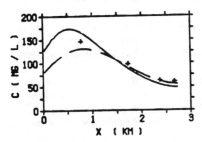

Figure 4: Cl$^-$-concentrations downstream for α_L = 300m in model without background level of Cl$^-$-concentration

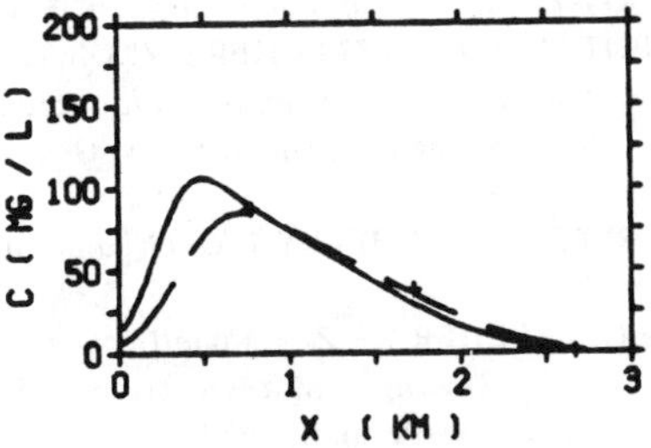

Figure 5: Cl$^-$-concentrations downstream for α_L =50m in model with background level of 60 mg/l Cl$^-$

Decay

Constants of decay for radionuclids are known. Parameters of chemical decay, if known, can be treated in the same way.

IV. Conclusions

Most of the hydromechanical parameters in a model for the distribution of contaminants can only be determined by calibration with measured concentrations. Therefore the quality of these measurements is most important. This means in particular:

1. Measurements have to be made at different locations
  1.1 downstream from a contaminating source (for the calibration of longitudinal dispersion lengths)
  1.2 aside from downstream flow (for the calibration of transversal dispersion lengths)
  1.3 upstream from a contaminating source (for the calibration of background-levels of concentration)

2. Measurements have to be made at different depths
   (to check if the transversal dispersion length from 1.2 is valid for pollutant distribution in depth as well - only in case when flow is not dependant on depth)

3. Measurements have to be made at different times

3.1 Timeperiods between measurements have to be small enough to catch the effect of unsteady disturbances

3.2 Timeperiod for all measurements has to be large enough to monitor the distribution development

4. Measurements have to be made for different components

4.1 A tracer substance should be measured or at least one substance out of set 1 according to classification (4) (for calibration of transport parameters)

4.2 The geochemical environment – pH, Eh, natural components, should be known (for determination of retardation factors out of the geochemical model)

BIBLIOGRAPHY

BEAR,J.- *Dynamics of fluids in porous media*, New York, Am.Elsevier, 1972

BOWEN,R.- *Groundwater*, London, Elsevier, 1986

BÜTOW,E./HOLZBECHER,E./KOß,V.- *Approach to model the transport of leachates from a landfill site including geochemical processes* Symposium on 'Contaminant Transport in Groundwater', IAHR-Proc.3, Rotterdam, 1989, pp183–190

FREEZE,R.A./CHERRY,J.A.- *Groundwater*, Englewood Cliffs (USA), Prentice-Hall, 1979

HOLZBECHER,E.- *Zur Modellierung von Ausbreitungsvorgängen im Grundwasser*, VII. Tagung 'Mathematische Simulation der Grundwasserentnahmen', Janowice (Poland), 1988

KINZELBACH,W.- *Groundwater modelling*, Amsterdam, Elsevier, 1986

KOß,V.- *Modeling of nickel sorption and speciation in a natural sediment groundwater system*, Mater.Res.Soc.Proc., in press

SCHEIDEGGER,A.E.- *General theory of dispersion in porous media*, J.Geophys.Res. 66, pp.3273–3278

SCHWEINGRUBER,M.- *Users guide for extended MINEQL EIR Version – standard subroutine data library package*, EIR-TKM-45-82-38

VERRUIT,A.- *Theory of groundwater flow*, London, 1970

WESTALL,J.C./ZACHARY,J.L./MOREL,F.M.M.- *Technical note No 8*, Water Qual. Lab., Dept.of Civil Eng., Cambridge (USA), 1976

Chapter 26

ELECTROCHEMICAL TREATMENT OF HEAVY METALS FROM ELECTROPLATING WASTEWATERS

C P C Poon (Water Resources Center, University of Rhode Island, USA)

ABSTRACT

A laboratory-scale electrochemical reactor using a platinized columbium screen anode and a stainless steel screen cathode, generating chemicals from a rock salt solution to precipitate heavy metals as a floating scum, is described. Removal of Cd, Cu, and Ni varies from 0.27 to 109.67 g/kwh with the effluent quality meeting the USEPA pretreatment effluent limitation. Favorable operating conditions are lower current density, higher initial metal concentration and deeper wastewater column. The collected scum is dissolved in sulfuric acid from which the concentrated metal is plated out onto a specific metal cathode in a separate reactor.

Key words: electrochemical, cyanide, cadmium, copper, nickel, electroplating wastewater, scum

1-INTRODUCTION

Large quantities of heavy metals and cyanide are discharged into sewer systems that receive very little treatment prior to their discharge. In-plant conventional treatments by chemical precipitation require large quantities of chemicals and produce metal sludges which are economically and socially unacceptable. Under the 1976 Resources Conservation and Recovery Act (RCRA) and the 1984 Hazardous and Solid Waste Amendments Act, the minimization of sludge production and recovery of metals are strongly encouraged. The associated costs of sludge disposal on hazardous waste disposal sites as well as the unpredictable price of possible future liabilities add to the incentive of in-plant recovery and reuse.

Electrochemical method of metal recovery requires a relatively high metal concentration in the liquid to be economically feasible. This study uses a three-step approach, (1) electroflotation for metal separation from the liquid in the form of a scum, (2) acid dissolution of the metal scum, and (3) electrolytic recovery of metal from the metal acid solution. This treatment/recovery system can be applied equally well to high or low concentration metal bearing wastewater.

A bench-top reactor of plexiglass material is built with an outside dimension of 45 cm x 16 cm x 30cm high. It has a maximum working volume of 12.0 liters. The reactor is equipped with a 33 x 14 cm platinized columbium screen anode and a 34 x 14 cm stainless steel screen cathode. Submerged inlets are provided separately for rock salt solution and for the wastewater at one end of the reactor. A single outlet for the spent rock salt solution and multiple outlets at various elevations for the treated wastewater are provided at the other end of the reactor. The multiple outlets controls the depth of the wastewater column in the reactor for the experiment. Variable speed pumps are used for wastewater and rock salt solution supply from their respective reservoirs. Various pumping rates are used to control the hydraulic residence time of the wastewater in the reactor. The spent rock salt solution is returned to its reservoir for recycling in order to minimize its disposal. Flowmeters and direct current power supply are installed to complete the reactor. Separate control of current and voltage are provided from the direct current power supply unit.

Many electroplating wastewaters contain cyanide. Traditionally alkaline chlorination is used for the treatment of cyanide-bearing wastes. The method involves the addition of chlorine gas to a metal plating wastewater at high pH. Sufficient alkalinity is added, e.g., $Ca(OH)_2$ or $NaOH$, prior to chlorination in order to raise the pH to 10 or above, thus ensuring the complete oxidation of cyanide. Violent agitation must accompany the chlorination in order to prevent the cyanide salt of sodium or calcium from precipitating out prior to oxidation. The presence of other metals may also interfere with cyanide oxidation because of the formation of metal cyanide complexes. Extended chlorination may be necessary under these conditions. Sodium or calcium hypochlorite may be used in chlorination to replace gas according to MARIN et al. (1979). Another known method of cyanide treatment is to use hydrogen peroxide which oxidizes cyanide to cyanate CNO. A pH of 8.5 to 10 is best for this reaction. The reaction rate is greatly increased by the presence of traces of catalytic metals such as copper. In the Soviet block countries, electrolysis has been used for cyanide removal. An ample amount of sodium chloride is added to the cyanide bearing waste before the solution is passed through an electrolytic cell. Hypochlorite is generated within the cell as a mixed solution is electrolyzed and the reaction takes place in the receiving tanks which follow.

In the present study, the rock salt solution is electrolyzed below a column of metal plating wastewater containing cyanide and a heavy metal. The chemicals generated from the rock salt solution electrolysis rise in the reactor and bring reactions which complete the treatment of cyanide and heavy metals. Both hypochlorite and chlorine gas are produced which oxidize CN to nitrogen and carbon dioxide.

Electrolysis of rock salt solution also generate hydroxides which raise the pH in the wastewater column. The metal liberated from the cyanide-metal complex through cyanide oxidation is then precipitated as a result of the low solubility in the highly alkaline condition.

2-EXPERIMENTS

The independent variables in the treatment process include (a) current density, (b) depth of wastewater column, (c) electrode spacing, (d) initial metal concentration in wastewater, and (e) initial pH of wastewater. Dependent variables are effluent metal concentration and power consumption. The effluent metal concentrations are set by the United States Environmental Protection Agency (USEPA) in the pretreatment effluent limitation. The 1987 USEPA pretreatment effluent limitations are Total Cd=0.11 mg/l, Total Cu=1.20 mg/l, Total Ni=1.62 mg/l, and Total CN=0.58 mg/l.

Preliminary experiments show that the electrode spacing within the range of experimental values of 1.0 to 3.0 cm has negligible effect on the process performance. Therefore, a fixed space of 2.0 cm is used throughout the experiment. The initial metal concentration for different metal species in the wastewater samples is quite different and is therefor used as one independent variable in the experimental design. In addition, current density, depth of wastewater column, and initial pH of the wastewater are the other operational variables in the process. The statistical design of the experiment, therefore, uses a three-level factorial design with four variables, each using a high, medium, and low value. The high, medium, and low values chosen for these four variables depend on the metal species and the source of plating wastewater obtained for this study. Several series of experiments are conducted. In each series the effect of each control variable is investigated while the others are held constant. Metal removal is measured for the determination of grams of metal removed per unit power consumption.

The lower portion of the reactor is filled with rock salt solution (33 gram/l) to a level approximately 2 cm above the anode. Wastewater is then added to a designed depth. Power is switched on to achieve the designed current density. Pumps for continuous supply of rock salt solution and wastewater are activated to provide the predetermined flow rates. The mean hydraulic residence times for various metal removal are predetermined from preliminary experiments in batch testing. Sufficient residence time is allowed in each experiment to ensure an acceptable effluent meeting the USEPA effluent limitation.

Metal determinations are performed by Perkin Elmer Model 3030B Atomic Adsorption Spectrophotometer with HGA 600 Graphite Furnace. A strict cleaning procedure is followed to ensure a minimum of contamination. Plastic containers are used exclusively for sample gathering and analysis dilutions. The plastic containers are cleaned after each use by the same procedure. The containers are rinsed with tap water followed by soaking in concentrated nitric acid for 24 hours. Subsequently they are rinsed and then soaked in double distilled water for 24 hours, air dried, and stored in sealed plastic bags until use. All samples of wastewater from experiments are diluted to parts per billion concentrations for Atomic Adsorption analysis using a matrix water. Since rock salt solution is used in the process, a matrix water is formulated to minimize inconsistencies in solution composition, which reduces interference with the atomic adsorption process. The composition of this matrix water is 5 per cent filtered rock salt solution and 95 per cent double distilled water with 5.0 mg/l concentrated nitric acid and 1.0 mg/l hydrochloric acid for pH control. Following this procedure it is possible to measure the metal concentration accurately down to 0.05 parts per billion. Total cyanide is determined by distillation followed by titrimetric method using a silver-sensitive indicator, p-dimethylaminobenzal-rhodanine. Residual chlorine is determined by Iodometri Method I, and chloride by Argentometric Method. All procedures follow the STANDARD METHODS (1985).

3-EXPERIMENTAL RESULTS

Dragout solutions are obtained from several plating shops. They are diluted with tap water to simulate actual rinsing water for experiments. Because of the different starting metal concentrations, pH, and wastewater column depths, different flow rates and therefore different reaction times are required to obtain effluents meeting the USEPA pretreatment limitation. For the wastewater samples used in this study, the initial conditions are: Cd=16 mg/l (pH=9.1), Cu=65 mg/l (pH=2.4), Ni=92 mg/l (pH=3.2). Metals removed per unit power consumption (gram/kwh) are then compared for all experiments, with those giving the highest values defined as the most effective treatment. TABLE 1 presents the results of the most effective treatments from this study.

The results as indicated in TABLE 1 as well as the results from other experiments giving lower effective treatment suggest that a deeper wastewater column and a lower current density will be a better use of the power for metal removal. In other words the most favorable operating condition to obtain the most effective removal of metal is to use a treatment unit with deeper wastewater column and longer reaction time but lower power input. Initial pH and initial metal concentration also have significant effect on power consumption when initial metal concentration and initial pH are higher. However, both are not control variables in actual application.

All wastewater samples contain cyanide in the form of metal-cyanide complexes. Much of the alkaline generated by rock salt solution electrolysis is first consumed by cyanide oxidation. Only when this process is completed before the free metal can be precipitated.

A multiple regression analysis is carried out to fit performance equation to the experimental data. The following equations are for Cd, Cu, and Ni removals.

$$R \text{ cadmium} = 4200(P)^{-0.372}(D)^{0.340}(C)^{0.717}(H)^{-0.903}$$

$$R \text{ copper} = 20563(P)^{-0.695}(D)^{0.260}(C)^{0.418}(H)^{0.488}$$

$$R \text{ nickel} = 1.13 \times 10^{7}(P)^{-0.667}(D)^{0.110}(C)^{0.654}(H)^{-5.27}$$

in which R is the metal removal in mg/kwh, P is power input in watts, D is wastewater column depth in cm, C is initial metal concentration in mg/l, and H is initial pH of the

TABLE 1
Experimental condition and treatment results

Wastewater	Wastewater column depth (cm)	Electrode Spacing (cm)	Power Consumption (watt)	Metal removal per unit power consumption (g/kwh)
Cd	10		35	2.14
	25	2.0	15	6.20
	25		25	4.15
Cu	25		15	72.67
	10	2.0	15	55.35
	10		13	53.81
	10		35	47.43
	25		25	38.62
Ni	25		15	109.67
	10	2.0	15	97.55
	10		25	96.34
	10		25	90.27
	10		15	79.87

TABLE 2
Standard β values of treatment performace in the
electrochemical process for metal removal

Wastewater	Standard β value			
	Power input	Wastewater column depth	Initial metal concentration	Initial pH
Cadmium	-0.334	0.467	1.125	-0.275
Copper	-0.618	0.257	0.303	0.413
Nickel	-1.185	0.113	0.846	-0.322

wastewater. While the equations above are for treatment performance only, the relative importance of the independent variables on the treatment performance can be compared with their standard β values listed in TABLE 2.

The correlation coefficients for the performance equations above are 0.906, 0.883, and 0.962 respectively for cadmium, copper, and nickel. The β values in TABLE 2 suggest that consistently the process is most efficient in removing more metal per unit power consumption when a lower power input, and therefore a lower current density, is applied. Also the β values indicate deeper wastewater columns and higher initial metal concentrations having positive impacts on treatment efficiency of all metals.

3.1-Cadmium removal

Of all three metal species, Cd is the most difficult for removal. The wastewater sample contains 52 mg/l total cyanide thereby much power is consumed to liberate the metal from the metal-cyanide complex leading to cyanide oxidation. It is observed that in batch process, almost thirty minutes are required to raise sufficient chlorine and hypochlorite for cyanide destruction to such extent that cadmium precipitation begins to take place. The low cadmium concentration (the lowest initial metal concentration among all three metal species in this study) is also responsible for the low removal efficiency. In addition, the effluent concentration as required is far lower than Ni and Cu. All these factors contribute to the low treatment efficiency from 2.14 to 6.20 g/kwh removal. Initial pH has relatively the least effect on treatment efficiency for Cd.

3.2-Copper removal

Copper can be removed much more efficiently than Cd by the electrochemical process. As much as 72.7 g/kwh can be obtained. Lower cyanide content of 18 mg/l total cyanide and higher initial copper concentration of 65 mg/l have the positive impact on the treatment. Relatively speaking, power input has a greater impact on treatment efficiency than initial pH which in turn has a greater impact than initial copper content and wastewater column depth.

3.3-Nickel removal

With a high initial concentration of 92 mg/l and a low cyanide content of 7.6 mg/l, nickel removal efficiency is the highest at 109.7 g/kwh. The higher effluent limitation at 1.62 mg/l also is responsible for the lower power consumption. The impacts of current density and initial metal concentration far exceed those of the initial pH and wastewater column depth.

3.4-Acid dissolution of scum and metal recovery

Some initial attempts have been made to dissolve the metal scum by using 0.25N sulfuric acid followed by plating out the metal in a reactor. Using a platinum clad columbium screen as the anode (6 cm x 7 cm) and a specific metal plate of the same size as the cathode, the metal from the concentrated metal-acid solution is plated out onto the cathode, e.g. copper is plated out onto a copper plate. A corrugated stainless steel plate (6cm x 10 cm) also has been used as the cathode for plating. The reactor is an open tank 7.5-cm in diameter, 12-cm in depth, holding a working volume of 0.35-l of acidified metal solution. A spinning magnetic bar provides the mixing of the solution in the plating process, with a velocity gradient value (G) ranging from 180 to 500 sec^{-1}! Only batch experiments have been conducted to date. Continuous flow experiments will begin soon with an objective to determine the limiting current density and current efficiency for each metal species.

4-CONCLUSION

A compact treatment reactor using rock salt solution as the anolyte can be used successfully in removing Cd, Cu, Ni and CN from plating wastewaters to meet effluent standards. Control variables important to the process are wastewater column depth and current density. A 25-cm wastewater column depth and a current density of 6.3mA/cm^2 is the best operating condition resulting in the successful removal of metal and CN with the highest amount of metal removal per unit kwh power consumption basis. The scum as a product of the treatment process can be dissolved in acid from which the metal can be plated out for recovery.

ACKNOWLEDGEMENTS

This study was supported in part by the U.S. Department of Interior and by the Department of Environmental Management of the State of Rhode Island.

BIBLIOGRAPHY

APHA, AWWA, WPCF.-Standard Methods for the Examination of Water and Wastewater, 16th edition, 1985.

MARIN, S., et al.,-"Methods for Neutralizing toxic Electroplating Rinse Water-Part 1", Industrial Wastes, May/June, 1979. pp.50-52

Chapter 27

MODELLING THE IMPACT OF A CHEMICAL MANUFACTURING FACILITY ON A MAJOR AQUIFER

M Loxham, W Visser (Delft Geotechnics, The Netherlands)

ABSTRACT

A systems analysis approach to calculating the impact of (petro)chemical facilities on a surrounding aquifer is described and illustrated. It is shown that for commonly expected release scenarios of contaminants, considerable difficulty can be anticipated in meeting the attenuation factors required by modern environmental legislation.

Key words: pollution, industry, aquifer.

INTRODUCTION

The appreciation that groundwater forms an essential link in many environmental chains has only recently become fully apparent. In particular attention has only to often been focussed upon the quality of groundwater for the immediate purpose of drinking water production without examining the wider implications of a general degradation of the groundwater as a major element in what has become to be known as the urban ecological system.

Real urban systems are characterised by a mixed and intensive land use where industrial activity, past and present, is contained within the same watershed or hydrogeological system as residential areas, small scale crop production, and limited domestic or industrial extraction points. These activites are intimately connected and for the purposes of the study reported here attention will be focussed on the impact of (petro)chemical industrial complexes on the local groundwater quality. The methods illustrated in this paper are applicable however to a wide range of local pollution sources.

Large scale petrochemical complexes tend to be associated with the immediate hinterlend of large urban deltas such as New Jersey, La Harvre, Singapore and Rotterdam. Typical distances between the complexes and the surrounding urban centers are between one and ten kilometers and groundwater velocities are in the order of one to ten meters per year, indicating that the problem time scales for significant contaminant transport are between ten and one thousand years. As most of these complexes have been constructed in the last fifty years, the urban communities surrounding them can expect to be confronted with the consequences of any historical releases in the near future and the action time scales for remedial actions focus on the coming fifty years.

Classical risk analysis techniques call for the identification of a source of contaminants, characterised by a emission strength of a particular chemical, a target where that chemical will have an impact and whose sensitivity to the impact is often defined in terms of a maximum allowable concentration, and a pathway connecting these elements, which in this case is associated with convection via the groundwater. Any analysis of the problem has to address the future behaviour of the chemicals of concern both at the source and on the pathway.

The pathway can be divided into three regimes that are often associated with hydrogeological units in the system. These are the local, the near and the far field regimes. The local field is usually dominated by the unsaturated zone in the immediate vicinity of the site, whilst the far field is the aquifer proper. The near field contains the immediate surrounding of the factory and is usually a saturated flow regime. In this paper attention will be given to this last component of the overall problem.

In what follows, a simple model for the near and local fields of a petrochemical complex will be described. The objective of the model is to generate source terms that can be used as input into the more conventional convection-dispersion based models for the aquifer itself.

MODEL STRUCTURE

The releases of process materials into the environment by an operating chemical manufacturing facility are controlled and reduced to the lowest possible level that can be technically achieved. None the less incidents do occur from time to time but is impossible to predict in any certain way the extent and nature of the release. Furthermore, the local and near field geohydrological regime under a factory is complex and more often than not dominated by the operation of the facility itself. Leaking drains, pavements, multiple piles, sand cunets, local dewatering well points, not to mention the highly heterogeneous soil system, make a description of the site in classical geohydrological modelling terms impossible. Counteractions to mitigate the effects of any release also have a strong impact on the groundwater flow patterns and thus the expected release scenarios to the far field.

These considerations make it impracticable to construct a determinate contaminant transport model of the near and local fields and recourse has to be made to less precise modelling techniques. Fortunately the accuracy to which answers have to be sought in order to define the impacts and countermeasures of the site, is also limited and these simplified approaches are have often proved highly fruitful. For the purposes of this study, a systems analysis model, identified as the Refinery Contamination model will be used.

The conceptual basis of the model is to identify possible migration pathways though the near and local fields and to subdivide these into transport elements and barrier elements. The former are then implimented as mixing cell elements and the latter as time-lag and bypass elements. These elements are then interconnected in a network reflecting the convection processes in the system.

The construction of the network is based on intuition and experience considering the, often limited, available information of the site. Calibration of the model then has to be sought either against existing data, which is rare, or against constructed benchmarks using simplified geohydrological schematisations to generate travel time based transfer functions which are then assumed to reflect the Greens function of the system. Typically the benchmark will consider the migration of a single conservative tracer through the system and the matched network will be implimented to calculate effects such as biodecay, adsorption-desorption phenomina, multiple interacting species chemical transport etc, for which parameter studies using more rigorous models would be prohibitively expensive and time consuming.

Automatic network generators are available and spreadsheet level implimentations are possible. It is anticipated that this type of model will be widely available in user orientated interactive modes in the near future.

APPLICATIONS

In order to illustrate the use of this approach to the problem, consider the facility illustrated in figure 1 which has been taken from a study in the Rotterdam Europort area.

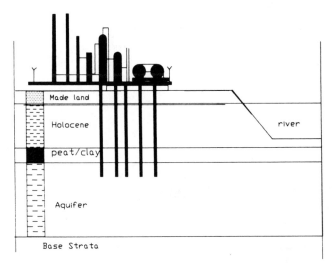

Figure 1

The factory is built on made land above a sandy-silt holocene profile. The made land has a depth of 5m and the holocene of 15m. Below this profile is a clay-peat horizon with a low permeability and high adsorption capacity of 1m thickness, which may or may not be continuous and in any case has been perforated by piles and other subsurface structures as shown. The aquifer proper is below this clay-peat layer.

Two targets can be identified in this problem, the aquifer and the adjoining surface waters in the harbour. Only the former is considered here.

Releases such as those of neat hydrocarbon liquids or aqueous solutions to the groundwater are typical of industrial facilities. The migration pathways are shown in figure 2. where it can be seen that infiltration, which can leach out the oil from any floating layer, and transport contaminants moves down and across through the holocene horizon.

As there is a significant leakage into the aquifer through the clay layer, contaminants will also be found there in due time.

In environmental terms, although oils and other so-called immiscible solvents have but a limited solubility in water, the concentrations are well above current environmental standards. Dilution factors of 1000 to 100000 are required to avoid impacts in the aquifer.

The objective of the modelling exercise is to calculate the expected time dependent concentrations and fluxes into the aquifer. In order to do this the model illustrated in figure 3. was used.

As can be seen the model divides the holocene and made land zones into six mixing cell elements of which two (1 & 2) are associated with the local field and the remaining four with the near field. This division has

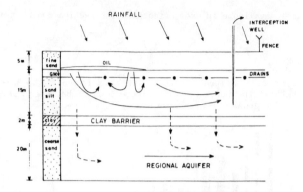

Figure 2

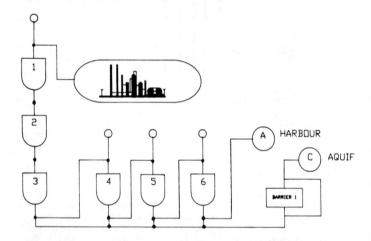

Figure 3

been made to reflect the changed conditions directly under the release point. The clay-peat layer has been modelled as a barrier element.

The network has been calibrated against a benchmark generated by a finite difference calculation of the travel time distribution in the flow field to the underside of the clay layer. A maximum deviation between model and benchmark of 10% was achieved over the predictions for the first centuary after the release.

As pointed out above, various release scenarios from the facility are possible. Consider firstly a simple release of hydrocarbon contaminated sewer water of unit concentration for a period of one year. (In practical terms this could represent a release of 400 cubic meters of water with a concentration of 600 ppm contaminant. The concentrations have been normalised here to allow the determination of the attenuation factors, these are the ratio of the source to target concentrations). Neither adsorption nor biodecay effects have been taken into account in setting the parameters for the calculation.

The development of the concentration input to the aquifer is shown in figure 4. for two effective permeabilities of the clay layer.

In one case the leakage across this layer is 65% of the infiltration and in the second only 20%. This latter value was found in a study of the water balance of a large Rotterdam refinery. As can be seen there is significant delay in the arrival times of the contaminants in the aquifer,

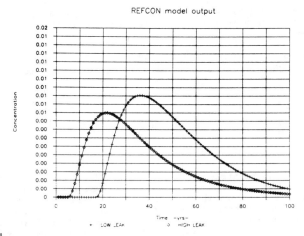

Figure 4

in this case 20 or 40 years. This result is especially interesting in that the facilities considered were constructed some fourty years ago. The attenuation factors are in this case simple dilution factors and amount to roughly 100 in both cases whereas, of course, the fluxes differ by a factor of 3. As most environmental legislation is formulated in terms of concentrations rather than fluxes, and as groundwater standards usually require dilution factors very much higher the 100, these results indicate that the facility management will be confronted with serious problems in the near future.

It can be argued that in the upper, oxygen richer zones of the profile, biological attenuation effects will play a significant role. This is illustrated in figure 5. where the biodecay half-lives have been taken as 2 years in the unsaturated, and as 5 years in the upper saturated zones under the spill. (the base case parameters unless otherwise noted).

These values are optimistic for unenhanced biodecay coefficients. The results indicate that there is indeed a substantial reduction in the level of impact of the release on the input into the aquifer. However, even with this level of bio-activity, an attenuation factor of only slightly above 300 can be achieved.

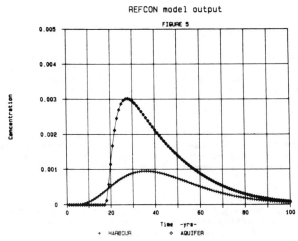

Figure 5.

In order to achieve an attenuation factor above 1000, biodecay enhancement to deep in the profile is required as the following table illustrates :-

Depth in profile -m-	Half life -yrs-	Peak Attenuation factor
5	5	125
12	5	230
20	5	450
20	2	1500
20	1	3500

It can also be argued that adsorption effects can help to mitigate the impact of the pollutant on the aquifer. The adsorption can take place in either the holocene zones or in the clay and peat layers. It is expected that for hydrocarbons these effects will be small and in any case concentrated in the clay layers. In figure 6, the effect of attributing a distribution coefficient of 2 ml/g and 4.5 ml/g to the clay layer or to the holocene is shown.

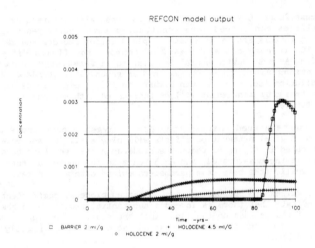

Figure 6

The effects are to significantly delay the breakthrough, but not to seriously change the attenuation factors.

However a combination of biodecay and adsorption can have major mitigating effects. The effective attenuation factors achieved by combinations of the above adsorption in the upper zones and the base case biodecay parameters are 1500 and 2500. It would appear that a combination of bio-enhancement and adsorption improvement could prove a very fruitful line of attack on these problems.

Unfortunately the clay layer is rarely intact and in any case its integrity can be seriously impaired by the construction activities for the facility itself. This effect can be modelled by invoking the by-pass parameter on the barrier element. The results of this are shown in figure 7.

As was to be expected, it can be seen that only small leakages around the barrier can have major effects on the peak attenuation factors, and that the results will become apparent on a much earlier time scale than otherwise expected. Similar effects can be expected for back-pumping clean-up strategies if the well points are spaced too widely apart.

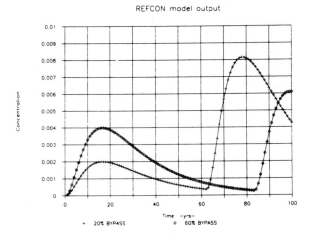

Figure 7

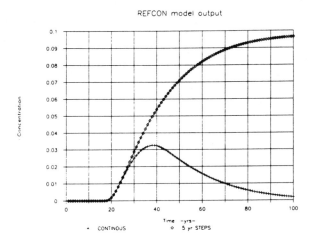

Figure 8

The release scenario considered up to now is a very optimistic one and it is rarely so that a major petrochemical facility can respond so quickly to a pollution incident. Typical releases involve neat hydrocarbons to the groundwater which either float there or become distributed in the unsaturated zone by fluctuations in the groundwater level. The hydrocarbon then forms an essentially perminant source of pollution. This situation is illustrated in figure 8 which should be compared to figure 4 above. The long term releases show that the effective attenuation is small and the impact on the aquifer correspondingly serious.

In figure 8 is also shown the results of a typical clean-up strategy favoured by many facility managements, that is excavation of the accessable soils exposed when process units have to be replaced. The clean-up of the site has been simulated as 4 stages each of 25% at five year intervals. It can be seen that whilst the results indicate that this strategy is an improvement, it is less than perfect.

Of especial interest in the light of these results is the possibility of a combination of attenuation mechanisms. Given the fact that these facilities have released contaminants in the past, and are likely, albeit less frequently, to do so in the future, the question arises as to the degree of adsorption and bio-activity that will be required to form an effective barrier to unacceptable impacts on the aquifer.

Much work requires to be done in this area, but the results of this study would seem to indicate that only a combination of deep profile enhanced bioactivity combined with adsorption reinforcement can achieve the required attenuation levels. Future technical research should be directed towards this end.

CONCLUSIONS

The results of this study have demonstrated the usefulness of system analysis type models for assessing the loading terms from industrial releases to aquifers. These can then be used with the conventional convection-dispersion based codes to model the far field impact of such a facility.

They also illustrate that it will be extremely difficult to ensure that these impacts, measured in terms of local and near field attenuation factors, can be reduced to acceptable levels in terms of current legislatitive frameworks.

Chapter 28

A PROBABILISTIC MODEL FOR URBAN POLLUTANT LOAD CONTROL THROUGH DETENTION STORAGE

R Segarra-Garcia (University of Puerto Rica)

ABSTRACT

A model has been developed that determines the optimal values of the storage capacity and treatment rate for a stormwater detention unit in terms of pollutant trap efficiency. Runoff event variables are treated as random variables in a lumped storage model of the detention unit. The derived probability distributions are utilized to obtain expressions for the pollutant trap efficiency under uniform loading conditions in terms of the distribution of overflows, and for non-uniform loading in terms of the standard first-order load equation. For a specified trap efficiency level the optimal design values of the storage capacity and runoff treatment rate are obtained.

Key words: stormwater detention, storage, probabilistic model, pollutant load control, optimal storage, stormwater treatment.

1 - INTRODUCTION

Urban stormwater management is receiving increasing attention within the water resources field as urban runoff is the transport mode of a wide variety of substances made available on the surface through human activity.

With increased urbanization and growth of industry the environmental problems associated with heavily developed and industrialized catchments have become numerous and of a greatly increased complexity. It has become evident that runoff is a source of pollutants comparable at times to other major sources of waste flow (JEWELL, 1980), particularly with regards to metals, solids (including organic), and coliforms.

A most promising planning alternative to control runoff pollution is the storage/release system. In this system a separate stormsewer conveys the pollutant-laden runoff to a storage unit from which water is pumped at a certain rate to a treatment plant.

Because the storm arrival process is, fundamentally, a random process, the runoff storage problem can be studied from the statistical point of view. This paper presents a statistical formulation of the problem coupled to a physically-based formulation of storage.

2 - FORMULATION OF THE STATISTICAL MODEL

2.1 - The basic storage model

The statistical model for pollutant load control has two formulations. One corresponds to uniform loading conditions based on the storage equation, and the other to non-linear loading based on a first-order washoff model. The basic storage model is presented in this section.

The urban catchment is conceptualized as shown in Figure 1. Storms arrive at the urban catchment where they produce runoff events. The stormsewer system collects the runoff that its inlet capacity allows and conveys it to the storage unit. If the volume exceeds the storage capacity the excess overflows to receiving waters. Continuously, water is being pumped from storage to a treatment plant. The mass balance at the storage unit yields the following equation for the value of the available storage at the end of the nth runoff event:

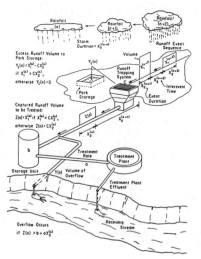

Figure 1. The Urban Stormwater System

$$S(n) = Min\{[Min(S(n-1) + aX_3, b) + aX_2 - Min(X_1, cX_2)], b\} \quad (1)$$

$$S(n-1) = \begin{cases} S(n-1) & \text{if } S(n-1) > 0 \\ 0 & \text{if } S(n-1) \leq 0 \end{cases} \quad (2)$$

where:

$S(n)$ — available storage (empty space) at end of nth runoff event (in. or cm);
$S(n-1)$ — available storage at end of $(n-1)^{th}$ event;
X_1 — runoff volume of nth event (in. or cm);
X_2' — effective rainfall duration of nth event (hr);
X_2 — runoff duration of nth event (hr);
X_3 — time between end of $(n-1)^{th}$ event and start of nth event (hr);
$Y(n)$ — overflow volume at n^{th} event (in. or cm);
Y_I — runoff not captured by inlets;
a — treatment rate (in/hr or cm/hr);
b — storage unit design capacity (in. or cm);
c — a measure of the surface runoff-trapping capacity of the inlet system (in/hr or cm/hr).

All volumes are expressed in terms of basin inches or centimeters, and the process is assumed stationary so that no indexing with respect to events is utilized.

In the storage equation the term Min $(S(n-1)+aX_3, b)$ represents the storage available at the start of the nth event, aX_2 is the volume pumped to the treatment plant during the event duration, and $Min(X_1,cX_2)$ is the incoming runoff volume, which may be controlled by the system runoff-trapping capacity. The storage space available at the end of the event is limited by the physical capacity b, which is the upper bound of the storage equation. The equation is unbounded in the negative direction, and it is seen from Equation (1) that negative storage occurs when the incoming runoff exceeds the available empty space in the unit. This negative storage corresponds to the overflow volume from the storage unit.

The event parameter variables X_1, X_2, and X_3 are assumed independent exponential variables. These are the independent process variables, while the storage and overflow variables are the dependent variables. The treatment rate and storage capacity are parameters.

Utilizing the technique of the derived distribution of functions of random variables the distribution of the available storage capacity $S(n)$ is obtained. With the probability density function of $S(n)$ the distribution of the overflow variable $Y(n)$ can also be obtained because storage overflow corresponds to negative storage. The relationship between overflow and storage is given by:

$$Y(n) = \begin{cases} - S(n) & \text{if } S(n) < 0 \\ 0 & \text{if } S(n) \geq 0 \end{cases} \qquad (3)$$

The complete derivation of the distributions can be found in SEGARRA-GARCIA and LOGANATHAN (1989). The expressions obtained are analytical, but conditioned on the specified value of the previous storage capacity $S(n-1)$. Thus a markovian formulation is obtained for the state transition probabilities.

2.2 - The uniform load model

The probability distribution function of the storage overflow is used to obtain the long-term runoff capture efficiency of the detention unit. Assuming uniform pollutant concentration over the event duration, the long-term pollutant capture efficiency is given by:

$$\mu = 1 - C_{av} E[Y]/C_{av} E[Z] \qquad (4)$$

where:

μ	-	pollutant trap efficiency;
C_{av}	-	uniform pollutant concentration;
$E[Y]$	-	expected value of storage overflow volume;
$E[Z]$	-	expected value of incoming runoff volume, with $Z = Min(X_1,cX_2)$.

It is evident that the pollutant trap efficiency for uniform concentration is also the hydraulic trap efficiency. The concept is similar to the concept of flow capture efficiency employed by DI TORO AND SMALL (1979), HYDROSCIENCE (1979), and LOGANATHAN et al. (1985).

The expressions for trap efficiency serve as the basis for estimating the storage capacity and treatment rate necessary to obtain a specified level of system performance. The locus of storage capacity and treatment rate combinations that produce a set level μ_o of capture efficiency is called the storage/treatment isoquant. By setting $\mu = \mu_o$ the isoquant is defined by finding the roots of the equation

$$h(b,\theta) = 0 \qquad (5)$$

where:

h( ) - the isoquant as a homogeneous equation;
b   - the storage design capacity - the desired root;
θ   - the fixed parameter set describing the hydrologic
       process (mean duration, depth, and time between
       events), storage parameters, and treatment rate
       (varied for additional roots).

For some situations it is possible to obtain explicit expressions for an isoquant. The derived expressions are found in SEGARRA-GARCIA and LOGANATHAN (1989).

Since uniform concentration conditions are only approximate in terms of real world situations, a non-linear loading formulation was also pursued.

2.3 - The non-linear load model

The non-linear load model is based on the so-called first flush condition. This assumes that a large percentage of the accumulated surface load is washed away during the first stages of the event. The model assumes the form of a first-order process in which the rate of change of pollutant washoff is proportional to the amount of pollutant remaining at time t. The integrated form of the model is given by:

$$
L = \begin{cases} P_o\{1 - \exp[-K(T + at_o)]\} & \text{if } Z \geq T + at_o \quad \text{(a)} \\[2mm] P_o[1 - \exp(-KZ)] & \text{if } Z < T + at_o \quad \text{(b)} \end{cases} \tag{6}
$$

where:

L - pollutant load trapped by storage unit (lb or kg);
P_o - initial accumulated surface load (lb or kg);
K - pollutant first-order washoff rate (in.$^{-1}$ or cm^{-1});
T - storage space available at start of nth event, given by $T = Min[S(n-1) + aX_3, b]$ (in. or cm);
Z - runoff volume, $Z = Min(X_1, cX_2)$ (in. or cm);
a - treatment rate (in/hr or cm/hr);
t_o - time required to fill available storage (hr).

Equation (6a) corresponds to the overflow situation wherein only the pollutant load corresponding to the hydraulic load that fills the available storage is retained. The situation in (6b) corresponds to that of no overflow and the trapping of the totality of the pollutant load. The time required to fill the available storage is used to obtain the additional storage capacity that is produced while the unit is being filled up. This time is defined as $t_o = T/(i_r - a)$. Here i_r is defined as the mean rate at which the unit fills, taken as a constant to simplify the analysis, and obtained as the larger of the system runoff trapping rate and the conditional expectation of the mean runoff intensity. This definition is an approximation that simplifies the analysis.

Utilizing the load equation and the probability density functions of the random independent variables the pollutant trap efficiency is obtained from:

$$
\sigma = E[L]/E[L_t] \tag{7}
$$

where:

σ - non-linear pollutant trap efficiency;
$E[L]$ - expected value of load trapped by storage unit;
$E[L_t]$ - expected value of total event load.

Equation (7) defines the storage/treatment isoquant for pollutant load control. As for the previous model the value of the storage capacity is obtained as the root of the homogeneous equation in terms of the specified value of the treatment rate and all other hydrologic and system parameters, including the rate K, which must be established for the particular pollutant under consideration. The resulting expressions are found in SEGARRA-GARCIA and LOGANATHAN (1989).

   With the expressions for the isoquants the optimization problem is formulated.

2.4 - Detention storage optimization

   The optimization formulation obtains the best value of the storage capacity and treatment rate in terms of some criteria.   Storage/treatment isoquants are treated as production functions.   A production function is a basic representation for the transformation of resources to products (DE NEUFVILLE and STAFFORD, 1971; FERGUSON, 1975). It also represents the maximum output attainable through a production process by any set of inputs.   The optimization formulation, as related to production function theory is given by

$$\text{Minimize } C = g(a, b) \tag{8}$$

$$\text{subject to } \mu = f(a, b)$$
$$(\text{or } \sigma = f(a, b))$$

$$\mu \geq \mu_o$$
$$(\text{or } \sigma \geq \sigma_o)$$

$$a, b \geq 0$$

where:
   C - total cost, dollars;
   $g(a,b)$ - treatment and storage cost function;
   $f(a,b)$ - production function, or storage/treatment isoquant;
   μ_o, σ_o - specified trap efficiency levels for uniform and non-linear pollutant loading, respectively.
   A major advantage of the procedure discussed here is that the constraints in the optimization model are analytical functions, which facilitates the solution of the non-linear optimization problem.   Past efforts have relied on graphical optimization techniques (NIX, 1988).
   The following section presents an application of the model.

3 - APPLICATION AND CONCLUSION

3.1 - Application to an urban catchment

   The statistical model has been applied succesfully to three catchments for which simulation studies were available. An application is illustrated for a catchment in the city of Minneapolis, Minnesota.   A detailed simulation study for a storage unit in this catchment was conducted by NIX (1982) using hourly precipitation records from a representative weather year in the Storm Water Management Model, version III (SWMM-III), of the U.S. Environmental Protection Agency.   The statistical model is compared to the results of the simulation study in the absence of realworld data for this type of facility.
   For this 640 acre catchment the mean runoff volume is 0.088 inches (2.24 mm), the mean runoff duration 7.3 hours, and the interevent time 94 hours.   The pollutant first-order washoff rate is given by $K=4.6$ in.$^{-1}$ (1.8 cm^{-1}).
   Figure 2 compares the results of the probabilistic model to those of the simulation study in terms of the storage/treatment isoquants.   The isoquants correspond to non-linear loading.   The parameter δ represents the percent of previously available storage, and indicates the storage condition at the end of the previous event.   The most critical is the previously full tank condition.   It is expected that the extreme conditions of previously empty or full storage yield isoquants that straddle the simulation results.   This is observed for the higher trap efficiencies

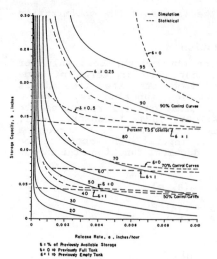

Figure 2 . Pollutant Control Isoquants for Minneapolis Catchment (Non-linear Load)

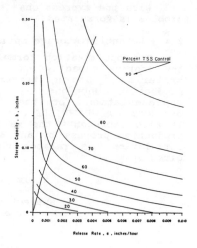

Figure 3 . Expansion Path for Pollutant Load Control, for the Minneapolis Catchment (Uniform Load).

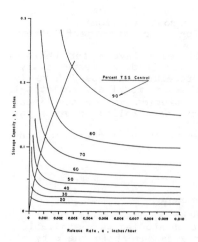

Figure 4 . Expansion Path for Pollutant Load Control, for the Minneapolis Catchment (Non-linear Load)

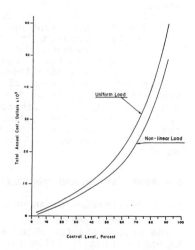

Figure 5 . Total Annual Costs for Levels of Runoff and Pollutant Control, for the Minneapolis Catchment.

but not for some of the lower ones. This is due in part to the fact that a slightly different release strategy than that of the statistical model was employed in the simulation. Also, the simulation was conducted with the same historical storm sequence for all variations of storage capacity and treatment rate. The statistical model allows for variations in storm intensity, which does not allow the isoquants to drop off as rapidly with increasing treatment rate as suggested by the simulation study.

Figure 3 shows the expansion path for uniform load Total Suspended Solids (TSS) control. The expansion path is the curve representing the optimal combination of storage capacity and treatment rate. It is defined by solving the optimization problem of Equation (8) for each of the trap efficiency or control levels shown in the Figure. This yields the optimal value of storage capacity and treatment rate necessary to obtain that level at the least cost. The curve joining these points on each isoquant is the expansion path. Figure 4 shows the expansion path for non-linear TSS loading.

Figure 5 shows the total annual cost for both load mechanisms as a function of the control level. The control level represents the percentage of the pollutant load that is intercepted by storage and eventually treated. Non-linear costs are lower for the same control level because a greater amount of sediment can be borne by the same runoff volume for the non-linear condition as can be for the uniform loading condition. This reduces storage requirements. The uniform load model is also the runoff volume control model, so a comparison can be made between storage requirements for volume control and for pollutant load control under the non-linear mechanism.

The cost function for the present application is linear but, given the analytical nature of the formulation, non-linear cost functions can be used as well, and no graphical solution is required.

3.2 - Conclusion

The statistical model is found to be a useful tool for stormwater detention planning, as it does not require extensive simulation to assess the response surface of storage/treatment rate interactions. The use of analytical constraints in the optimization problem removes the need for graphical optimization procedures. The study of the probabilistic nature of urban pollutant generation and transport mechanisms is an expanding area of research.

LIST OF SYMBOLS

a	- treatment rate (in/hr or cm/hr).
b	- storage unit design capacity (in. or cm).
c	- runoff trapping capacity of inlets (in/hr or cm/hr).
$C=g(a,b)$	- total cost function.
C_{av}	- uniform pollutant concentration.
$E[\ ]$	- the expectation operator.
$f(a,b)$	- production function (storage/treatment isoquant).
$h(b,\theta)=0$	- isoquant, expressed as homogeneous equation.
i_r	- rate at which storage unit fills during an event.
K	- first-order washoff rate (in.$^{-1}$ or cm^{-1}).
L	- pollutant load trapped by storage (lb or kg).
L_t	- total pollutant load from the nth runoff event.
n	- index representing the nth runoff event.
P_o	- initial accumulated surface load (lb or kg).
$S(n)$	- storage space at end of nth event (in. or cm).
$S(n-1)$	- storage space at end of $(n-1)^{th}$ event.
T	- storage space at start of nth event.
t_o	- time required to fill available storage (hr).
X_1	- runoff volume of nth event (in. or cm).
X_2	- runoff duration of nth event (hours).
$X_2{}'$	- effective rainfall duration of nth event (hr).
X_3	- time between end of the $(n-1)^{th}$ event and start of nth event (hr).
$Y(n)$	- storage overflow at n^{th} event (in. or cm).
Y_I	- runoff volume not captured by inlets (in or cm).
Z	- actual volume of runoff arriving at storage unit, $Z = Min(X_1, cX_2)$.
δ	- percent of previous available storage.
μ	- pollutant trap efficiency.
μ_o	- a fixed trap efficiency level
σ	- non-linear pollutant trap efficiency.
σ_o	- a fixed pollutant trap efficiency level.
θ	- a fixed parameter set describing the hydrologic process (mean duration, depth, and time between events), the storage parameters, and the treatment rate.

BIBLIOGRAPHY

DE NEUFVILLE, R.; STAFFORD J.H. - Systems Analysis for Engineers and Managers.McGraw Hill, 1971.

DITORO, D.M.; SMALL, M.J. - "Stormwater Interception and Storage".Journal of Environmental Engineering, ASCE,105, EE1, Feb. 1979, pp. 43-54.

FERGUSON, C.E. - The Neoclassical Theory of Production and Distribution.Cambridge University Press, 1975.

HYDROSCIENCE, INC. - "A Statistical Method for Assessment of Urban Runoff".Environmental Protection Agency, EPA440/3-79-023, May, 1979.

JEWELL, T.K. - Urban Stormwater Pollutant Loadings.Ph.D. Dissertation, Univ. of Massachusetts, 1980.

LOGANATHAN, G.V.; DELLEUR, J.W.; SEGARRA, R. - "Planning Detention Storage for Stormwater Management".Journal of Water Resources Planning and Management, 111,4, Oct. 1985,pp. 382-398.

NIX, S.J. - Analysis of Storage-Release Systems in Urban Stormwater Quality Management.Ph.D. Dissertation, University of Florida, 1982.

NIX, S.J.; HEANY, J.P. - "Optimization of Storm Water Storage-Release Strategies".Water Resources Research, 24(11), Nov. 1988,pp. 1831-1838.

SEGARRA-GARCIA, R.; LOGANATHAN G.V. - A Planning Model for the Control and Treatment of Stormwater Runoff through Detention Storage.Final technical report to U.S. Dept. of the Interior, University of Puerto Rico Water Resources Research Institute, Jan. 1989.

PART V

Agricultural pollution

Chapter 29

RESEARCH NEEDED TO REGULATE PESTICIDE CONTAMINATION OF GROUNDWATER

G Chesters, J M Harkin (Water Resources Center, Wisconsin-Madison University, USA), J B Wood (Wood Communications Group, Wisconsin, USA)

ABSTRACT

The policy issues embraced here are of major concern to the general public and their political representatives. Plentiful supplies of high-quality water must be provided for all uses. Future decisions on groundwater use and preservation will have profound impacts on land use and national economics. A research program to assist in regulation of pesticide contamination must consist of a mix of fundamental and applied projects, and of short- and long-term theoretical, laboratory and field investigations. The research should be accompanied by a strong technology transfer/information dissemination effort and a training scheme capable of producing a new generation of high-quality professionals. An adequate data base backed by well-designed sampling and monitoring protocols is essential to the development of predictive mathematical models capable of delineating those lands likely to be sources of groundwater contamination. Proposed control measures should be site- and compound-specific. The goals of the research should be to describe the chemical, physical and biological processes controlling transport, transformations, distribution and fate of pesticides and to evaluate their toxicological significance. Eventually, these findings will allow formulation of a rational protocol for protecting water resources founded on fundamental scientific principles and sound economic and policy considerations.

Key words: groundwater, pesticides, research, contamination.

During the past decade water resources research programs have increasingly focused on groundwater. It is now obvious that serious gaps exist in our knowledge of: a. The characteristics of the quality and quantity of the groundwater resource; b. ways in which groundwater becomes contaminated; c. techniques to minimize impacts of contamination; d. the realistic magnitude of public health problems that arise from contaminated drinking water; and e. cost-effective management methods to protect groundwater from contamination. The research needs discussed will provide scientific information to minimize water quality deterioration resulting from pesticide use and misuse. The eventual aim is to ensure a rational protocol to establish water quality standards for ground and drinking waters.

1 – PUBLIC POLICY ISSUES

The public policy issues embraced in this document are of major concern to the general public and their political representatives. Nations must provide a plentiful supply of high-quality water for all uses. Future decisions on groundwater use and its preservation are likely to have profound impacts on land use and national economies. Vital questions include: 1. Where can agriculture be conducted and how will farmland be managed? 2. How can predictions be made to determine which lands--on a site-specific, chemical-specific basis--are overly susceptible to movement of pesticides to groundwater? 3. What health risks can be tolerated in establishing drinking water and other groundwater standards? 4. How can land (agriculture, landfill sites, urban) be managed to minimize use of pesticides and protect water resources from degradation? These questions tie in closely to the aspirations defined for "Sustainable Agriculture" programs.

These are but four of the highly germane questions which will face decision-makers between now and the turn of the century. Research has been identified as an activity desperately needed to complement the monitoring, inventory and regulatory programs presently being conducted in many parts of the world. Because researchers must be given time to assemble data and evaluate their findings, it is essential that this urgent task start immediately if research is to play its appropriate role in decisionmaking.

National, regional and local governments must harness and coordinate research personnel in a way that maximizes total output from their investments in solving groundwater problems. The prioritization of issues needing research should be updated continually by a strong oversight advisory group.

Groundwater research should pursue three early initiatives: 1. Evaluate existing research information and repackage it in a format and language readily understandable for use in policy evaluations; 2. establish priorities for research utilizing the inputs of top decision-makers; some decisions have to be made quickly and decisionmaking deadlines should be an important consideration in setting priorities; and 3. evaluate information distribution systems to provide decision-makers with the best techniques for quickly accessing relevant information and policy alternatives.

For the purpose of this discussion, categories of research are identified and discussed under nine headings, with high priority research groundwater research from a characterization of the quality and quantity of the resource to delineation of the economic consequences of particular policy alternatives. Information dissemination/technology transfer and formal education programs also are discussed. The research seeks to develop an understanding of: The movement and transformations of pesticides in the environment; methods to minimize their impacts by the use of conversion methodologies; their toxicological significance; and innovative management practices to reduce the use and transport of pesticides. A quality-assured data base available to researchers and policy analysts is essential.

2 - RESEARCH NEEDS

2.1 - Prediction of environmental behavior of pesticides from chemical properties

Experience has shown that it is too expensive and time-consuming to study each individual pesticide in the variety of soils, landscapes, climates, crop rotations in which it is used. Short cuts must be sought. Fortunately, the transport, fate, and effects of pesticides in the environment are linked to their molecular structures. This interdependence exists because chemical reactivity, mobility, and toxicity are controlled by chemical properties determined by molecular structure. Quantitative structure-activity relationships form linkages to chemical properties that provide a basis for predicting the fate and effects of chemicals in the environment. The overall number of chemicals of environmental concern is immense ($>50,000$); the number and classes of pesticides are far fewer (ca. 300 active ingredients). However, new pesticides are being introduced and there is gathering concern over pesticide metabolites in soils and waters. Assessment of chemical properties by testing individual chemicals is impractical. Rapid, cost-effective methods for predicting chemical properties are vital. Most rapid and least costly are those methods requiring only information on chemical structure.

While related to the chemical properties of the compound, pesticide behavior is also determined by environmental interactions. For example, sorption to soil and sediment components modify transport rates, and susceptibility to chemical or microbial degradation may be retarded or enhanced by surface reactions and the presence of other compounds. Thus, an understanding of the linkages between chemical properties and the interactions and transformations occurring in the environment is essential in predicting pesticide residue transport and fate. Information on chemical properties is vital to environmental research on transport, transformations, and toxicology of pesticides. Acquisition of available information on chemical properties, reliable methods for measuring and predicting them, and valid models for predicting fate and effects using this information are urgently needed.

Research goals are to: 1. Develop and evaluate methods for predicting chemical properties of pesticides; 2. establish and predict relationships between chemical properties and the interactions and transformations of pesticides; and 3. incorporate predictive relationships into models of transport, transformations, fate and toxicology of pesticides in aqueous environments.

Chemical properties cannot be incorporated into models of transport, fate, and effects without taking into account the chemical property-dependent interactions occurring in natural systems. These interactions include partitioning between air-water, air-soil/sediment, and water-soil/sediment phases, and chemical and biochemical transformations. Predictive methods for some physicochemical interactions have been developed. Several important questions to be addressed include: 1. The compound and particle-specific properties controlling adsorption reactions; 2. the dependence of water-sediment/soil partition coefficients on water:solid ratio; and 3. the kinetics and mechanisms of adsorption/desorption reactions. In this area, a major goal is development of readily measured parameters of the solid and aqueous phase which can be combined with information on chemical properties of the pesticide to achieve accurate predictions.

2.2 - Pesticide transformations

The fate of toxic organic chemicals is governed largely by the rate and extent to which they degrade in the environment. Consequently, determining degradability of pesticides to nontoxic metabolites is of major importance in evaluating the environmental hazards associated with their use. Transformations of pesticides may occur through biological and/or nonbiological mechanisms. Some pesticides undergo relatively rapid chemical degradation mainly through hydrolysis, oxidation or reduction while others undergo photodegradation or conjugation. Some chemical and photolytic reaction rates are enhanced by sorption of the chemicals on particle surfaces. However, microbial metabolism is considered to be the major pathway of degradation in soil and aquatic environments. Its efficiency depends on the composition of the microbial population as affected by soil, limnological, and hydrogeological factors. Similar types of metabolic processes may be found in soils, sediments, and groundwater but degradation rates and products vary widely because of differences in environmental conditions.

Pesticides differ in their resistance to microbial degradation. Some are highly resistant, others are easily biodegraded. This varying degree of resistance is attributed to: 1. The chemicals' inherent resistance or susceptiblity to microbial attack; and 2. the biodegradable chemicals' becoming largely or totally unavailable for microbial degradation due to environmental factors and interactions. For example, some biodegradable compounds become resistant when adsorbed by clay minerals and organic constituents of soils and sediments, especially at sites where microorganisms cannot penetrate.

Little is known about biodegradation of organic compounds in the soil's subsurface (vadose zone) and in groundwater (phreatic zone). Most information on pesticide degradation has been accumulated for surface soils and only recently has interest been shown in the fate of pesticides in the vadose zone and groundwater. Potential for biodegradation of pesticides below the root zone exists, but the rate at which it proceeds is unknown. The subsurface environment differs considerably from microbial habitats found in surface soils and waters. Little is known about the effect of subsurface conditions on biological activity and the biotic transformation of organic contaminants.

To clarify the fate of pesticides, studies on their transformations in soils, sediments, and groundwater must include: 1. Elucidation of degradation pathways and metabolites; 2. determination of degradation rates of parent compounds and metabolites; and 3. identification of environmental factors which control degradation rates and pathways. It is essential to identify major metabolites in various environments so that their toxicological significance can be assessed. Degradation products of pesticides may be as toxic, more toxic or less toxic than the parent, and each metabolite may have a different effect on human health. Analytical methodologies for identifying parent compounds and metabolites and confirming their composition are desperately needed, as well as cost-effective assays for them in complex media (soils, sediments, and groundwater). Until recently, in most degradation studies, only the parent compound was considered when determining degradation rates because principal concern was with avoidance of phytotoxic carryover. Unless all the important toxic products which appear in the degradation pathway are taken into consideration, this approach likely leads to underestimation of the residence times of hazardous residues in the environment. Accurate degradation pathways and rate data are needed to develop dependable mathematical models for predicting transport and fate of pesticides and their metabolites in surface and groundwater.

2.3 - Pesticide transport

Pesticide transport of pesticides from point of formation, emission, application, or storage into surface and groundwaters and their reactions en route are key aspects of the toxic pollutant problem. A thorough understanding of the transport-reaction chain is essential for effective and economic control of chemical pollutants. Overall objectives are: 1. To establish the factors controlling water and pesticide movement in and through the root and vadose zones of soils to groundwater; and 2. to measure and model the transport and dispersion of pesticides in groundwater.

Although numerical models have been developed to predict contaminant transport in the vadose zone and groundwater, indications are that the models are considerably more reliable in predicting water flow than in their ability to simulate the transport of organic contaminants. Present models are inadequate in simulating dispersion and the interactions and transformations affecting contaminant solute transport in soil and groundwater systems so that prediction of distribution and concentration is extremely inaccurate. Problems arise in part from dealing with aqueous-solid phase interactions, particularly adsorption, and transformations occurring through chemical and biological reactions during transport. Some models have incorporated dispersion, adsorption-desorption, and first-order degradation rate information on the associated equilibrium and kinetic parameters applicable under field conditions, but to date the models are not sufficiently refined to allow site-specific decisions to be made about the susceptibility of particular soils to contaminate groundwater. A three-tier approach should be used in advancing the capability for assessing transport: 1. Data should be obtained from representative field sites to provide ground-truth information on transport; 2. the rate and extent of transformations and adsorption-desorption should be determined under conditions representative of those encountered in the field; and 3. transport models should be modified and/or developed to accommodate information on flow path, interactions with solids, and transformations occurring during transport.

A quantitative understanding of the influence of solid-solution interactions and transformations is perhaps the greatest impediment to predicting transport and distribution of pesticides in soil and ground-water systems. For organic compounds, adsorption-desorption reactions represent the main form of solid-solution interaction. To model the attenuation of pesticides, information on adsorption-desorption kinetics and equilibrium relationships is essential. Another key process affecting transport involves chemical and microbial transformations occurring in the system. A principal goal is to predict degradation based on structure-susceptibility relationships and incorporate this information into transport models.

In addition to pesticide transport from surface soils and waste disposal sites, the mobility of pesticides in sediments associated with surface waters may also represent a problem. Surface and groundwater systems are interconnected, and polluted sites are potential sources of groundwater and subsequent surface water contamination through groundwater discharge. Consequently, research on the factors controlling mobility of contaminants associated with bottom sediments is needed, including dispersion into surface water through diffusion and particle-mediated transport and advective transport into aquifers. In addition to the chemical and biological interactions affecting transport, important problems must be addressed concerning the movement of water in the unsaturated and saturated zones including the influence of heterogeneity in porous media, modeling unsaturated flow and gas-phase transport, and assessing important physical parameters, including hydraulic conductivity.

Mathematical models should be tested to evaluate movement of landfill leachate and leaching of pesticides through soils, describe their fate in the root and vadose zones and estimate their potential for predicting groundwater quality for a variety of management schemes.

2.4 - Innovative management practices

2.4.1 - Agricultural management practices

Continued field-scale investigations are needed to develop so-called "best" management practices for agriculture. In the past decade numerous investigations have been made on no-till and conservation tillage practices. It is widely reported that most of agriculture will be under some form of conservation tillage by the turn of the century. Studies on minimum tillage practices have been designed in large measure to reduce fuel consumption, runoff, and soil erosion; certainly these practices reduce the threat of nonpoint pollution to surface waters. However, the impact of the processes on transmission of pesticidal and fertilizer chemicals to groundwater is little understood. Some investigations show that leaving stubble and other plant residues at the soil surface can enhance some disease vectors, particularly fungal diseases. Also, increased use of herbicides is often associated with these cultivation practices. After a few years of minimum tillage, earthworm populations rise dramatically and worm holes short-circuit water and associated pollutants to greater depths in the soil profile. By reducing runoff, particularly during heavy storms, more water is available for transport through the soil to carry pollutants to groundwater. Research on water movement in soils at the field-scale level is urgently needed particularly with reference to macropore infiltration, e.g., down cracks or worm holes in fine-textured soils.

Farmers who use pesticides must understand that it is not essential to preempt intrusion or remove every vestige of weeds, nematodes, insects, etc. from a field to assure high crop yields. Prior to the use of herbicides, the mechanism for curtailing growth of grasses in corn fields was to allow rapid growth of the corn to shade out the grasses. Investigations need to be made to see if production can be maintained while using very low amounts of herbicide in the early phases of growth and then allowing the crop canopy to eliminate weeds by shading.

Some simple methods that might be investigated for retaining pesticides in the root zone are to modify the surface hydraulics of the soil so that under rainfall or irrigation conditions excess water is diverted around the main sites of chemical placement. Only enough water need reach the chemical to allow it to perform its pesticidal function. Minor surface landforming on susceptible soils (particularly sandy soils) can provide this water diversion with little mechanical energy input. A second possibility is to localize placement of pesticides to reduce the amount of water coming in contact with them. Successful procedures have been developed to do this with fertilizers (particularly nitrate). Slow-release formulations allow plants to take up plant nutrients as they are released. A similar concept might be applied to pesticides, especially those taken up systemically. Investigations are needed for particular plants to determine if these compounds are transmitted by the xylem or phloem, to determine how localized the pesticide placement can be. A third possibility is that some of the new biodegradable plastics could be used to form temporary barriers around pesticides to shield them from percolating waters. Some, but not excessive redesign, of farm equipment may be needed to apply some of these proposed techniques.

2.4.2 - Waste containment technology

A major problem in industrialized societies is the proper, safe and long-term handling of wastes. Waste disposal in landfills impacts everyone because all are direct or indirect waste generators and many suffer from the economic and health impacts of improper waste disposal. The obvious, but unrealistic, solution is not to generate waste materials. Clearly, waste minimization, reduction and beneficiation should be actively pursued. However, complete elimination is simply impossible. Therefore, construction of secure waste disposal sites with containment and collection of waste effluent that cannot be treated or until it can be treated becomes an important issue in evolving innovative management tactics. Waste containment systems currently include bottom liners, sideslope liners, covers and vertical barriers, and utilize earthen materials and geosynthetics. The purpose of these waste barriers is not merely control of water seepage. They must also restrict the escape of hazardous substances such as pesticides or volatile organic compounds in solution or in the vapor phase. Laboratory and field experience shows that some chemicals can affect the structural integrity of the most widely used liner material—soil. Hydraulic conductivity and its susceptibility to change with time or exposure to chemicals are factors impacting use of clay in waste containment barriers. The current data base comprises mainly test results for simple monochemical systems and test programs that have extended over short time periods relative to the active life and post-closure security periods for most waste containment and storage facilities. Even in field situations involving little or no fluid flow, chemical flux through a barrier may be completely controlled by chemical concentration gradients. The complexities of diffusion and its importance as both a transport and an attentuation mechanism have been recognized. An acute shortage exists of well-documented, long-term field performance with respect to suitability of barriers to waste leakage.

It is now clear that waste containment technology is in dire need of innovative approaches to bring about another generation of barriers. Economic and construction factors are often limiting in this respect. The feasibility of using improved designs and new materials as a substitute or amendment for soils needs to be investigated. For instance, pozzolanic fly ash is a relatively inexpensive material shown to have favorable hydraulic characteristics and interesting properties for removing certain contaminants. Additionally, its attractive physical properties such as resistance to damage by freeze/thaw and wet/dry exposures and much superior strength and mechanical characteristics suggest innovative developments of barriers (both as liners and covers) using pozzolanic fly ash as a substitute for clay or more likely as an additive to clay or as a material interlayered with clay. For instance, the superior strength of pozzolanic fly ash could allow construction of steeper sideslope liners resulting in increased waste disposal volumes over the same area of disposal. The constructibility of liners, their cost-effectiveness, their interaction with mixed chemical systems, their performance in field circumstances, and the effects of time are significant research elements. Furthermore, development of technologies for rapid detection of leakage and use of technologies such as reverse or electro-osmosis to concentrate contaminants or control solute transport across barriers are items of immediate concern.

2.5 - Pesticide reduction and conversion

Methods for cleanup of hazardous organic wastes suffer from many problems, making management a pressing scientific and social issue. Annually, many new pesticides are synthesized and new manufacturing processes come on line, potentially contributing to water pollution. Techniques such as landfilling for disposal of byproducts, expended major products, and waste products formed in the manufacture of pesticides are inadequate. Analytical advances allow detection of lower and lower trace levels of pollutants in water, but treatment methods remain archaic. The need exists for new, efficient and economic methods for reduction and conversion of wastes containing pesticides and their byproducts. Two possible innovative strategies for degradation of pesticides in aquatic systems are: 1. Photodegradation catalyzed by hydrous oxide surfaces; and 2. development of mixed microbial populations capable of degrading target compounds under natural or modified environmental conditions.

The photocatalytic approach involves the use of energy from light (possibly sunlight) to promote oxidation of organic compounds on the surfaces of oxide particles. The metal oxides serve first as adsorbants to concentrate the organic compounds and then as catalysts to degrade them. Ceramic membranes constructed of hydrous oxides may serve the same purpose and provide a means for separating the products from the liquid waste. If research indicates that pesticidal wastes should be treated using this approach, the final research phase should involve pilot plant-scale testing of the method. Technology transfer should involve communicating the results to government agencies and the industrial sector interested in process development.

Bioremediation, the controlled biodegradation of toxic chemicals, has been used for many years to remove petroleum contaminants from soil. Recent biotechnology advances in microbial ecology and genetic engineering suggest that bioremediation has a broader applicability and may provide a practical, cost-effective system for clean-up of a variety of organic contaminants from soils, sediments, and groundwaters. A problem encountered in microbial degradation of organic contaminants is the formation of highly recalcitrant intermediates. In fact, biological conversion of pesticides and other organic pollutants to metabolic dead-end products is a common phenomenon. For many organisms, the phenomenon arises as a result of cometabolism, involving low-specificity enzymes designed to attack and mineralize naturally occurring analogs of contaminants by normal pathways. Furthermore, since degradation of pollutants through cometabolism is not growth-linked, cometabolism requires additional substrates to build and maintain an active microbial biomass at a sufficient level. An alternative to the use of pure cultures is to exploit the diversification achievable with microbial consortia to accomplish complete mineralization of pollutants. Construction of consortia to exclude a cometabolic component may be achievable by maintaining pollutant concentrations at very low levels.

Objectives of bioremedation are: 1. To construct stable consortia of soil or aquifer microorganisms capable of complete transformation of such target compounds as pesticides to microbial biomass and/or inorganic end products; 2. to characterize the factors controlling the kinetics of degradation to environmentally acceptable levels by successful consortia; and 3. to evaluate the potential of using the consortia for field-scale treatment of pesticides in wastes. Pesticide-contaminated soils or sediments should be used as starter sources for developing pollutant-degrading microbial consortia. For successful consortia, the effects of environmental factors should be evaluated. Loading rate limits should be established, and the production of microbial inocula should be scaled up to allow development of procedures for evaluation of conditions necessary to establish consortia in soils of diverse composition. In the final stage, successful systems should be evaluated under field conditions.

2.6 - Characterization of groundwater flow systems

The primary objectives of hydrogeology-related research are to improve understanding of: 1. The geologic factors which affect ground-water movement; 2. groundwater flow systems which dictate the direction,

depth and rate of flow and the area or point of discharge to the surface;
3. rock/water/contaminant interactions which directly affect the
chemical/biological quality of groundwater; and 4. the boundary conditions
which determine the relationships between aquifers and between drainage
basins. The importance of these objectives are discussed in section 2.3
on pesticide transport.

Applied hydrogeologic research helps provide solutions to immediate
problems associated with groundwater-management and groundwater-quality
problems. Such problems continually arise as national, regional and local
regulatory officials and the public seek an increased understanding of
groundwater movement and groundwater contamination and demand better
methods for groundwater monitoring and the protection of groundwater
quality. For example, current regulatory and management concerns include
the effects on groundwater of hazardous waste disposal and agricultural
practices, as well as groundwater quality protection in areas of fractured
bedrock and thin or coarse-textured soils. In response, applied research
studies have been initiated on groundwater movement in clays, groundwater-
capture-zone analysis, and groundwater monitoring in fractured rocks.

2.7 - Toxicological assessment of pesticides in groundwater

The United States Federal Water Quality Act of 1987 for the first
time addressed the issue of toxicants in surface water at the national
level and required that all states--under the guidance of the U.S.-
Environmental Protection Agency--create standards or procedures for
deriving criteria for individual toxic substances defining maximum
permissible ambient concentrations in various categories of surface water,
depending on the use of the particular body of water. Furthermore, the
same legislation required states to promulgate rules and regulations to
ensure that appropriate ambient water quality criteria for toxic
substances be maintained. To this end, the states should regulate the
input of individual toxic compounds from industrial and municipal sewage
treatment plants by means of National Pollutant Discharge Elimination
System (NPDES) permits. Thus, it allows states to create standards to
safeguard against acute and chronic toxicity in receiving waters towards
fish, other aquatic species, and even aquatic plants important for
fish/aquatic organism habitat, as well as waterfowl and wildlife relying
on the affected water for drinking and as sources of food. Beyond this,
criteria must establish adequate protection for humans and domesticated
animals using the water as sources of drinking water or fish. The
contaminant levels in water for such uses must be low enough to preclude
unnecessary risk of mutagenic, teratogenic, carcinogenic, neurological or
immunosuppressive effects. Derivation of responsible criteria and control
measures, which guarantee adequate levels of human and animal health and
environmental protection, without undue social and economic stress on
dischargers and the communities or industries they service, becomes an
extremely complicated task requiring input from many scientific and
technological disciplines.

At the same time, several states and the U.S.-EPA are trying to
develop programs and strategies to cope with the emerging issue of
groundwater contamination by toxic compounds particularly pesticides and
volatile organic compounds such as soil fumigants. As a consequence of
measures to eliminate waterborne toxics in discharges to surface water,
larger and larger amounts of more and more hazardous waste materials
including pesticides are being managed by disposal on land, increasing the
risk of degradation of groundwater quality as traces of these toxics leach
through the soil with percolating rainwater. This situation compounds the
widespread groundwater problems already extant with improperly
constructed, abandoned refuse, solid waste, and chemical dumps, leaking
underground storage tanks, field and chemical spills and agrichemical
usage. Regrettably, most instances of groundwater pollution from such
sources arose due to lack of foresight in application of appropriate
regulations and scientific/technological options for management of the
polluting materials. The seriousness of such groundwater contamination
lies in the fact that groundwater provides for the drinking and household
water needs of approximately half of the United States population, as well

as supplies for many industries, agriculture/irrigation and livestock operations.

Since new protocols have been developed at the federal level and previous standard-setting mechanisms have changed, the time is opportune to reevaluate the scientific basis for establishing groundwater standards. Criteria for the quality of groundwater must center mainly around its suitability for human and domestic animal consumption. Most toxicants of concern here are different from those in wastewater discharges to surface water, and their persistence and distribution also are quite different. Thus, although expertise from several disciplines is required to derive appropriate groundwater quality standards, this expertise is different from that required to address surface water quality issues.

The concentrations of toxicants in groundwater are likely to be lower than those in surface water, but their persistence will be longer and their variety greater. Questions of long-term multiple exposures are therefore more important, especially the likelihood of subtle, chronic effects. Health risks from low-level chronic exposure to toxicants are a function of toxic potency, exposure level, and exposure frequency and duration for each individual substance or combination of substances. For reliable assessment of risk, compounds of concern such as pesticides, must be identified and their sources, distributions, fates, toxicities and pharmacokinetic behaviors derived from known toxicological data. The resultant information must be evaluated by means of an appropriate, biologically based mathematical model in order to derive guidelines for adequately cautious standards.

Objectives for research in this arena should include: 1. Selection of appropriate compounds or compound mixtures for study; 2. identification of known problem sites or potential experimental sites; 3. assemblage and evaluation of existing toxicological data to improve extrapolation of known effects on laboratory animals exposed to high doses to humans exposed for protracted periods to low levels of groundwater contaminants; and 4. apply the best risk assessment models to predict expected increased incidences of pathological effects. The research requires interaction with hydrogeologists to predict dispersion, diffusion and attenuation of pesticides in groundwater and with agronomists, horticulturalists, and agricultural and civil engineers to conceive and test improved management practices to reduce future contamination or remedy existing contamination. For instance, the nature, distribution, persistence, and fate of trace residues of the commonly used pesticidal chemicals (e.g., triazines and N-chloroacetyl herbicides, thiophosphate and thiocarbamate fungicides, carbamate and organophosphate insecticides), volatile organic solvents, and their metabolites should be examined and information on the toxicity and disposition of the parent compounds and major breakdown products assembled. If these compounds are, for example, carcinogenic, potency factors and acceptable daily intakes should be calculated using appropriate levels (10^{-5} or 10^{-6}) for excess cancer risk.

2.8 - Pesticide sampling, monitoring and data base

To conduct research on groundwater quantity, quality and management a well formatted data base is required. A program should be developed and managed to accommodate existing and additional data necessary for a complete description of the extent and quality of the groundwater resource under investigation. The data base should be well-documented and quality-assured since it must serve as the foundation for developing, validating and using mathematical models to delineate areas which are sensitive zones of potential groundwater contamination by pesticides. A detailed description of the contents of the data base including all mechanisms for assurance of analytical quality should be prepared. The base should permit easy access and be compatible with existing data bases and with mathematical models being generated and mounted locally, regionally and nationally. The base would be most useful if it were uniformly formatted with neighboring regions or countries to permit information retrieval on a broad scale. The inter-compatibility and formatting of the data should be reviewed by groundwater modeling groups. Initially, a fully documented data book with abstracts should be prepared to inform potential users of data base capability.   The data base would reveal information gaps and research needs.   These consultations would bring out potential data base applications   and   evaluation   of   existing   data   will   determine   the completeness and accuracy with which these applications could be met.

Data collection should emphasize parameters important to evaluation of pesticide and hazardous waste impacts on groundwater.   Nonetheless, the data base should have capability and flexibility for addressing a diversity of problems and be applicable to different water and land resource uses and management.   The data system must be sufficiently flexible to acccommodate regionally variable information and multifactoral analysis and be compatible with other systems.   This analytical capability is essential if data are to be used for identifying policy options and delineating problem solutions.   Data in the system should accommodate local data entries and yet be compatible with national government information on demographics and climate.   The system should contain geological, hydrological, water quality, and water- and land-use data.

Devising a groundwater policy which treats all users similarly and is applicable to a large region is a difficult but not insurmountable task. Using available scientific knowledge and groundwater monitoring information, a government or academic entity could select contaminant-sensitive areas for investigation and analyze five to eight likely groundwater contaminants chosen because they display a diversity in use pattern and location; chemical properties; point, nonpoint and multiple sources; transport in the environment; chemical and biological degradability; and potential health effects.   Such a planned strategy is better than conducting multiple analyses on a grab bag of samples.   Analysis is expensive and thousands of "no detects" at $100/parameter are not warranted.   Appropriate "indicator parameter" choices might include pesticides such as aldicarb (highly water soluble) and atrazine (strongly adsorbed), a volatile synthetic organic chemical of high usage such as trichloroethylene, a typical aromatic petroleum product such as benzene, and nitrate.

These examples illustrate the principle that a small number of compounds representing a wide diversity of contaminant types can be investigated without producing intolerable strain on a region's financial and analytical capability.

From recent experiences with pesticide monitoring of groundwater it is clear that in qualitative and quantitative terms, analyses from different laboratories are inconsistent and show poor replicability. Comparability and precision of quantitative analyses using different methodologies on the same sample leave much to be desired.   The need for adequate quality assurance and quality control and extensive confirmatory (qualitative) analysis greatly increases the cost of monitoring programs. However, it is essential to confirm groundwater contamination findings before data are released to the public.   For the sake of credibility, groundwater samplers and analysts must utilize reliable methods to assure the validity of data and establish reliable chains of custody, secure storage and other quality assurance measures.   Thus, on a regional basis creation of a centrally-located quality assurance program is urgently required.

2.9 - Economic and policy alternatives

A nation depends on its groundwater resource in numerous and incalculable ways.   Identifying the myriad economic and significant public policy issues that emerge as this resource becomes threatened by toxic materials is a local and national imperative.   The economic implications of pollution are staggering.   Nielsen and Lee (1987) of the Economic Research Service of the U.S. Department of Agriculture placed the cost of a national, one-shot monitoring effort to evaluate groundwater contamination at $2.1 billion.   The economic significance of groundwater contamination can better be understood in looking at some individual components of the problem:

1.   Sampling and monitoring soils and groundwater are extremely expensive.   The average price of a good, reliable pesticide or volatile

organic compound analysis is $100. At the regional level, and in areas of high contamination even at the local level, this involves millions of dollars. Added to sampling is the cost of well drilling. Deep sampling can cost $1,000/day for equipment and might require a whole day to sample one location.

2. The number, variety, and degree of pollution sources are pervasive throughout society. Groundwater protection in the final analysis requires education programs directed not only at thousands of farmers, but also at the urban populations. Such educational programs are costly because they must address themselves to individuals or small groups.

3. High costs are associated with development of regulatory schemes. To set responsible standards, scientific inputs from laboratory and sampling programs must be merged with toxicological information for humans and farm and domesticated animals. All toxicological investigations are expensive and time-consuming. Added to this are investments in public hearings, writing legislation, etc.

4. Once a regulatory scheme is in place it is of no value if it is unenforceable. For the case of 40,000 farmers in Wisconsin using a single compound, e.g., atrazine, as a herbicide for corn, even at the local level the cost of a regulatory scheme based on a set of limited variables becomes prohibitive unless creative self-regulatory mechanisms can be established. Regulation in general requires inventories of farmers, location of farms, chemicals purchased and used, waste disposal practices in place, numbers of manufacturers and distributors, and chemical formulations used. Substantial staff additions will be required by government agencies to implement such regulations.

5. If regulation results in bans, moratoria, or limited use of some chemicals, a commensurate reduction in agricultural output will likely ensue. Net profitability of farms is further affected if such regulations are more stringent in one region or country than another.

6. Once groundwater quality is impaired, environmental and ecological damages are enormous. For instance, in the United States 97% of all rural and a significant proportion of municipal water supplies rely on groundwater for drinking water. Groundwater pollution, even in small areas has serious consequence. The costs of adding filtration systems to household faucets, purchasing bottled water, drilling wells to greater depths to reach unpolluted aquifers or of developing new supplies at distant locations are often staggering.

7. A realistic assessment of the cost of groundwater pollution should include the medical treatment costs of intoxicated humans and animals exposed to pollutants in their water supplies. As in most evaluations regarding human health, be it in terms of medical treatment costs, lost productivity, losses in quality of life experiences or shortened life-expectancy, determination of meaningful measures becomes overwhelming in complexity. A reasonable surrogate is the cost of restoring the water supply by removal of the contaminant on the assumption that this cost will always be a lower bound on the true damage caused if action were not taken.

8. Even if some costs are impossible to determine, the economic marketplace responds to them. In areas where groundwater has been contaminated, sharp reductions in assessed property values have resulted. Equity considerations might suggest legislated compensatory schemes for this value reduction or at the least for the replacement of contaminated wells. Even with such policy instruments it is improbable that property values will ever fully recover, causing a reduction in private wealth and in tax bases at all levels--local, regional or national.

Research is needed to determine some of the important economic costs delineated above, along with associated legal and intergovernmental policy

questions. Incidental or site-specific water contamination should be
addressed locally. On a site-specific basis, estimation of costs of
economic relocation, costs of alternative water supplies, and costs of
remedial alternatives should be studied. A strategically important
question is whether and how present institutions at all levels of government must change to develop workable, cost-efficient and implementable
plans for clean-up of contaminated groundwater.

Policy investigations must focus on developing new mechanisms of
financing plausible policy options. This requires economic analysis of
costs and benefits, especially as related to health issues. Once new,
workable institutional arrangements are identified, it is necessary to
determine with what rapidity existing structures can accommodate changing
intergovernmental ties. Economic evaluations of pollution impacts and
costs of remedial actions are vital. Given the large social costs,
measurement difficulties and human dislocations involved, indications are
that management should largely be geared to finding protection strategies.
To do this, appropriate intergovernmental financial control mechanisms
must be developed and identification of appropriate cost-sharing
approaches among levels of government and the private sector must be
found. In assessing policy alternatives, legal issues need to be
addressed as well.

In a more specific context, policy research should focus on
development and testing of innovative management practices and policy
instruments as they relate to source reduction and, as appropriate,
transport mechanisms. In each instance economic evaluations are
critical. To facilitate such inquiry, scientific research results should
be translated into policy parameters applicable to state and local
government intervention. These investigations should be actively linked
to proposed technology transfer and outreach activities. Analyses of
existing programs will guide new policy implications and impact
assessment.

3 - TECHNOLOGY TRANSFER, INFORMATION DISSEMINATION

Research information is being gathered faster than our institutions
are able to utilize it. This situation arises in part from an inability
to communicate information fast enough and in a sufficiently comprehen-
sible format for the public and its leaders to utilize the knowledge to
its best advantage in decision-making. Furthermore, a great amount of
information gathered and compiled is replicative of other efforts. An
efficient system of information and technology transfer is desperately
needed in the area of water resources management. Such a program must
serve a more useful purpose than merely collecting and compiling data; it
must utilize the experience and wisdom which have been accumulated to
evaluate the information available and propose rational alternative
solutions to major resource management questions of regional and national
significance. The advantages and disadvantages of policy alternatives
must be discussed and communicated and the constraints placed on certain
solutions must be delineated for different locations. Only in this way
can an informed public be utilized to impact on the decision-making
process, and decision-makers obtain the best information to allow them to
make rational judgments.

Research is needed to develop a scheme of information and technology
transfer on pesticidal impacts on groundwater quality and to facilitate
its dissemination to a diversity of groups. The scheme should be
structured to allow an advisory council to define and prioritize the
important pesticide problems needing evaluation. Specifically it is
necessary to: 1. Utilize an advisory council--in conjunction with program
staff--to define and prioritize those resource areas requiring re-
evaluation by providing and disseminating alternative management schemes;
2. mount an information retrieval system on water resource management
including scientific, technical, health, legal, social, economic and
political considerations; 3. identify information gaps and research needs
and conduct the research necessary to allow rational development of
resource management alternatives; 4. delineate alternative land
use/management schemes and discuss the advantages and disadvantages of

each; and 5. train students, technicians and leaders in the important elements of the research, information and technology transfer programs.

At a minimum, an effective technology transfer program involves retrieving all available information on a given topic, evaluating that information, utilizing the information to develop a series of potential solutions, and conveying those alternatives to decision-makers. Research is the mechanism for filling basic information gaps. A general manual of information dissemination methodologies should be prepared which can be appended as necessary with specific details describing a particular problem. Information recipients include the general public, educators, legislators, administrators, planners, and technical staffs in government agencies and industry. Each of these recipients has a different level of technical expertise and each has a different reason for needing the information. Some types of materials and the intended audiences are:

1. Press releases stating the background of the problem, reasons for current concern, and a summary of alternative solutions.

2. Brochures or leaflets with brief general summaries of problems and alternative solutions by region.

3. Slide sets, films or videotapes portraying the local area related to a particular problem and its significance.

4. Semi-technical reports outlining problems and alternative solutions by region.

5. Graphs, charts, and maps displaying potential impact of alternative solutions.

6. Technical details for implementing a particular alternative.

7. Cost-benefit analyses for alternative solutions.

Audiences for items 1, 2 and 3 are the general public and elementary and secondary school teachers. For items 3, 4, 5 and 7, principal audiences are legislators, administrators, planners, researchers, and university faculty. The principal outlet for item 6 is disseminators and technicians.

Concomitant with the development of a technology transfer program, a training scheme should be instituted to make the greatest use of the broad scope of expertise developed by both the research and technology transfer program.

4 - EDUCATION

Contamination of surface and groundwater by toxic organic compounds will be the principal issue in water resources management for the remainder of this century. The problem of groundwater contamination received little attention for many years, especially while more visible environmental management problems such as point source discharges of organic wastes captured the public's attention. As more compounds found their way into water systems, and more were identified as being dangerous to ecosystems and especially to humans, and as analytical techniques improved to demonstrate their presence, a strong political response has emerged to "do something about it!" University faculties should be tapped to form the interdisciplinary research teams required to meet the objectives delineated in this paper. The major research activities should be framed with a primary mission to train scientists and other professionals and to conduct relevant research. The need for and scarcity of scientists trained to deal with toxic pollutants in surface and groundwater in all levels of government agencies, research institutes, firms in the private sector, and universities is amply demonstrated by the many position openings currently being advertised. Even though many universities have comprehensive degree programs in place, a critical and well documented need exists to provide funding opportunities to students for training in water resource fields.

A wide range of specialized degree programs are available for undergraduate and graduate students in such water resources-related areas as limnology; hydrogeology; soil science/agronomy; water and environmental sciences; civil, chemical and environmental engineering; geology; pharmacology; preventive medicine; meteorology; mathematics; chemistry; physics; microbiology; policy studies; economics; business; political science; law and sociology.

A significant broadening in interdisciplinary education can be achieved through employment of undergraduate students as hourly workers or interns on research projects. These involvements bring extremely valuable insights to young students who have never confronted the complexity of environmental issues in the real world and who may redirect their energies and future professional involvements towards groundwater problems. Graduate students working toward advanced degrees and post-doctoral researchers become part of the critical pool of professionals to fill positions in universities, government, industry, and private and public research institutes. Students at all degree levels should be oriented towards career positions, especially those in the public sector where the need is greatest.

A significant educational activity should also center around the various outreach and technology transfer activities. Faculty and advanced graduate students should participate in short courses, institutes, and conferences directed at translating significant research findings to the community of users.

5 - SUMMARY AND RECOMMENDATIONS

The research discussed here should provide scientific information to minimize water quality deterioration from toxic chemical contaminants including such agricultural and industrial chemicals as pesticides. By evaluating the toxicological significance of pesticides, an eventual aim will be to provide a rational protocol to establish water quality standards for ground- and drinking-water.

This information should be acquired in part by extensive field-scale research geared to maximizing implementation of control strategies aided by a program of standard setting which adequately protects the citizen's health and minimizes environmental deterioration without negatively impacting a nation's economy.

Science should drive the standard-setting and implementation decisions. The interconnectability of surface and groundwater should be recognized so that solutions are not fashioned that protect water quality in one sector by degrading it in another. Areas of extreme importance in evaluating the fate and transport of pesticides are: 1. Chemical and biological transformations of pesticides in soils and groundwaters; 2. transport through the root and vadose zones of soils into groundwater; and 3. fate and dispersion of chemicals in the groundwater system *per se*. These transport phenomena must eventually provide us with mathematical models which are sufficiently accurate in their predictions to allow decisions to be made about appropriate and inappropriate land uses on a site-specific basis. Ideally, the decision unit for the models should be the size of a typical farm field. For pesticides we need to know their rates of degradation, the metabolites formed, and the factors controlling the rates. The toxicological significance of pesticides and their metabolites which impair water quality must be fully evaluated if wise decisions on standards are to be made.

Based on the United States concept of so-called "best management practices," innovative research is desperately needed to protect groundwater from pesticides. "Sustainable agriculture" is a worthy goal if it involves maintaining agricultural productivity and sustaining farm profitability while using lesser amounts of chemicals. Research should be directed toward reducing pesticide usage by significant amounts. New methods are needed to protect groundwater by retaining pesticides in the root zone. As a complementary program, groundwater should be protected by improved methods of hazardous waste disposal and by developing new methods of "waste reduction" including non-incineration techniques.   Research findings must eventually lead to implementable, cost-effective control measures.

From this research base policy alternatives should be forged whose advantages and disadvantages should be evaluated to provide assistance to decision-makers.   Plans should be made to hold workshops for officials from legislative, executive and administrative, and judicial branches of governments to debate the rationale of the several and separate alternatives. Complementing this program should be an outreach effort to: 1. Translate research findings into hard copy and media reports and oral presentations for a diversity of audiences; 2. conduct meetings to inform the public of what is being done and 3. evaluate problems and suggestions with regard to maintaining water quality.

The goals of the research program are to describe the chemical, physical, and biological processes controlling the transport, transformations, distribution and fate of pesticides in aquatic systems and to evaluate their toxicological significance.   The eventual aim is to provide a rational protocol for protecting surface-, ground-, and drinking-waters founded on fundamental scientific principles and rational economic and policy analyses.   Emphasis should be on development of inexpensive predictive techniques based on the chemical structure of the organic compounds and the relationships of their structures to transport, transformations, and toxicology.

The total program can be divided into nine major research program areas, including: 1. Estimation of chemical properties and behavior of pesticides; 2. transformations of pesticides in soils, and groundwaters; 3. investigations of transport, fate, and dispersion of pesticides in soils and groundwaters; 4. pollutant reduction and detoxification methods, including chemical and biological techniques; 5. team evaluations for ambient water quality health criteria; 6. innovative management practices; 7. description of the groundwater resource; 8. monitoring and sampling protocols; 9. economic assessments and policy alternatives.

This paper recommends collaboration between universities, government agencies having responsibilities for hazardous waste disposal and groundwater protection and the private sector.   It also proposes integration of multidisciplinary basic and applied sciences in a research effort to resolve top-priority water quality problems using a unified team approach.   It also proposes to include toxicological evaluation teams in the program who will help assure that standards for contaminated water are promulgated using the best science available.

ACKNOWLEDGEMENTS

During the preparation of this paper the authors have had conversations with scores of people. Particularly we wish to acknowledge the contributions of Professors Marc A. Anderson, David E. Armstrong, Kenneth R. Bradbury, Tuncer B. Edil, Colin R. Jefcoate, Erhardt F. Joeres and Champ B. Tanner of the University of Wisconsin-Madison as well as Orlo R. Ehart of the Wisconsin Department of Agriculture, Trade and Consumer Protection, Michael R. Schmoller of the Wisconsin Department of Natural Resources, and William P. O'Connor of the law firm of Wheeler, Van Sickle, Anderson of Madison, Wisconsin.

BIBLIOGRAPHY

NIELSON, E; LEE, L. - "Potential Groundwater Contamination from Agricultural Fields: A National Perspective, Economic Research Service, U.S. Department of Agriculture, Washington, D.C., 1987.

Chapter 30

THE EFFECT OF PATTERNS OF USE ON THE INGRESS OF PESTICIDES TO GROUNDWATER

R J Hance (Consultant, Oxford, UK), J A Guth (Ciba-Geigy, Basel, Switzerland)

ABSTRACT

The way a pesticide is used, its chemical properties together with local climatic, edaphic and hydrological factors largely determine the likelihood that it will reach ground water. The ways in which these factors operate are examined in order to identify ways in which the risk of ground water contamination can be minimised. The use of controlled release formulations does not seem promising at the moment and modifications to soil management practices and the use of undersown cover plants are unlikely to make a major contribution. Reduced rates of application, restricting the time of application to periods of high soil microbial activity and when evapotranspiration exceeds precipitation together with limitations on areas where pesticides are used are likely to be effective. This is supported by early indications from minimization strategies in practice

Key words: pesticides, application timing, use restrictions, minimization.

INTRODUCTION

For many years there was little concern about the likelihood that pesticides might reach groundwater in significant quantities from normal agricultural use, although accidental releases or improper disposal were recognised as providing local dangers. The reason for this complacency was probably the assumption that organic materials are removed from solution by physical, chemical and biological action as water percolates through the soil. In addition the application rates of pesticides, even herbicides used for long term industrial weed control, are low compared with the quantities of organic substances released by grazing animals onto pasture which traditionally have not been regarded as major hazards. The assumption seems basically still to be sound as recent estimates by BOESTEN (1987) and PIONKE et al (1988) put the fraction of a pesticide application that reaches ground water at less than 1%, usually less than 0.1%.

However, perceptions, at least in Europe, have been changed by the Council of European Communities (1980) directive which arbitrarily sets a maximum admissible concentration in drinking water of $0.1 \mu g/l$ for any one pesticide, with a total for all pesticides that may be present of $0.5 \, g/l$. There are reports that such levels have occasionally been exceeded in ground water, although the validity and reliability of measurements made at these concentrations raises questions that cannot be discussed here. For the purposes of this paper it is merely accepted that in some cases, for reasons that may or may not be justified scientifically, it might be necessary to modify the way a pesticide is used in order to reduce the quantity that may reach ground water.

The factors which control the quantity of a chemical that moves to groundwater include those determined by hydrogeology, those which derive fron the properties of the chemical and those which are affected by the way the chemical is used. These factors interact so there is scope for modifying the influences of the first two groups by manipulating the third.

HYDROGEOLOGICAL FACTORS

The hydrogeology of a region determines the vulnerability to contamination of the local ground water. The principle factors include: the depth of the water table; the ease of recharge of the aquifer; the porosity and material of the aquifer; the physical, chemical and biological properties of the soil (if present); topography.

ALLER et al (1985) produced a scoring system based on these factors, the DRASTIC index which attempts to give a numerical indication of the vulnerability of an area to ground water contamination. Their main object seems to have been the assessment of waste disposal sites and the approach does not seem to have been formally developed by those concerned with pesticides. This is unfortunate, particularly as the policy in some countries of defining groundwater protection zones implies the use of such principles. At the very least, the use of some sort of formal vulnerability assessment would improve the efficiency of ground water sampling strategies as pointed out by RAO et al (1988). There thus appear to be good reasons to believe that the development of an index of the DRASTIC type validated for pesticide usages, could make an important contribution to removing anxieties about the potential pollution of groundwater by pesticides as it would assist manufacturers and registration authorities to restrict use to regions of low vulnerability.

The processes that occur in the soil can be included in the group of hydrogeological factors and they offer some scope both for the identification of risks and for manipulation. The extreme situation occurs where soil is absent for example on railway tracks and on some industrial sites. Under these circumstances there is little or no opportunity for a pesticide to be broken down by microbiological processes, the area is likely to be free-draining and herbicides may be used at rates 5 or more times higher than those used in agriculture. It is interesting that CROLL (1985) reported concentrations of atrazine, of 0.2-0.5 g/l in shallow ground water in eastern England, a region where the main crops are wheat, barley and sugar beet so atrazine has no significant agricultural use. He concluded that the source must have been railways and industrial sites. Restricting the recommended uses of atrazine to agricultural situations, therefore, should greatly reduce the chances of it being found in groundwater.

Pesticides used in agriculture and horticulture are mostly dissipated by soil processes so the possibility must be considered that rates of loss could be accelerated by soil management. Apart from leaching, which we are trying to avoid, a pesticide is lost by evaporation, or by chemical, photochemical or biological transformation. Evaporation is normally something to avoid since it represents a total waste as far as the grower is concerned and the more volatile pesticides are either formulated to reduce such losses (for example as granules), or they are mixed into the soil soon after application. It is difficult to modify chemical processes in the soil except by changing pH, which also affects micro-biological activity, and in any case it is normally the requirement of the crop which determines the acid/base status of the soil. Photochemical reactions have so far not been manipulated in practice, although in principle it is possible to encourage them by the inclusion of photosensitizers in the formulation.

For most pesticides, transformation by the activity of the soil microbial population is the most important process of loss. Biological activity is, of course, greatly influenced by environmental conditions, particularly soil moisture and temperature and useful predictions of persistence in the field can be made on the basis of laboratory observations of their effects on breakdown rates (see NASH, 1988 and WALKER, 1989 for recent reviews). However, changing soil temperature or water conditions to affect field persistence of pesticides is not a practical proposition in agriculture although it may be possible on waste disposal sites. Nutrient availability should also have an effect and the addition of organic and inorganic nutrients, by stimulating microbial activity, would be expected to enhance pesticide degradation. The evidence (WALKER, 1980) is that this does not always occur and there is no general pattern that emerges. In any case, again it is the needs of the crop that will be paramount.

Another possibility would be to undersow the crop with a ground cover plant in order to increase the proportion of the soil influenced by roots (the rhizosphere) which is regarded as being of most significance as far as soil microbiology is concerned (GREAVES & MALKOMES, 1980). Unfortunately experimental results comparing cropped and fallow situations are inconsistent and not encouraging, probably because the effect of crop cover on degradation depends on how it affects soil temperature and moisture which will vary between soils and seasons (HURLE & WALKER, 1980). However, in greenhouse studies where such variations would be small, MUDD et al (1983) found that the rate of disappearance of isoproturon was not apparently affected by the presence of wheat plants and SEIBERT et al (1981) reported maize did not affect atrazine dissipation. It is interesting that the latter authors noted that after harvest of the maize tops, atrazine degradation was increased which indicates that the decomposing roots developed a microbial population that could degrade atrazine. These studies all compared cropped with uncropped soil so do not provide direct evidence of the likely effect of undersowing the crop plant with a plant specifically chosen to encourage microbial activity. The results are, however, sufficiently disparate to show that the effect of a cover plant cannot be predicted with existing knowledge.

It seems that increased general microbial activity is not necessarily related to the rate of decomposition of a particular pesticide. This was clearly shown in work by DOYLE et al (1978) in which both sewage sludge and dairy manure increased microbial activity but the former inhibited the degradation of three herbicides while the latter accelerated it.

The other aspect of soil management that might affect persistence and also the movement of a pesticide in the soil is the cultivation that is used. As already mentioned, mixing a pesticide in the soil can reduce losses in the vapour phase. Cultivation might have two other effects on persistence. Firstly, by moving the pesticide into the soil it reduces the fraction near the surface which could become dry and hence microbially inactive, but on the other hand it will increase the fraction in the cooler layers of the soil. Thus the outcome could be an increase or a decrease in persistence. Secondly, mixing reduces the concentration of a pesticide in the soil and there are many reports (see HURLE & WALKER 1980) that degradation proceeds more quickly at low than at high initial concentrations. In practice it seems that cultivations generally have little effect (HANCE & COTTERILL, 1984) unless they move the pesticide into the subsoil where breakdown tends to be slower. They may be effective in reducing the phytotoxicity of a carried-over herbicide residue but this is largely because of the simple dilution effect reducing the dose the plant receives.

There is little beyond speculation to give an indication of the effect of cultivation on water and solute movement. HANCE & COTTERILL, (1984) noted that there is evidence that the surface soil is more compact and with lower porosity after a period of direct drilling than when the soil is ploughed and cultivated but that ploughing disrupts the continuity of the pore system so that water usually penetrates to the lower layers more easily in direct drilled land. With present knowledge there seems to be no basis for proposing a system of cultivation to reduce leaching to ground water although if the major route of groundwater contamination by pesticides is through preferential paths like cracks and channels (BOESTEN, 1987; RAO, et al 1988), then any system of cultivation is likely to lead to less contamination than zero tillage. However, the effect of cultivation on erosion and run-off to surface water would seem to be a much more serious consideration (LEONARD, 1988).

For completeness it is necessary to mention irrigation as a management practice that may be relevent. Little specific attention seems to have been paid to this in the context of pesticide movement to groundwater. Perhaps this is because in many circumstances, irrigation is applied to correct a natural deficiency in rainfall so there is unlikely to be a significant excess over evapotranspiration and the management of other solutes, notably salt, is a more pressing problem.

The conclusion from this section is that although hydrogeologic (including soil) processes are important there is little that can be done to manipulate them but by taking proper account of them pesticide use could be restricted to the less vulnerable situations.

PROPERTIES OF THE CHEMICAL

The properties of the pesticide chemical itself determine its susceptibility to the various processes of decomposition and also its propensity to move with the mass flow of water. For this reason they are considered very carefully in registration evaluations as indicators of the hazard to groundwater posed by a compound, although it is possible to argue that too much attention is given to this aspect compared with the other factors involved (HANCE & GUTH, 1989).

There is little scope for altering the intrinsic behaviour of a pesticide without the likelihood of compromising its pesticidal properties. However, the way in which a compound is formulated for application does raise possibilities of controlling the rate at which it is released.

It is usual to formulate pesticides with surface active agents of various sorts for a number of reasons including improving retention on plant or other surface and aiding penetration into the target organism. Also, because many pesticides are of relatively low water solubility, surfactants are used to stabilise emulsions or wettable powders so that the product can be sprayed using water as the carrier. There has therefore been interest from time to time on the effect of additives included in formulations on adsorption and leaching of the active pesticide ingredient. By and large the results have been inconsistent (CALVET, 1980) and the effects relatively small and there certainly seems to have been no major attempt to influence leaching behaviour by the use of surface active agents.

There are, however, other formulations which attempt to limit the rate at which the active ingredient is released. The initial incentive to devise such formulations was to extend the period of activity of substances which are of low persistence because of their volatility or rapid degradation. This in itself could limit the risk of ground water contamination by increasing the range of uses of otherwise transient compounds. However, controlled release formulations could in principle be used to reduce the amount of a relatively persistent compound that is required (WILKINS, 1983).

One difficulty to be overcome is that the simpler, cheaper methods using matrices which deliver the active ingredient by diffusion or erosion have release rates that are a function of the square root of time, whereas in order to reduce the quantity of pesticide needed for long term control a constant rate of release is required. Microcapsules which release their contents by osmotic rupture have the potential to give a nearly constant rate if a mixture is used containing capsules with a range of osmotic potentials so that groups of them rupture sequentially but so far most microcapsule systems in commercial use have release rates that are proportional to the square root of time.

Research into controlled release formulations has been pursued optimistically for several years and patents abound but commercial success seems largely to have been restricted to the more toxic (to mammals) insecticides, insect pheromones and aquatic molluscicides. A successful formulation designed specifically to reduce the risk of ground water pollution has yet to emerge.

THE WAY THE PESTICIDE IS USED

Although the opportunities to reduce the possible hazards to ground water from pesticides by soil management and by changing the formulation are limited there is considerable scope to improve the situation by paying attention to application rates and timing.

Application rate

Intuitively it would be expected that the quantity of pesticide (if any) reaching ground water would be closely related to the rate of application. For example, bromacil applied to citrus at 8kg/ha would surely be more likely to appear in groundwater at than metsulfuron applied to wheat at 4g/ha. Of more practical interest is whether reducing the application rate of, for example atrazine, from 2kg/ha to 1kg/ha will halve the chances of finding a measurable quantity in groundwater.

Direct measurements on this point are difficult to find. WEHTJE et al (1984) reported that the amount of atrazine reaching a sampling point 0.2m deep was directly proportional to application rate. Otherwise there is only the indirect evidence that estimates (some of them rather subjective) of the fraction of an application rate leaching below sampling depth summarised by BOESTEN (1987) are low and show no obvious differences that could be ascribed to application rate. From a theoretical viewpoint, the currently available modelling procedures whether based on mechanistic or stochastic principles do not take account of application rate (RAO et al, 1988) and it is certainly difficult to conceive how the low quantities (in relation to the size of the soil matrix) of pesticides that constitute a normal application could overload the physical processes in the soil. Thus it seems reasonable to assume that the quantity of a pesticide reaching a particular depth in the soil or sub-soil will be closely related to the quantity applied to the surface.

There are, however, grounds for believing that the proportion of an application which reaches groundwater may be smaller as dose rate is decreased. As mentioned earlier when discussing the effects of cutivation HURLE & WALKER (1980) quoted a number of laboratory and some field studies in which rates of loss decreased as initial concentration increased. The evidence is not totally convincing, particularly that from the field, but where differences in rate have been observed it is the lower initial concentration that shows the fastest rate of dissipation.

Timing of application

As mentioned earlier, the rate of breakdown of a pesticide varies with soil moisture and temperature but there is little that can be done to control them. However, because both vary with the season the time of the year at which a pesticide is applied affects its persistence and hence the quantity that may reach groundwater. WALKER (1989) calculated degradation rates for a hypothetical pesticide at different temperatures and soil moisture contents. Fig 1 shows a family of curves from which it is clear that (under U K conditions) application in the period from June

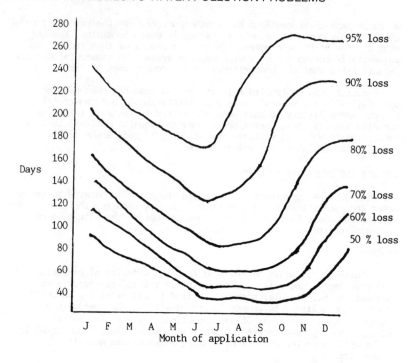

Fig. 1 Calculated times to loss of different proportions of an applied pesticide with a half life of 30 days at 20°C under UK conditions (WALKER, 1989)

to September leaves the lowest residues because degradation rates are then at their most rapid.

A second consideration is the seasonal distribution of rainfall which in most temperate regions is lower in the summer than in spring and autumn. Also during the main period of crop growth evapo-transpiration usually exceeds precipitation so the net water movement is upwards.

Therefore, combining these two factors so that pesticide application is restricted to the period of most rapid breakdown and lowest leaching, the likelihood of groundwater contamination is at its minimum.

DISCUSSION

The conclusion from these considerations is that the possibility of pesticides reaching groundwater can be reduced by lowering rates of application, restricting the period during which they may be applied, refraining from using them in vulnerable areas and (possibly) by the use of controlled release formulations. Many of these limitatioms were applied to the use of aldicarb in Wisconsin and preliminary indications were that aldicarb levels in groundwater are declining (HARKIN et al, 1984). A similar strategy is now in action for atrazine in Switzerland and The Federal Republic of Germany but results are not yet available.

There are of course cost penalties to pay for reducing the likelihood of groundwater contamination. Restricting the area in which pesticides may be used may make some farms unprofitable so they will be abandoned. Because weeds can be fire hazards and hamper industrial operations in mant ways, the industrial user could resort totally untested methods simply because they fall outside the scope of pesticide legislation. Reducing application rates and limiting application periods is likely to reduce reliability of pest control and create problems for the growers even if they do not totally destroy profitability.

The balance of the cost of reduced agricultural productivity against the perceived benefits of reduced water pollution will vary from country to country and will reflect political, ecological and economic pressures. Until recently the main consideration has been efficiency in terms of agricultural production or industrial site maintenance and current recommendations for use are based on this. If the ground rules change then pesticide use will adapt to them but there will be a quantifiable cost.

BIBLIOGRAPHY

ALLER, L.; BENNETT, T; LEHR, J.H.; PETTY R.J. - "A Standardised System for Evaluating Groundwater Pollution Potential Using Hydrogeologic Settings". US Environmental Protection Agency, Washington, D.C., 1985

BOESTEN, J.J.T.I. - "Leaching of Herbicides to Ground Water: a Review of Important Factors and of Available Measurements", Proceedings British Crop Protection Conference - Weeds 1987, pp 559-568

CALVET, R. - "Adsorption-Desorption Phenomena", in Interactions Between Herbicides and the Soil, edited by R J Hance, Oxford (UK), Academic Press, 1980, pp.1-30

CROLL, B.T. - "The Effects of the Agricultural Use of Pesticides on Fresh Water", in The Effects of Land Use on Fresh Waters, edited by J.F. de L.G. Solbe, Ellis Horwood, 1985, pp 201-209.

DOYLE, R.C.; KAUFMAN, D.D.; BURT, G.W. - "Effect of Dairy Manure and Sewage Sludge on ^{14}C-Pesticide Degradation in Soil", Journal of Agricultural and Food Chemistry, **26**, 1978 pp 987-989.

GREAVES, M.P. ; MALKOMES, H.P. - "Effects on Soil Microflora", in Interactions Between Herbicides and the Soil, edited by R J Hance, Oxford (UK), Academic Press, 1980, pp 223-253

HANCE, R.J.; COTTERILL, E.G. - "Relationships between Soil Cultivation and Pesticide Performance', in Soils and Crop Protection Chemicals, edited by R J Hance, Oxford (UK), British Crop Protection Council, Monograph No 27, 1984, pp 65-74.

HANCE, R.J.; GUTH , J.A. - "Constraints on the Use of Models to Predict the Movement of Pesticides to Ground Water", Proceedings of Watershed '89 - The Future of Water Quality in Europe Pergamon, 1989, in press

HARKIN, J.M.; JONES F.A.; FATHULLA, R.; DZANTOR, E.K.; O'NEILL, E.J.; KROLL, D.G.; CHESTERS, G. - "Pesticides in Groundwater Beneath the Central Sand Plain of Wisconsin", Technical Report University of Wisconsin Water Resources Center No 84-01, 1984.

HURLE, K; WALKER, A. - "Persistence and its Prediction", in Interactions Between Herbicides and the Soil, edited by R J Hance, Oxford (UK), Academic Press, 1980, pp 83-122

LEONARD, R.A. - "Herbicides in Surface Waters", in Environmental Chemistry of Herbicides, Vol 1, edited by R. Grover, Regina (Canada), CRC Press, 1988, pp 45-87

MUDD, P.J.; HANCE,R.J.; WRIGHT S.J.L. - "The Persistence and Metabolism of Isoproturon in Soil", Weed Research, **23**, 1983, pp 239-246

NASH, R.G. - "Dissipation from Soil", in Environmental Chemistry of Herbicides, Vol 1, edited by R. Grover, Regina (Canada), CRC Press, 1988, pp 131-169

PIONKE, H.B.; GLOTFELTY, D.E.; LUCAS, A.D.; URBAN, J.B. - "Pesticide Contamination of Groundwaters in the Mahantago Creek Watershed, Pennsylvania, USA", Journal of Environmental Quality,17, 1988, pp 76-84

RAO, P.S.C.; JESSUP, R.E.; DAVIDSON, J.M.. - "Mass Flow and Dispersion", in Environmental Chemistry of Herbicides, edited by R Grover, Regina (Canada) CRC Press, 1988 pp 21- 43

SEIBERT, K.; FUHR, F.; CHENG, H.H. - "Experiments on the Degradation of Atrazine in the Maize Rhizosphere", Proceedings EWRS Symposium, Theory and Practice of the Use of Soil Applied Herbicides, Versailles, France, 1981, pp 137-146

WALKER, A. - "Factors Influencing the Variability in Pesticide Persistence in Soils", Aspects of Applied Biology 21, Comparing Laboratory and Field Pesticide Performance, 1989, pp 159-172

WEHTJE, G.; MIELKE, L.N.; LEAVITT, J.R.C.; SCHEPERS, J.S. - "Leaching of Atrazine in the Root Zone of an Alluvial Soil in Nebraska USA", Journal of Environmental Quality 13, 1984, 507-513

WILKINS. R.M. - "Controlled Release - Present and Future", Proceedings 10th International Plant Protection Congress, 1983, pp 554-559

Chapter 31

PLANNED STUDIES OF HERBICIDES IN GROUND AND SURFACE WATER IN THE MID-CONTINENTAL UNITED STATES

M R Burkart (US Geological, Survey, Iowa, USA), S A Ragone (US Geological Survey, Virginia, USA)

ABSTRACT

Studies have been planned to investigate the effects of natural and human factors on the occurrence of herbicides, such as atrazine, in ground and surface water in the mid-continental United States. To answer the question: what happens to a herbicide after its application? procedures have been developed to integrate information related to atrazine--a representative herbicide having widespread use in this agricultural region. This integration involves: research on natural processes; spatial data managed and analyzed in a geographic information system; and validation of digital models and analytical methods. Studies will range in scale from laboratory and field plots, to regional-scale investigations. The integration of information will enable identification of major processes affecting the transport and fate of atrazine and an understanding of the effects that spatial variability of environmental and anthropogenic factors have on atrazine in the hydrologic system. Processes to be studied include physical, chemical, and biological actions, such as chemical decomposition, that can affect the transformation, transport, and storage of atrazine. Factors, such as soil pH, are variables that can affect which processes are dominant. Processes are independent of scale, and factors can be scale-dependent. A research matrix has been developed that uses a mass-balance concept to account for the distribution of atrazine in the environment and to identify areas of needed research. A geographic information system that will permit analysis of spatial variability of important natural and human factors, such as atrazine application rates and depth to water, has been developed. The planned studies include the use of statistical methods to relate regional patterns of the occurrence of atrazine to a limited number of factors. Completion of the planned studies may take 5 to 10 years or more depending on the availability of information and the need for additional studies. The approach involves an iterative process that uses information from detailed studies, such as laboratory and field-plot research of natural processes, to direct the analysis of the regional distribution of factors. The results of regional analysis will be used to identify areas and parts of the hydrologic system where factors can be measured in more detail and verified. If verification of factors cannot be achieved, additional process research can be initiated in that area.

Key words: herbicides, ground water, surface water, geographic information system

INTRODUCTION

This is an inter-agency study which derives from the mutual interests of The Department of Agriculture, The Environmental Protection Agency and The Geological Survey. Mutual interests are; water quality, and fate and transport of herbicides in the environment. The three agencies bring together expertise in technical fields including agronomy, economics, food policy, water resources sciences and environmental regulation to increase our understanding of an issue of National and global concern.

The use of pesticides in the United States has steadily increased over the last three decades. Estimates of use obtained from statistical sampling of application rates indicate that between 300 and 400 metric tons of active ingredients are applied every year. The major increase in usage has been in herbicides. The eleven States included in this study use about 57 percent of all the pesticides in the Country. Three States, Iowa, Illinois and Indiana consume more that 26 percent of the Nation's pesticides.

The boundaries of the region of interest include counties within which more than 35 percent of the land area was planted in corn or soybeans during 1982, the latest published agricultural census year. The mid-continental United States could also be called the corn belt because more than 70 percent of land in many counties in the region is commonly planted in corn. Soybeans are rotated with corn, but generally on a 1 out of 3 or 5 year basis. Chemicals applied to both these crops include atrazine and other triazine compounds.

In the mid-continental United States, atrazine is the herbicide most frequently found in streams and aquifers. Over the last 10 years, atrazine has been detected at virtually every routinely sampled stream station. More recently, sampling in watersheds with limited atrazine use have yielded some sites where atrazine is not found in the streams. Atrazine is being detected in aquifers at an increasing frequency in the region. Some of this increase is due to improved laboratory detection methods (smaller detection limits) and some increase may be due to sampling locations selected to confirm a suspected presence. Normalizing for multiple detection limits results in an increase in detections with time as well. Many of these samples are from public supply wells.

One of the areas where atrazine is most extensively used coincides with the corn and soybean producing region. Rates of application to fields planted in corn range from 250 to 350 grams per hectare. The total amount of atrazine in this region is sufficiently large to provide the potential for occurrence in many parts of the environment. The application rate and the detection in ground and surface water makes atrazine an excellent compound to use to undertand the fate of pesticides in the environment.

A key question is: where are the herbicides? Some evidence suggests the 2-4 percent of atrazine applied to a basin may be transported by stream flow. Ground-water concentrations of atrazine are generally an order of magnitude less than streams. Does this mean the ground water contains less mass of atrazine than streams? Wide ranges of estimates are available for atrazine stored in the unsaturated zone. Limited sampling of precipitation has detected atrazine; how much can be accounted for in the atmosphere? Transformation products are hypothesized to be a major pathway for dissipation of atrazine, but methods of detection of these compounds have only recently become widely available. Little information is available on bioaccumulation of atrazine or residues of atrazine in crops. These and other questions are of concern to scientists attempting to provide information to decision makers regarding the continued use of atrazine and similar herbicides.

STUDY OBJECTIVES

The overall objective of this study is to develop appropriate methods, knowledge and information to understand how atrazine is distributed in the environment. This plan develops a protocol to; collect data, evaluate methods, and conduct research essential to determine the distribution of atrazine in the environment.

MASS BALANCE APPROACH TO INTEGRATING RESEARCH

A mass balance concept is being used to integrate research related to the diverse natural processes acting to distribute herbicides in the environment. The mass balance equation used to describe this concept states that the amount of herbicide (atrazine) applied to the land can be accounted for by summing the amount which is transformed, stored and transported out of any part of the environment. While this equation is not intended to be quantitatively solved, it does form the basis for considering the fate of herbicides.

The mass balance equation is combined with a compartmentalization of the environment to form a matrix of topics for which research, data and methods may be needed. The environmental compartments represent subdivisions of the hydrologic system because the plan is an attempt to deal specifically with water quality issues.

A critical need for success in meeting the objectives of the initiative is the integration of expertise from many disciplines and governmental and academic institutions. Three general technological categories of work are being organized. These include; research on processes related to hydrology, agronomy and water chemistry; acquisition, management and analysis of spatially distributed data in a GIS; and validation of models and analytical methods. The interactions among scientists working in each of these areas will involve iterations of exchange of information. An example of this exchange includes identification, by process research teams, of critical factors for which data or estimates will be needed. This list of factors will be followed by attempts by the spatial-data team to obtain the needed data or develop estimates from surrogate data which can be included in a GIS. Analysis of the spatial data generated by this process will be used to identify areas where additional research will yield results which can be used over large areas. Process research has already yielded a number of research models for simulating transport, transformation or storage of herbicides. The methods team will use data provided by the GIS to test or validate these research models in a variety of environmental and crop settings.

An example of one of the developing geographic data bases is the distribution of atrazine detections in wells. The frequency of detection of atrazine in wells has been determined for each of the major land resource areas. Land resource areas are classified by characteristics of soil, water, climate, and topography to determine suitability of the land for agriculture and other uses. These areas allow a summary of the distribution of atrazine in ground water related to several environmental characteristics.

The distribution of estimated atrazine use throughout the region is determined in geographic units of county. The application rate and the detection in ground and surface water makes atrazine an excellent compound to use to understand the fate of pesticides in the environment.

A combination of the atrazine detection by land resource areas and estimated atrazine application rates can be used to identify areas where extremes in the ranges of variability exist. For instance, the areas where frequency of detection is small can be located within areas with relatively large atrazine application rates. These will be interesting areas to pursue research on the environmental factors which retard movement of atrazine to aquifers. Also, the areas with relatively large detection rates and small application rates may be isolated to research the processes acting under conditions which appear to make aquifers particularly vulnerable to contamination.

SUMMARY

The mid-continental United States and atrazine, a frequently used corn herbicide, have been selected for studies of the fate and transport of herbicides in ground and surface water. The experience gained from using integrated technology and expertise from major Federal and State institutions needed to accomplish the objectives of this study may provide the prototype for solving questions related to other agricultural chemicals and perhaps other non-point sources of water resource contamination.

Chapter 32

CONTAMINATION OF THE MARESME AQUIFER BY AGRICULTURAL PRACTICES (BARCELONA, SPAIN)

J Guimerà, L Candela (Polytechnic University of Catalonia, Barcelona, Spain)

ABSTRACT.

Intensive agricultural activities on the Maresme region have been carried out for decades. The initial composition of groundwater - calcium bicarbonate type - has been shifted to calcium sulphate type. Sometimes even nitrate is the dominant anion. Excesive fertilizer amounts, jointly with local soil characteristics and hydrogeological features produce a high mineralization of resources by the recirculation of groundwater. The great thickness of the unsatured zone explains the low concentration levels of pesticides found in preliminary surveys in groundwater. General features are commented in this introductory paper.

Keywords: Nitrate, Sulphate, Groundwater Recirculation, Pesticides.

1- INTRODUCTION.

The Maresme coastal aquifer is located north of the city of Barcelona and it extends forming a 150 km^2 strip along the seaside, between the Besós and the Tordera rivers. The granitic Coastal Range constitutes the inner boundary of the aquifer. It is crossed by several recent alluvial aquifers related to small creeks (Fig.1). Further details can be found in COROMINAS and CUSTODIO, (1982).

The aquifer is formed by an alluvial fan complex which covers the weathered granite. Near the seaside the geology becomes more complex due to interlayered marine sediments. The thickness of the aquifer varies from 60 to 70 m. Although locally is a multilayered aquifer, regionally is unconfined.

Groundwater flow is regionally normal to the coast, but well pumpings produce local drawdown depresions that may attain some meters below the sea level. General features of the aquifer after COROMINAS and CUSTODIO (1982), are as follows: The rainfall mean value is 600 mm/year. The transmisivity varies from 20 m^2/day near the coast to 1000-2000 m^2/day in some areas. The mean value is about 200 m^2/day.The specific yield ranges between 0.1 and 0.25.The transit time through the unsatured zone obtained from tritium data varies between 1 and 2.5 year/m of vertical movement, CUSTODIO *et al.* (1985) and this agree with vertical flow calculations. Time of water residence in the aquifer ranges from 8 to 20 years, according to the thickness of the aquifer, the wate table depth and the recharge. The recharge originates by direct rain infiltration around 100 mm/year.

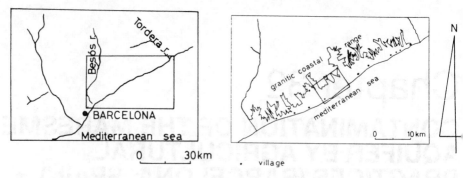

Fig. 1 - General setting. The rectangle shows the specific area under consideration in this paper.

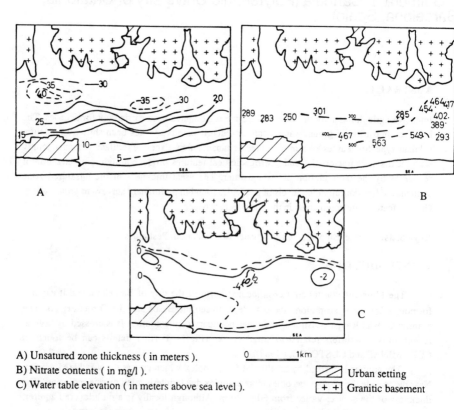

A) Unsatured zone thickness (in meters).
B) Nitrate contents (in mg/l).
C) Water table elevation (in meters above sea level).

Urban setting
Granitic basement

Fig. 2 - Maps of the most contaminated zone under consideration.

The most important use of water is for irrigation. Urban and industrial supplies also exist, although most of the water for those uses are imported from outside the area. Groundwater extraction is made all the year long, but pumpings are intensified in summertime.

Groundwater exploitation is carried out through partially penetrating, old shallow dug wells. Irrigation practices do not use any kind of drainage and the return of water used in agricultural activities produces an important water recirculation in the upper part of the aquifer - 1 to 10 m - variable from one well to another. The recirculation is loading the upper part of the aquifer with leached ions applied on the surface as agricultural fertilizers. Because the mixture of water is rather different from one well to another, specially depending on its penetration below the water table, conspicuous differences in water composition between neighbouring wells appear. According to this, chemical data from groundwater sampled 20 - 30 meters deep in bore-holes, show a lower concentration of nitrate ion than the shallow wells.

2- HYDROCHEMICAL FEATURES

A long term quality monitoring serie is available - from 1970 to 1988 -, but samplings use to refer different wells each year - only 3 among them are extensives all aquifer long -. So, only few local quality evolutions are available.

The initial groundwater composition, calcium bicarbonate type, changes conspicuously during its transit through the aquifer. The spatial distribution along a flow line is shown in a trilinear diagram (Fig.3). The calcium bicarbonate type, near the granitic range changes to sulphate-nitrate type on its downward movement in places where the crops are more intensives, revealing the agricultural origin of the contamination. Near the seaside the water can become sodium chloride type due to seawater intrusion.

 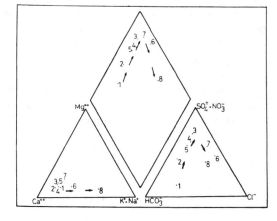

Wells sampled: 1 to 8 from the upper part (1) downward to seaside (8).

Fig. 3 - Trilinear diagram showing spatial evolution of groundwater quality. Concentration in % of meq/l.

In the agriculturally affected area, the major anion is sulphate eventhough nitrate concentrations attain 600 mg/l locally. Their origin is the down leaching of agrochemical products. The regresion analysis of populations of different years does not show large variations of the slope (Fig.4). Since the analysis does not come always from the same wells, these slope changes may be attributed neither to different agricultural practices nor to hydrochemical variations of the aquifer but to sampling spatial variations. Therefore, the effect of agricultural practices on groundwater chemistry changes is quite similar during this period of time - 1970 to 1984 -, and the increase of nitrate is paralleled by a sulphate increase too.

Chloride ion content in groundwater is important. The mean concentration varies between 100 and 1500 mg/l, eventhough 3000 mg/l may be reached locally, VILLARROYA (1986), associated to seawater intrusion and the recirculation of water, CHICOTE and MEDIAVILLA,

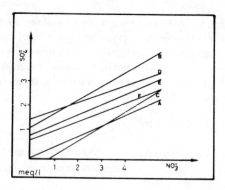

A - 1970 . R = 0.78 ; B - 1978 . R = 0.64 ; C - 1981 . R = 0.83 ;   D - 1983 . R = 0.74
E- 1984 (January). R = 0.89 ; F-1984 (September). R = 0.71

Fig.4 - Linear regresion analysis of $SO_4^=$ upon NO_3^- .

(1982). Since the Cl^- content is rather stationary, it has to be related to a recesion of recent intrusion. The linear regression of Na^+ and K^+ concentration upon Cl^- content does not show high correlation coefficient (R<0.5) and so, the origine of these alkaline ions could be explained as leached products of the agricultural practices. Indeed, in a multiple regression analysis of sulphate and potassium upon nitrate, the multiple correlation coefficient is 0.8. Alkaline relation Na^+/K^+ - in meq/l - ranges between 10 and 20.

Potassium is heavily added to the soil and so, large concentrations in groundwater would be expected. The mean values of K^+ contents are between 4 and 6 mg/l, so the absortion by plants and the retention in the unsatured profile has to be large, to dimish the presence of K^+ ions in the satured zone. Nevertheless, contents of 10 mg/l and even 70 mg/l of potassium ion have been reached locally, VILLARROYA (1986), COROMINAS and LOPEZ (1988). In theses cases, the potassium origin can be a mixture of seawater intrusion and irrigation return flow.

3- CONTAMINATION BY AGRICULTURAL ACTIVITIES

For the last two centuries, irrigated croplands have been the traditional use of land on the Maresme region and since 1940's intensive crops of flowers and vegetables are common - 3 or 4 harvest per year - . The majority of croplands are sandy soil where the silty clay fraction is less than 20% and permeability is around 0.1 m/day. Thus, a quick infiltration is expected. Until five or ten years ago, flooding was the typical irrigation method and the volume of water applied were between 9000 and 15000 m3/ha/year, COROMINAS and CUSTODIO (1982). The irrigation returns of water were expected to be around 30% of the applied doses. Recently, microsprinkler irrigation method have reduced the doses and consequently the excess irrigation flow . Fertilization practices apply up to 1000 kg N/ha/year of several agrochemicals (Table 1).

3.1 Inorganic solutes in groundwater

Presence of sulphate

Concentration of sulphate ion in non agriculturally affected waters varies from 50 to 100 mg/l. Downward flow, groundwater increases the sulphate content making this anion the major constituent in the coastal aquifer. Iron, copper and ammonium sulphates are used in crops and

FERTILIZER	QUANTITY (10^3 kg/year) (in 250 - 300 ha)	MAIN COMPOSITION (%)				NO_3^-	$SO_4^=$
		N	P	K	S	(in kg/year)	
Ammonium sulphate	190	21	-	-	24	173.	67.
Ammonium nitrate	146	17	-	-	-	25.	
"15 - 15 - 15"	125	15	15	15	-	19.	
"12 - 12 - 24"	80	12	12	24	-	10.	
Potassium nitrate	70	13	-	46	-	9.	
Potassium sulphate	70	-	-	45	18		13.
Ammonium nitrosulphate	64	26	-	-	-	73.	11.5
Calcium superphosphate	60	-	13	-	-		
Total NO_3^- , $SO_4^=$						309.	91.5
$SO_4^=/NO_3^-$ relation in weight (kg/kg)						0.29	
Total compounds	36	Total quantity 10^9 kg.					

Table 1. - Major fertilizers (Aproximately 80% of fertilizers used).Data from Public Agriculture Service of Mataró, Vilassar and Malgrat.

soil treatements. First data - 1970 and 1973 analysis - show concentrations of 200 - 300 mg/l of $SO4^=$; some analysis with more than 500 mg/l are related to salt water intrusion processes (relation $SO4^=/Cl^-$ < 0.5 in meq/l). After a period of 15 years, mean aquifer values is 600 mg/l $SO4^=$ .The concentration of sulphate ion is up to double or threefold the chloride content in the central part of the aquifer. The spatial variation of the relation $NO_3^-/SO_4^=$, - in meq/l - in the seaward direction, shows a light deacrease from the central area to the seaside. This phenomena is attributed to sulphate sea water providing, yet the agricultural practices near the sea, are the same as the rest of the aquifer. Thus, as reported in VILLARROYA (1986), a double source of sulphate ions exists: agrochemical leaching and seawater mixture.

Presence of nitrate

First chemical data available show nitrate concentrations between 100 and 300 mg/l. In 1978, values bigger than 500 mg/l were reported. As it is shown in Fig. 5 nitrate concentration in groundwater clearly shows an increasing trend. The monthly variation shows agreement between maximum concentration in dry seasons, and minimum concentration during wet seasons (Fig.6). A direct relation of nitrate concentration with electrical conductivity is not well stablished. Spatial variation of nitrate concentration shows an increase in flow direction. Usually, maximum values appear related to drawdown depressions. In general terms, fertilization practices are not concentrated during a season, so the non-point pollution of the aquifer is rather homogeneous, reaching maximum concentration in areas where the agricultural activities are intensive, and minimum in upper areas of the aquifer and where urban stablishement is concentrated, CUSTODIO (1982).
Table 1 shows that nitrogen is the most plentiful element in inorganic agrochemicals applied in the area. A rough estimation - more precise calculations are not available - shows that 200 tones of N/year are applied over an area of 250 - 300 ha.The oxidizing conditions of the unsatured zone allows that all the nitrogen leached down toward the water table, were oxidized to nitrate.
Two methods are used to estimate the down leaching of nitrate. Firstly, a relation of

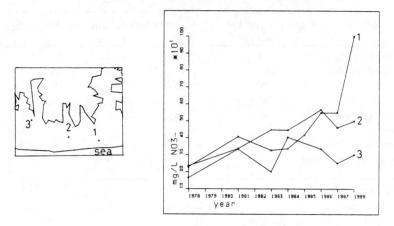

Fig. 5 - Nitrate content evolution in groundwater in the most contaminated zones. Each zone reports more than one sampled well.

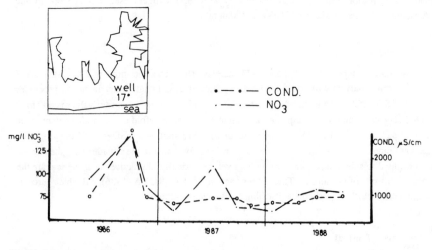

Fig.6 - Monthly variation of nitrate contents and electrical conductivity.(Data from Municipal Laboratory of Mataró. Well 17 (C.A.A.M.S.A.)

$SO_4^=$ /NO_3^- in kg/year of fertilizers applied upon the relation of $SO_4^=$ /NO_3^- in mg/l of sampled water is used. From Table 1, it is found that the total amount of $SO_4^=$ applied to the crops divided by the total quantity of nitrate applied as well, is 0.29. From Fig.4, one can see that the relation of $SO_4^=$ /NO_3^- for 1984 population of groundwater analysis corrected to mg/l is almost one. Since sulphate passes fully to the aquifer, it is found that nitrate loses by down leaching are aproximately 20-30%. This calculus agree with previous works of CUSTODIO, (1982). Secondly, another type of estimation is carried out. A nitrate increase ratio is necesary to make the following calculus. Since the penetration of wells is between 5 and 10 m below the water table, an assumption is done: The increase of concentration is located in the first meters of the aquifer, so the deeper parts are not contaminated by agricultural activities - wich is found in some deep bore-holes -. Thus, the calculation of the nitrate leaching, in kg of nitrate / year, is:

$$L_{NO_3^-} = V * \Delta NO_3^-$$

V = Total porosity (m) * well penetration (b) * surface (S)

ΔNO_3^- = nitrate increase ratio (kg NO_3^-/m^3/year)

Then, as the weighted mean value of nitrate applied is 4500 kg/ha/year, a simply estimation can be done for zone number 1 from fig. 5:

$$L_{NO_3^-} = 30000 \ m^3 * 0.04 \ kg \ NO_3^- / m^3/ \ year = 1200 \ kg \ NO_3^-/year$$

$$(m = 0.3 \ ; b = 10 \ m \ ; S = 1 \ ha \ ; \ \Delta NO_3^- = 0.04 \ kg \ NO_3^- / m^3/ \ year \ )$$

So 1200 kg NO_3^-/ha/year are the loses of the fertilization practices in the zone number 1, wich corresponds to 26% of the dose applied. Nevertheless, this method is conditioned by parameters usually poorly known, and only aproximative estimations can be done.

3.2 Organic compounds in groundwater

A general inventory of pesticides has been done. Over one thousand products are used in order to improve the crops over 300 ha, and the major constituents are grouped as follows: Insecticides 60 types; Fungicides 59; Herbicides 18; Others 32. Although only aproximated data has been obtained, the major active principles are distinguished. The main compounds are shown in table 2 where only those used in quantities greater than 1000 kg or 1000 l are considered. The main compounds are carbamates and thio carbamates, and given its difficulties to pass through the unsatured zone, HOUZIM et al (1986), they are not present in first analysis of groundwater.The organochlorinated compounds are also present, and its high half-life and the quick infiltration explain the low concentrations of the herbicides atrazine and simazine - $5 \cdot 10^{-9}$ g/l - in groundwater. However, these are preliminary results because only one analysis is available.

INSECTICIDES (kg)		FUNGICIDES(kg)		HERBICIDES(kg)		OTHERS(kg)	
Sodium metam	54000	Propineb	2500	Linuron	1000	Methil bro-	
Metomile	2400	Captan	2200	Paraquat	4800	mide	5000
Acephate	1000	Bitertamol	1800	Diquat	1700		
Phenamiphos	1000	Zineb	2000				
Aldicarb	1600	Methil tolclo-					
		phos	1000				
		Mancozeb	1100				

Table 2. - Main plaguicides applied (Total annual quantity per year of active principles used).

4- CONCLUSIONS.

- Since the loses of fertilizers are not large, recirculation of groundwater is the most important process related to the mineralization increase in the upper part of the aquifer. So a groundwater layering is shown. However, a process of slow mixture of water in the vertical direction of the aquifer is present and provide contents around 80 mg/l of NO_3^- in piezometers 20 meters deep.

- Not much data are available about presence of pesticides in groundwater, but first ones reflect a small presence. The thickness of the unsatured zone is usually greater than 10 m; therefore, the concentration levels are dimished because of the sorption processes in the soil, and degradation of compounds. However, primarily results could not be representative of the entire aquifer situation and predictions for the following years can not yet be done.

REFERENCES.

COROMINAS, J.; CUSTODIO,E.- "Contaminación por nitratos e intrusión marina en el acuífero costero del Maresme. Barcelona " in <u>Jornadas sobre análisis y evolución de la contaminación de las aguas subterráneas en España,</u> C.I.H.S./A.I.H. Barcelona, España, 19 - 23 Oct. 1982 pp.537 - 552

COROMINAS,J.; LOPEZ,J.A. - "Interferencias y analogías entre la contaminación por abonos y por intrusión marina en acuíferos costeros del litoral catalán", in <u>Tecnología e intrusión de acuíferos costeros</u> Almuñécar, Granada, España. 15- 20 May 1988 pp. 1 - 10.

CHICOTE,A.A.; MEDIAVILLA,C. - "Estudio de la contaminación agrícola y marina del acuífero costero entre Masnou y Premiá". in <u>Jornadas sobre análisis y evolución de la contaminación de las aguas subterráneas en España,</u> C.I.H.S./A.I.H. Barcelona, España, 19 - 23 Oct. 1982 pp.131-136.

CUSTODIO, E. - "Nitrate buid-up in Catalonia coastal aquifers". in "<u>Impact of Agricultural Activities on Groundwater</u>". International Symposium A.I.H., Prague Czechoslovakia, 1982. pp. 171 - 181

CUSTODIO,E. ; VILLARROYA,M.; IRIBAR,V; PELAEZ,M.D. - "Plan Hidrológico del Pirineo Oriental. Sector del Maresme" C.H.P.O. - M.O.P.U. 1985

HOUZIM,V.; VÁVRA,J.; FUKSA,J.; PEKNY,V.; VRBA,J.; STIBRAL,J. - "Impact of fertilizers and pesticides on groundwater quality" in *Impact of Agricultural Activities on Groundwater.* Vrba, J. ; Romijn, E. (Eds.) International Contributions to Hydrogeology UNESCO - IAH - IUGS 1986, vol.5 ch.5 .

VILLARROYA, M. - " Estudio Hidrogeológico del acuífero costero del Maresme Sur ". Univ. Barcelona. Fac. Geología. Unpublished Doctoral Thesis, 1986.

Chapter 33

EFFECTS ON CROPS OF THE IRRIGATION WITH PRIMARY AND SECONDARY EFFLUENTS

M H F Marecos Do Monte (National Laboratory for Civil Engineering, Lisbon, Portugal)

ABSTRACT

The results of two years of experiments about irrigation of sorghum, maize and sunflower with municipal wastewater treated by primary sedimentation and high-rate biofiltration are described. Control plots were irrigated with potable water and given commercial fertilizers. Identical crop yields were obtained for the three treatments, which leads to the conclusion that the nitrogen content of wastewater can substitute the nitrogen from commercial fertilizers, subsequent fertilizer savings ranging from US$67/ha up to Us$184/ha. No significant crop composition changes were noticed.

Key words: Reuse; wastewater; irrigation; crop; sorghum; maize; sunflower; yield.

1 - INTRODUCTION

Irrigation is the only possibility of taking full advantage of agricultural soils in areas where rainfall is insufficient for all or part of the year. Generally groundwater is not readily available under such circumstances and therefore virtually no crops can be grown during the dry season. This is the situation in pratically all mainland Portugal, as well as many other areas in the world, particularly in developing countries. The use of treated municipal wastewater for crop irrigation can be an important component of integrated water resources management in those areas.

A five year water reuse research project was initiated by LNEC in collaboration with LQARS and other organizations in 1985 in order to assess the associated health effects (pollution and soil and crop contamination) and crop yields and qualities. The objective of this study was to obtain experimental data which could be used for the development of national guidelines for the use of treated municipal wastewater for crop irrigation.

In this poster we present the results obtained during the first two years of the project concerning the chemical composition and contamination of the crops.

2 - EXPERIMENTAL METHODS

Municipal wastewater from Évora (Upper Alentejo), Portugal, after treatment by primary sedimentation and high-rate biofiltration was used to irrigate three different crops - a forage crop (sorghum), a cereal (maize) and an oil-bearing crop (sunflower). These crops were selected for the important agricultural and economical benefits that they can bring to the region.

Three treatments - primary and secondary effluents and potable water with commercial fertilizers - were tested in quadriplicate in 36 plots (Fig.1).

Furrow irrigation was used in order to minimize crop contamination and to prevent cross contamination between adjacent plots.

Prior to the commencement of irrigation, the local climate, soil composition and effluent quality were determined. The control (potable water) plots were fertilized with commercial fertilizers, according to the productive capacity of the soil and its potencial

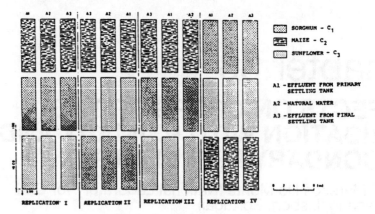

SORGHUM - C_1
MAIZE - C_2
SUNFLOWER - C_3

A1 - EFFLUENT FROM PRIMARY SETTLING TANK
A2 - NATURAL WATER
A3 - EFFLUENT FROM FINAL SETTLING TANK

REPLICATION I   REPLICATION II   REPLICATION III   REPLICATION IV

Fig. 1 - Layout of experiment

fertility. No organic fertilizer (manure) was added so that the contribution made by the waste water to the organic matter content of the soil could be determined.

The plots were irrigated on 14 occasions in 1985 and 11 occasions in 1986 at a total rate of 500 mm/yr.

The sewage effluents were sampled for chemical analysis at every irrigation session. Composite samples were taken at 15 minute intervals during the actual periods of irrigation. Wastewater samples were analysed for salinity (electrical conductivity, total dissolved solids and principal ions, such as calcium, magnesium, sodium, carbonate, bicarbonate, chloride and sulphate), macronutrients (nitrate, ammonia, organic nitrogen, potassium and total phosphorus) boron, pH, As and heavy metals (Cd, Cr, Co, Cu, Fe, Mo, Ni, Pb and Zn).

3 - RESULTS

3.1 - Water quality for irrigation

TABLE 1
Chemical composition of treated wastewaters used for irrigation

NUTRIENTS (mg/l)	PRIMARY EFFLUENT		SECONDARY EFFLUENT		SALINITY	PRIMARY EFFLUENT		SECONDARY EFFLUENT		TRACE ELEMENTS (µg/l)	PRIMARY EFFLUENT		SECONDARY EFFLUENT	
	85	86	85	86		85	86	85	86		85	86	85	86
NO3 - N	20	0.9	1.3	4.4	Ev (m mho/cm)	1.32	1.15	1.36	1.07					
NH3 - N	18.7	20.1	16.0	12.5	TDS (mg/l)	665.5	978.7	645.7	1073.1	As	3.8	7.9	4.4	7.2
Org. - N	28.2	13.5	16.8	9.0						Cd	<2	<2-145	<2	8.1
Total - N	48.9	34.5	34.1	25.9	CATIONS AND ANIONS (mg/l)					Cr	<20		<20	
K	31.4	22.8	32.1	23.9						Co	<30		<30	
Total - P	ND	12.7	ND	12.8						Cu	<10-100	150	<10-100	151
					Ca ++	54.6	33.6	55.6	33.1	Fe	60	377	<20-70	269
ORGANIC MATTER (mg O2/l)					Mg ++	34.5	15.7	34.9	15.9					
					Na +	119.6	83.6	128.9	89.7	Mo	<2-3.1	<2-23	<2-2.7	<2-3.4
BOD5	194.9	156.5	82.1	60.5	CO3 =	ND	ND	ND	ND	Ni	169.6	66.8	190	8.1
COD	401.7	330.5	211.7	243.5	HCO3 -	421.3	372.6	303.5	291.6	Pb	35.5	<20-26.2	24.3	<20-54
					Cl -	155.0	139.2	155.5	142.6	Zn	20	110.9	30	125
SUSPENDED SOLIDS (mg/l)					SO4 =	57.9	ND	71.9	ND					
										MISCELLANEOUS				
					ND - Not determined					CN -	ND	ND	ND	ND
					(*) - The values of 1985 are the mean of 14 composite samples					pH	6.4	7.4	6.6	7.4
Total	89.7	81.8	32.1	50.1	(**) - The values of 1986 are the mean of 11 composite samples					B 3+	1.10	0.61	1.2	0.65
Fixed	18.1	72.8	3.8	43.2						adj R Na	6.2	5.54	5.9	6.1
Volatile	71.6	9.0	28.3	7.0										

TABLE 2
Microbiological quality of treated wastewater

PATHOGEN	PRIMARY EFFLUENT		SECONDARY EFFLUENT	
INDICATOR	1985	1986	1985	1986
FC/100 ml	$2,73 \times 10^6$	$1,34 \times 10^7$	$3,77 \times 10^5$	$3,6 \times 10^4$
FS/100 ml	$4,25 \times 10^6$	$1,24 \times 10^6$	$8,67 \times 10^5$	$6,8 \times 10^4$
Helminths	(a)	ND	ND	ND

(a) Not detected in 2/3 of samples
ND Not detected in any sample

3.2 - Crop yields

Crop yields for 1985 and 1986 are shown in Table 3. The yields of crops irrigated with primary and secondary effluents were very similar to those of the crops irrigated with potable water and treated with commercial fertilizers. Furthermore the observed yields were very close to the theoretically expected yields; the lower yield obtained with sorghum can be explained by the fact that it was harvested only twice instead of the usual four times (because it was sown late in the season), and the 1986 sunflower yield was artificially low as a most of the crop was eaten by birds. On the other hand, the increased maize yield observed in 1986 may be due to the effect of mineralization of the remaining organic nitrogen in the sewage effluent applied during the previous year.

TABLE 3
Crop yield

	Crop yield** (t/ha)					
	Sorghum		Maize		Sunflower	
	1985	1986	1985	1986	1985	1986
Primary effluent	8,7	7,1	8,9	11,1	2,2	0,7
Secondary effluent	8,6	7,7	8,6	11,7	2,3	0,6
Potable water	9,1	7,4	8,1	10,7	1,9	0,5
Expected yield** (t/ha)	12.5		8.1		2.0	

* Grain with 15,5% moisture
** On a dry weight basis

The similar crop yields observed with the three treatment indicates that the nitrogen content of the sewage effluents has an equal fertilizing value to that of the commercial fertilizers. This represents a real saving in commercial fertilizers if treated wastewater is used for irrigation, as shown in Table 4.

TABLE 4
Savings in commercial fertilizers
resulting from wastewater irrigation

Savings in commercial fertilizers (US$/ha)	CROP		
	Sorghum	Maize	Sunflower
	184	172	79

3.3 - Crop quality

3.3.1 - Chemical composition

The composition of shorgum, maize and sunflower, was determined from samples composed of 12 plants selected at random from four plots irrigated with each of the three waters. No significant differences were detected in the composition of sorghum, and the slightly lower values found at the second harvesting are considered normal.

Although the composition of maize seems not to be affected by the different irrigation waters, there was a decrease in the concentration of elements in the second year, except for B, Ni and Cd which increased. The tendency for some elements to increase and for others to decrease in concentration requires confirmation in the remaining years of the project. Sunflower irrigated with primary and secondary effluents also has a composition similar to that irrigated with potable water. The general conclusion is that no adverse effect on crop composition results from irrigation with primary and secondary effluents.

3.3.2 - Crop contamination

The results showed that the contamination levels on the crops irrigated with primary and secondary effluents are very close to those found on the same crop irrigated with potable water. The higher values of faecal coliforms found on sunflowers irrigated with wastewater are most probably due to microorganisms carried on insects such as ants.

Chapter 34

FERTILIZATION POWER INCREASE OF SLUDGE

M C Marquez, C Costa, J L Martin, J Catalan (University of Salamanca, Spain)

ABSTRACT
    A method to increase the nutrient percentage in the sludges produced in a waste water treatement system is described. To improved the concentration of nitrogen and potassium in the sludges, these have been treated by a solution of ammonia and potassium nitrats in a stirred tank, and some variables with influence on the process have been studied. The obtained results have indicated an estimable increase in the fertilization power of the sludge.

Key words: wastewater sludges application, fertilizers.

INTRODUCTION
    The legislation requirements about effluent purification have obliged to develop a technology more efficient and cheap. This has solved the water pollution problem but it has produced a second problem very important: the sludge elimination.
    The sludges can be unloaded in tips, incinerated or applied to agriculture. The use of one or other method depends on the location of the waste water treatement system. At present, the sludge application to agriculture is acquiring boom in many european countries. In fact, a 35% of the sludges obtained in Europe are now employed in the agriculture , SCHELTINGA (1987).
    The importance of the sludges as fertilizer consists in their contribution to the humidic materials, however their contribution to the nutrients is very small. Thus, if the percentage of nitrogen and potassium in the sludges is increased, then a fertilizer shall be obtained with better characteristics than fertilizers now in the market, since the same solid would supply the soil an organic and inorganic fertilizer.

EXPERIMENTAL
    The ammonia and potassium nitrate retention capacity has been determined on the sludge originated in an anaerobic digester for 21 days.
    In this work, the sludge was subjected to some standard conditions: heating at 105°C for 24 hours (to eliminate humidity and pathogenic agents), grinding and sieving (to obtain a representative sample).
    The standard sample treatement was a static treatement and it was based on the contact between the solid and the corresponding cation solution (ammonia or potassium) into a stirred tank. The solid weight and the solution volume ratio was 1:10. After a

considerate contact time, the solid was filtered and washed until
absence of nitrates. The amounts of ammonia and potassium contained
in the sludges were analyzed by using Kjeldahl method for the ammonia,
KOLTHOFF et al. (1976), and atomic absorption for the potassium.

Four variables has been studied: agitation speed, solid-liquid
contact time, cation solution concentration and pH.

Obtained results have shown that the sludge retention capacity
depends on the studied variables. The better retention was reached
for 500 r.p.m. of agitation speed, 10 hours of contact time,
5M ammonia nitrate concentration and 3M potassium nitrate
concentration (value depending on salt solubility at room reaction
temperature) and pH between 0 and 7.

CONCLUSIONS

At these conditions the obtained ammonia and potassium
percentages in the treated sludges have been, respectivily, a 42%
and a 123% higher than ammonia and potassium percentages in the
untreated sludge, which shows the obtained sludge improvement
using the previous suggested treatement method.

BIBLIOGRAPHY

KOLTHOFF, I.M.; SANDEL, E.B.; MEEHAN, E.J.; BRUCKENSTEIN,
S.- Análisis Químico Cuantitativo. Buenos Aires (Argentina), Nigar
Ed., 1976, pp. 817-820.
SCHELTINGA, H.M.J.- "Sludge in Agriculture: the European Approach".
Wat. Sci. Tech., 19 (8), 1987, pp. 9-18.

PART VI

Measurement and data

Chapter 35

A NORTH SEA COMPUTATIONAL FRAMEWORK FOR ENVIRONMENTAL AND MANAGEMENT STUDIES: AN APPLICATION FOR EUTROPHICATION AND NUTRIENT CYCLES

C G Glas, T A Nauta (Delft Hydraulics, The Netherlands)

ABSTRACT

The status to date is presented of a set of coupled mathematical modules for the hydrodynamics, transport and chemical/biological processes determining the nutrient cycles in the North Sea. The ongoing development of this framework is part of concrete projects in the field of policy assessment, as well as part of strategic instrumentation within a five year research program in the Netherlands called MANS (Management Analysis North Sea).

Results of dynamical model simulations are presented, using input scenario's for nutrients. The models simulate the dispersion and relevant chemical and biological processes that determine the water quality and primary productivity in the North Sea. Future developments are discussed.

Keywords: North Sea, management, modelling, nutrient cycles, eutrophication

1 - INTRODUCTION

The present levels of nutrients entering the North Sea, through rivers and atmospheric deposition, are well above the natural background level. (RIJKSWATERSTAAT,1986). The anthropogenic fraction in the total inputs of nutrients to the North Sea is presently in the order of 25-30% (Figure 1).

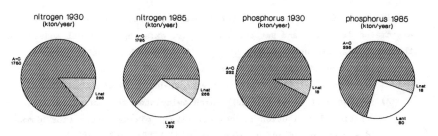

A+C = Atlantic + Channel
Lnat = land sources natural
Lant = land sources anthropogenic

Figure 1: Nutrient inputs (kton/year) into the North Sea in 1930 and 1985

A recent overview of observed effects in the North Sea, which have been attributed tot eutrophication, is given by DUURSMA et al. (1987). They include: an increases in phytoplankton biomass, nuisance blooms and changes in benthic communities. Less clear are the links with fish production or summer oxygen deficiencies in parts of the stratified German Bight.

During the second International Conference on the protection of the North Sea (1987) it was decided to achieve a substantial reduction (in the order of 50%) in inputs of nutrients to vulnerable areas by 1995 (ANONYMOUS, 1988).

Mathematical modelling plays an important role in the analysis of causal relationships between nutrient input strategies and eutrophication phenomena. It is a means of integrating scientific knowledge as well as a tool for the evaluation of managerial policies.

The model results presented in this paper have been achieved by application of the model DYNAMO, which was developed in the framework of two major projects in the Netherlands, namely: the MANS-project "Management Analysis North Sea" (RIJKSWATERSTAAT and SIBAS, 1988) and the scientific preparation of the "Third National Dutch Policy Note on Water Management". The study was commissioned by the Dutch Department of Public Works (RIJKSWATERSTAAT).

2 - METHODS

The model that is referred to as DYNAMO, comprises of:
1) a steady state description of the flow field in the North Sea,
2) a steady state description of the mass transport (water, solutes and solids),
3) a dynamical description of nutrient kinetics (pools and fluxes).

Although this paper deals with nutrient processes only, the computational framework has general applicability for fresh water or marine systems, and for all constituents of which the processes that determine the pools and fluxes can be mathematically expressed.

2.1 - Modelling of flow fields and mass transport

In order to simulate the yearly mean transport of water masses and associated dissolved and particulate matter, existing numerical models for the North Sea were used (POSTMA et al., 1987; DE RUIJTER et al., 1987).
   1) a hydrodynamical model (GENO) which simulates the vertically averaged residual flow field, using south-westerly winds of 4.5 m/s;
   2) a transport model (DELWAQ) which calculates the horizontal concentration distribution of conservatively mixing matter (for a model description and users guide see DELFT HYDRAULICS, 1989a).

With these models, simulations can be made of the dynamic or steady-state concentration distribution of the relevant nutrients and of all other state variables which are described by the biological process module.

Figure 2 gives the model area and inflowing rivers. The area was divided into some 1400 computational elements (16x16 km).

Figure 2:   Model schematization of the North Sea. River inputs are
             indicated. Locations I and II are referred to in section 4.

2.2 - Modelling of nutrient processes

General backgrounds

The model DYNAMO simulates the time and space dependant distribution of the major nutrient pools in the North Sea.

The following pools were considered:

dissolved   : NH4, NO3, ortho-PO4, SiO4 $(gN,P,Si/m^3)$

living particulate: diatoms, and so called 'other phytoplankton'
                     (gC/m^3)

dead particulate : detritus-N,P,Si in the water column $(gN,gP,gSi/m^3)$
                    detritus-N,P,Si in the seabed $(gN,gP,gSi/m^2)$

The following fluxes were considered (in $gN,P,Si,C/m^2$):
The numbers refer to figures 3 and 4.

- uptake of nutrients by phytoplankton   (1,2)
- respiration by phytoplankton
- sedimentation of diatoms   (10)
- mortality of phytoplankton   (9a,9b)
- sedimentation of detritus   (11)
- resuspension of detritus   (12)
- mineralization of detritus in the water column and
  in the seabed   (3,4)
- nitrification in the water column and in the seabed   (5,6)
- denitrification in the water column and in the seabed   (7,8)

The mathematics of the nutrient processes in DYNAMO

A full record of all mathematical formulations is given in DELFT HYDRAULICS (1989b). The description here is restricted to the basic elements of the algorithm.

The dynamics of both phytoplankton groups are described by the following classical mass-balance equation:

$$\frac{dA_i}{dt} = (P_i - K_i) A_i \qquad\qquad [1]$$

in which:

 i = phytoplankton species group 1 or 2
 K_i = specific loss rate (1/d)
 A_i = biomass of the phytoplankton group (g/m^3)
 P_i = specific production rate (1/d)

The specific production rate is determined by:

$$P_i = E_i * f_i \text{ (nut)} * f_i \text{ (temp)} * P_{i,max} - R_i \qquad\qquad [2]$$

in which:

 E_i = light efficiency (-)
 f_i (nut)   = nutrient limitation factor (-)
 f_i (temp) = temperature coefficient of growth (-)
 $P_{i,max}$ = maximum specific production rate at 20°C (1/d)
 R_i = specific respiration rate (1/d)

The average efficiency per day is calculated by intergrating the efficiency values over depth and during the whole day assuming a constant daylight intensity. The available light, of which the light efficiency is a depth integrated, function, is determined by Lambert-Beer's law.

The effects of nutrient limitation are expressed in the form of Monod-equations (see for a review MORISSON et al.,1987).

$$f_i \text{(nut)} = \frac{C_s}{C_s + K_s} \qquad\qquad [3]$$

in which:

 C_s = concentration of the limiting nutrient (g/m^3)
 K_s = critical Monod-value (g/m^3)

Phytoplankton respiration an mortality are defined as temperature dependant functions. Presently no explicit formulation is given for grazing by heterotrophic consumers.

Mass balance of anorganic nutrient pools

Figure 3 gives the model cycle of DYNAMO for nitrogen. Details of the formulations are given in DELFT HYDRAULICS (1989b). Figure 4 gives the cycle for phosphorus as it is described in the model. The silicate is identical, with one difference that only diatoms utilize silicate. From the Figures 3 and 4 it can be seen that detritus, and in fact phytoplankton too, are quantities derived from the nutrient mass-balance equations. The stochiometric coefficients which are used for the N,P,Si to C (carbon) conversion have been kept constant throughout the seasons.

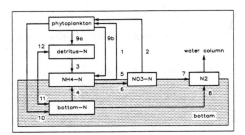

Figure 3: Nitrogen cycle
in the model.
For numbers see
text

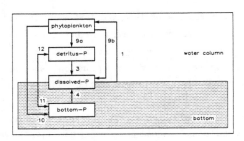

Figure 4: Phosphorus cycle
in the model.
For numbers see
text

3 - INPUT DATA FOR THE MODEL

All model steering has been based on either measured time series (i.e. light irradiance) or by defining 'smoothened' forcing-function (e.g. day length, temperature, resuspension etc.) For details about steering data and forcing functions reference is made to DELFT HYDRAULICS (1989b).

The model calibration and validation was performed by using measured nutrient inputs (1980 and 1985) transiting the Netherlands. For details of the emission definition and model simulation of rivers and lakes, the reader is referred to the contribution of STANS and GROOT (1989) in this volume. Inputs from other sources (UK, FRG etc.) have been provided by RIJKSWATERSTAAT. The following management strategies have been considered:

1) the present (1985) situation;

2) a 50% emission reduction aim of land based human nutrient sources in compliance with the North Sea Action Plan;

3) a 90% emission reduction aim of land based human nutrient sources in order to approximate natural input levels;

4) a 1930 emission level (assumed natural input levels).

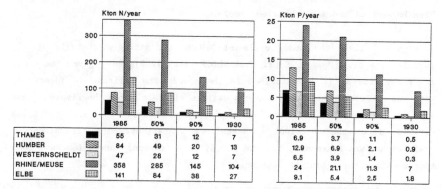

		1985	50%	90%	1930		1985	50%	90%	1930
THAMES	■	55	31	12	7		6.9	3.7	1.1	0.5
HUMBER	▨	84	49	20	13		12.9	6.9	2.1	0.9
WESTERNSCHELDT	☐	47	28	12	7		6.5	3.9	1.4	0.3
RHINE/MEUSE	▨	358	285	145	104		24	21.1	11.3	7
ELBE	▦	141	84	38	27		9.1	5.4	2.5	1.8

Figure 5: Nutrient inputs into the North Sea. For a description of management
strategies see text above and STANS and GROOT (1989) in this volume.

The annually averaged results for the major river inputs are given in
Figure 5.

4 - SIMULATION RESULTS AND DISCUSSION

For the validation of the model, the results of 1980 and 1985 simulations
were compared with available data from biweekly monitoring in Dutch coastal
waters by RIJKSWATERSTAAT. In the framework of a management oriented model,
three parameters here receive attention, namely phytoplankton species
composition, phytoplankton biomass and (potential) production limitation
factors.

The general dynamics of the annual phytoplankton species composition fits
well with field observations for the North Sea presented by BROCKMANN (1985).
In late winter (day 49) the diatom bloom starts in shallow coastal waters and
the Dogger Bank area as can be seen in Figure 6a. Non-diatom species are then
still very low in biomass (Figure 6b). By the end of spring (day 126) this
situation has reversed as can be seen in Figures 7a and 7b. The simulated
succession and biomass levels after the start of the spring bloom depends very
much on the area. Examples are given for one coastal location I (Figure 8a)
and one offshore location II (Figure 8b). For locations see figure 2.

Comparison of these results with field data is hampered by two facts; no
routine monitoring of species composition is available in absolute quantitative figures and secondly, no biomass estimate is available other than that in
units chlorophyll, for which the C/Chl-a ratio has high variability in nature,
and (due to lack of knowledge and consistent data) none in the model.

Using the DYNAMO model, extensive analyses have been made of the effects of
the 50% and 90% nutrient emission reduction strategies in comparison with 1985
and 1930 results. For Dutch coastal waters (location I in Figure 2) the model
simulates a doubling of total phytoplankton biomass during the spring bloom
between 1930 and 1985 (Figure 9). This is consistent with field observations
by CADEE (1986). A 50% reduction of emissions has a less than 50% effect on
the reduction of biomass. This is the joint result of 1) mixing eutrophic

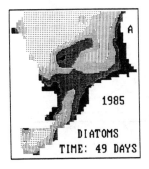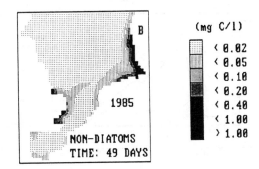

Figure 6a,b: Diatoms and non-diatoms simulated on day 49 for 1985 nutrient
               inputs to the North Sea

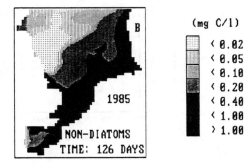

Figure 7a,b: Diatoms and non-diatoms simulated on day 126 for 1985 nutrient
               inputs to the North Sea

riverine inputs with background concentrations of inflowing Atlantic water
through the English Channel, and 2) the fact that phytoplankton in eutrophic
coastal waters is presently mostly energy (light) limited instead of nutrient
limited.

From the calculated concentrations of available dissolved nitrogen and
phosphorus, it seems that a strategy aimed at reducing P-loads with preference
has better impact in reducing blooms than a N-reduction scheme alone.

Taking into account that the seabed is still in equilibrium with the
present high nutrient loads, it will take in the order of at least 3 years before an abatement strategy will show the full intended benefit according to
the model results. It is estimated that in order to reduce the risks of reoc-
curring adverse phytoplankton blooms and associated detrimental water quality
(anoxia) to acceptable levels, a 50% reduction of the calculated spring
biomass should be attained. This can be realized by a combination of a minimum
anthropogenic emission-reduction of 50-75% for N and 75% for P.

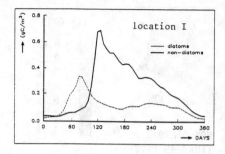 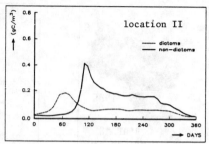

Figure 8a,b: Phytoplankton succession in Dutch coastal waters for 1985
              nutrient inputs. The locations I and II are indicated in fig. 2

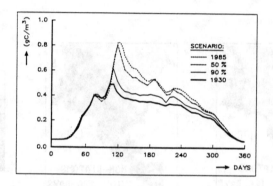

Figure 9   Total phytoplankton biomass in location I (figure 2) for
              different abatement strategies and the natural situation in 1930

5 - FUTURE RESEARCH

Presently, an number of modifications are being made to the model
framework. In 1989 and 1990 much attention will be devoted to incorporating
dynamical features of transport and temperature induced stratification in the
central and northern parts of the North Sea. With respect to the nutrient
cycling processes, the emphasis in 1989 will be on the collection of field
data and physiological data of various phytoplankton species. The aim will be
to develop a general watercolumn model (ECOLUMN) with a high degree of
deterministic formulations. This model will be applicable as an integrated
part of the present model framework, for which the model DYNAMO will provide
the nutrient boundary conditions.

6 - ACKNOWLEDGEMENT

The authors especially wish to thank drs. Olivier Klepper, formerly of
RIJKSWATERSTAAT, for his contribution to the first version of DYNAMO. Many

other people have played a crucial role in the genesis of the present integrated computational framework for the simulation of nutrient cycles in the North Sea. For this, we owe them gratitude. We wish to mention the following contributors in alphabetical order: drs. Janet van Buuren, drs. Louis Dederen, drs. Anton van der Giessen, dr. Liesbeth de Groodt, ir. Henk de Kruik, ir. Arjen Markus, ir. Leo Postma, drs. Ies de Vries. Also we would like to acknowledge the assistance of the people who helped us prepare this manuscript: Renée van der Beek, Rolf van Buren, Frank van Stralen and Engelbert Vennix.

7 - LITERATURE CITED

ANONYMOUS (1988)
Second international conference on the protection of the North Sea.
London, 24-25 November 1987.
Ministerial declaration.
Department of the Environment, London, April 1988, 73 pp.

BROCKMANN, U. (1985)
Bestandsaufnahme von Naehrsalzen in der Nordsee.
Forschungsbericht Bundesministerium fuer Forschung und Technologie, BMFT FB, MFU 0541, 172 pp.

CADEE, G.C., 1986.
Increased phytoplankton primary production in the Marsdiep area (Western Dutch Wadden Sea.
Neth. J. Sea Res. 20: 285-290.

DELFT HYDRAULICS (1989a)
DELWAQ-manual.
Report DELFT HYDRAULICS version 3.0 .

DELFT HYDRAULICS (1989b)
Modelling of nutrient cycles in the North Sea: EQUIPMONS and DYNAMO (in Dutch).
Draft report DELFT HYDRAULICS T234.01/T500.03, February 1989, part I: text, part II tables and figures.

DUURSMA, E.K., BEUKEMA, J.J., CADEE, G.C., LINDEBOOM, H.J. and DE WILDE, P.A.W.J. (1987)
Assessment of Environmental Impact of Nutrients.
International Conference on environmental protection of the North Sea, session two: nutrients, paper 10, 15 p.

MORRISON, K.A., THERIEN, N. and MARCOS, B. 1987.
Comparison of six models for nutrient limitations on phytoplankton growth.
Can. J. Fish. Aquat. Sci.44:1278-1288.

POSTMA, L., DE KOK, J.M., MARKUS, A.A., VAN PAGEE, J.A. (1987)
Long term and seasonal water quality modelling of the North Sea and its coastal waters.
International Council for the Exploration of the Sea, ICES, Hydrography Committee C.M. 1987/C:39.

RUIJTER, W.P.M. DE, POSTMA, L., KOK, J.M. DE (1987)
Transportatlas of the southern North Sea.
Rijkswaterstaat and Delft Hydraulics, The Hague, 1987, 20 pp.

RIJKSWATERSTAAT (1986)
Water Quality Management Plan North Sea (in Dutch)
Waterkwaliteitsplan Noordzee.
Tweede Kamer der Staten-Generaal, Vergaderjaar 1986-1987, 17408 Harmonisatie Noordzeebeleid, Nr. 22, 125 pp.
Achtergronddocument 2b: de ecologie van de Noordzee: analyse.

RIJKSWATERSTAAT and SIBAS (1988)
Management Analysis North Sea - MANS
Report of the preparatory study in 1987 and working plan for the main study
1988-1992. Report by RIJKSWATERSTAAT and SIBAS, February 1988, 61 pp., 9
appendices (in Dutch).
Summary report. Report by RIJKSWATERSTAAT and SIBAS, February 1988, 24 pp.

STANS, J.C. and GROOT, S. (1989)
Integrated approach for water management in the Netherlands: the policy
analysis for the National Water Management Plan 1990.
This volume.

Chapter 36

DATA ACQUISITION OF THE HYDROLOGICAL SURVEY IN DENMARK

O Houmøller (Danish Land Development Service, Denmark)

ABSTRACT

The latest years increase in pollutin of rivers, lakes and coastal waters has lead to a increase in monitoring activities in Denmark.

The work done by the local authorities in monitoring waterquantity and waterquality is being coordinated by the ministry of the environment and governmental surveys like the hydrological survey.

As of today there are approximately 450 gauging stations in operation in rivers and lakes

Dataprocessing is largly carried out by the hydrological survey, on computersystems developed by the hydrological survey, as consulting to the local counties.

1 - GENERAL

In Denmark the first environmental legislation was passed in 1974, necessitated by the increasing pollution of our surrounding environment, and expressing an increasing interest from the population on these issues.

In the 1970's the attention was paid esspecially to the use of waterressources and major single waste water discharges.

The public and private watersupply is based almost only on groundwaterressources, as 99% of water for household purposes is supplied from subsurface reservoirs and the last 1% is supplied from lakes. Also the industry is mainly supplied with groundwater.

Water for agricultural needs (irrigation etc.) was up to the 70's supplied from rivers and lakes, however a redistribution is ongoing towards groundwater use.

The environmetal consequences was obvious in the streams where the discharge was drastically reduced in the summer period, resulting in deterioration of the fauna and flora in the streams.

Thus the environrnmental surveillance to a considerable extent, focused on the quantity of water ressources rather than quality.

In the 1980's the interest is centred on the transport of nutrients to streams, lakes, groundwater and open waters. Approximately 50% of waste water from industry and households is carried directly to inlets and coastal waters, and 50% is discharged into streams and lakes.

The use of fertilizer in agriculture is as well suspected of polluting the environment with nutrients. In the last decade, increasing problems with nitrogen and phosporus in our lakes and inner waters, has been recorded, as well as the contents of nitrogen in subsurface water is alarming high. This has been followed by extended activities such as water quality surveillance and monitoring of polluting elements in the environment.

In 1987 the Danish Parliament passed a long-term environmental programme to improve the serious environmental conditions, throug strong measures against industry, agriculture and households; the objective is to reduce the admission of nitrogen and phoshorns with 50% and 80% respectively. To achieve these ambitious aims an investment of 12 billions D.kr. is planned.

Parallel to this, a large surveillance programme is initiated by the Government, in order to follow tendencies of development in the environmental condition, and to substantiate the effect of the activities, outlined in the former mentioned environmental programme.

2 - ORGANISATION OF SURVEILLANCE

The superior responsibility for the environmental surveillance lies with the ministry of Environment.

The monitoring activities in streams, lakes, groundwater and coastal waters lies with the local authorities (county authority); in addition, monitoring in open waters, groundwater and in the atmosphere is accoumplished by various State-institutions.

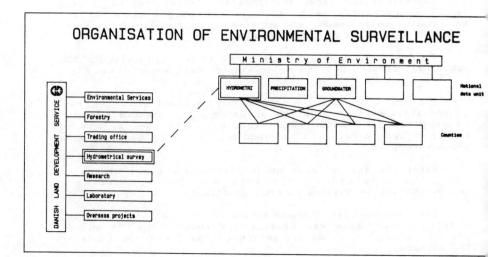

ORGANISATION OF ENVIRONMENTAL SURVEILLANCE

To ensure a properly and proffesional level in the surveillance programme, and homogenity in the choice of methods, the Ministry of Environment has appointed a number of so-called National Data Units. A National Data Unit is an institution having special expertise within a certain field of the surveillance programme, and is as such imposed to carry out development of methods and publishing of guidance manuals, and to hold advisory capacity.

A number of National Data Units has been appointed, covering various technical fields e.g.:

Groundwater: Danish Geological Survey

Hydrometri: Hydrometrical Survey

Freshwater: Danish Environmental Survey

Precipitation: Danish Meteorological Institut

All Data Units (all state-institutions except one) accomplish their tasks as a part of general research work and contultancy projects.

3 - THE HYDROMETRICAL SURVEY

The Hydrometrical Survey is a department of Danish Land Development Service, a private organisation working as consultant to public and private clients.
The hydrometrical Survey was founded in 1917 as a service to other departments in the Danish Land Development Service, which at that time was involved in some major drainage projects. The main task was to supply discharge data from danish rivers for design purposes.

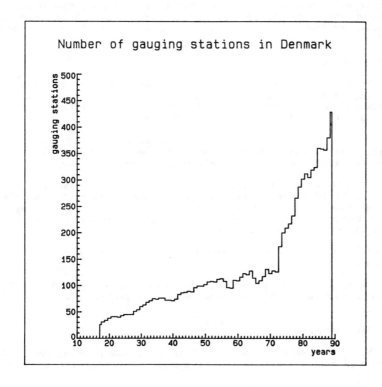

Since then the need for discharge data has increased and spread, and today the Hydrometrical Survey supply discharge data for environmental programmes, water ressource planning and water quality planning. Presently the Hydrometrical Survey runs a network of approximately 400 gauging stations, most of which an financed by local authorities and a minor part by the Government.

Besides working with datamonitoring, the Hydrometrical Survey acts as consultants for the State and county authorities on projects within hydrology, water ressources and transportation of dissolved solids.

In 1985 the Hydrometrical Survey was appointed "National Data Unit on Hydrometrical Data". As Data Unit the Hydrological Survey has assumed some responsibilities:

-   Running a National Hydrometrical network to supply general information on variation in discharge in danish rivers.

-   Running a National Hydrometrical Database available to public authorities, and to secure that hydrometrical data is stored and made available to future users.

-   Advise on hydrometrical procedures, including planning of network, training of staff in hydrometri and research in hydrometri and hydrologi.

4 - DATA AQUISITION

The collection of data is carried out by the Hydrometrical Survey mainly as consultants for the county authorities. The authorities often do a part of the field work at the gauging stations and with collection of watersamples, however data manipulation and calculations are centred at the Hydrometrical Survey.

4.1 - Field work

At 300 out of the 400 gauging Stations the continous data recording is done mechanically, while the rest of the Stations are equiped with eletronic dataloggers, some with the ability of being serviced from the office, via telephone.

In order to calibrate the relation between water level and disharge, 10-12 disharge measurements are made every year, the number necessitated by the growth of weed.

For the water quality surveillance, water samples are collected 18-32 times a year at approximately 200 stations. The water samples are analysed for a number of chemical elements, the concentration of wich forms an important part of nutrienttransport calculations.

Also the biological condition in streams and lakes is assessed.

The Groundwater quality is monitored in several hundreds of drillings, where water samples are collected 2-4 times af year.

4.2 - Data Storage

The recorded data is stored in a database developed and maintained by the Hydrometrical Survey. Waterlevel graphs on

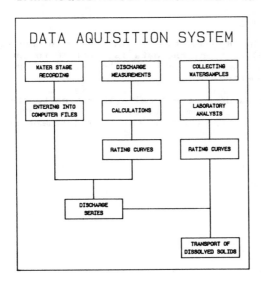

paper are digitized manually, and data from electronic dataloggers is transfered directly into the computer system via telephone or transfer unit.

When data has been stored in the National Data Unit's database, it is available for the local authorities either through telephone/terminal access or as plots and tables on paper.

4.3 - Data processing

There are certain corrections which must be applied to the recorded water levels before they are stored in the computer database. These include adjustments for incorrect time, errors in setting and horizontal and vertical corrections for paper shrinkage.

The first step in the calculation of disharge data, is to establish a characteristic water level-discharge relation (rating curve), based on a study of the discharge measurements. In this phase it is often profitable to use a hydraulic model to generate artificial rating curves, on the basis of measured cross sections and emperical hydraulic parametres.

Discharge data is now computed as the water levels is transformed through the rating curve.

When the rating curve is changing with time due to growth of vegetation, adjustments are made by applying corrected rating curves before entering the water levels. These corrections are determined from the discharge measurements.

In addition to the hydrometrical data processing computer programmes, a number of applications is developed at the Hydrometrical Survey, among others a software package to compute the transport of nutrients in rivers, and to identify the sources by character and by location. In the calculations, various models are applied, dependent on the type of relation between the concentration of dissolved solids in consideration and the runoff.

4.4 - Reporting

When the Hydrometrical Survey has accomplished the mentioned calculation, the data is reported to the authorities, usually as annual publications with tables, figures and maps, properly commented.

It is in the hands of the local authorities to use the supplied informations in the administration, and to report to a national collocation of which the Ministry of Environment is responsible.

Chapter 37

HYDRIDE GENERATION ATOMIC SPECTROMETRY IN THE ENVIRONMENT WATERS

C Baluja-Santos, A Gonzalez-Portal (University of Santiago de Compostela, Spain)

ABSTRACT

The aim of present review is to describe the hydride generation technique, to determine on the basis of ratio nal criteria, which of the various spectrometric methods (AAS, ICP, etc) is best fitted for use with hydride ge neration systems, and to compare with other techniques used in the water analysis. It is noteworthy that arsenic and selenium, the most sought-after of the elements inves tigated in environmental studies.

Key words: Hydride generation, atomic spectrometry, water analysis, arsenic, selenium.

1 - INTRODUCTION

The popularity of atomic absorption spectroscopy (AAS) for environ mental analysis and in particular for the analysis of water, is reflected by the large amount of significant research in this field in annual re- views Annual Report on Analytical Atomic Spectroscopy(1971-1984).

The hydride generation system for AAS was developed to meet the need for sensitive specific methods for the determination of the elements As, Sb,Bi,Ge,Pb,Se,Sn and Te which possess spectral resonance lines in the far ultraviolet region (below 200 nm). In 1967 the introduction the argon-hy- drogen flame for the determination of arsenic by AAS reduced absorption by flame to 15% a considerable improvement on the 62% absorption by air-acety lene flame. However, on applying the new method to the determination of trace quantities of As in real samples, considerable interference was encountered.It was to improve the signal-to-noise ratio of the old method without introducing interference problems that HOLAK(1969) conver ted the As of the samples into arsine, which be colected in U-tube immer sed in liquid nitrogen before using a strean of nitrogen to carry it to an air-acetylene flame.

HOLAK did not establish the limits of the new technique. The first to do so were DALTON and MALANOSKI(1971), who achieved a detection limit of o.1 μg for As on aspirating arsine into an argon-hydrogen flame directly (with no collector) and CHU et al.(1972) subsequently improved the sensi tivity for As to about 5 ng(theyfail to mention their detection limit) on remplacing flame by an electrically heated atoization tube with strean of argon as carrier. In 1973 the scope of the hydride generation -AAS me thod was broadened to the determination of Sb,Se and Bi,MANNING(1971) and by the use of sodium borohydride to that of Sn and Te, in all these cases an argon-hydrogen flame was employed. Finally, included Pb whithin the range of susceptible elements by using a spectral kind of quartz tube over an air-acetylene flame THOMPSON and THOMERSON(1974).

TABLE I. Trends in the development of atomic spectroscopy
techniques for waters analysis

	Paper on waters				
	1971–84			1984–89	
		G C			G C
ABSORPTION					
AAS	22			38	
ET-AAS	66			32	
HG	127	10		50	9
EMISSION					
AES	2			1	
d.c.P/arc,spark	6			2	
ICP / MIP	12			5	
HG	18	4		19	2

G C: Gas Chromatography,   AAS: Atomic Absorption Spectrometry
ET-AAS: Electrothermal atomic absorption Spectrometry
HG: Hydride Generation;   d.c.P: Direct Current Plasma
ICP: Inductively Coupled Plasma; MIP:Microwave Induced Plasma
AES: Atomic Emission Spectrometry

The practical analytical potencial of hydride generation systems for
AAS (HG-AAS) is apparent in previously published reviews, IHNAT and MILLER
(1977), ROBBINS and CARUSO(1979), GODDEN and THOMERSON(1980), IHNAT and
THOMPSON(1980), VERLINDEN et al.(1981), CAMPBELL(1984), BALUJA-SANTOS and
GONZALEZ-PORTAL(1986), QIN(1987); systematic studies of interferences on
atomization mechanisms, SMITH(1975), PIERCE and BROWN(1976), WELZ and
SCHUBERT-JACOBS(1986), WELZ(1986) and papers using the radio-tracer ^{75}Se in
the essential steps of the method as applied to environmental material (i.e.
waters) WELZ and MELCHER(1984), KRIVAN et al.(1985), JARON and JIJALKOWSKI
(1985), kinetic stability of gaseous hydrides (i.e. As, Sb,etc) FUJITA and
TAKADA(1986),optimization of the determination of Se for an ET-AAS heated
closed atom-cell method and a flame-heated opencell technique VERLINDEN
et al.(1980) or utilisant the chemometrics an optimization the factors governing the operation of the plasma and hydride generation PARKER et al.
(1985). TABLE I, summarise the development of atomic spectroscopy techniques for waters analysis, however lead with AAS and ET-AAS has been exclude.

A comprehensive review is given of the determination of As and Se by
the various atomic spectrometry methods and different type of waters (i.e.
natural waters, river, potable, drinking waters etc) is presented.

2 - FUNDAMENTALS

Covalent hydride are volatile compounds in which hydrogen is bound to
elements of Groups IIIb-VIb, the number of whose valency electrons equals
or exceeds that of their orbitals. So far, it has only been possible to
determine 8 of these elements by means of their hydride: As,Sb,Bi,Ge,Pb,Se,
Sn and Te. The spectrophotometric technique employed has generally been AAS
in any case the measured absorption or emission is proportional to the number of elemental atoms per unit cross-sectional area of optical axis at the
moment in which the determination is effected. The complete system may be
represented as a five-stage process, fig. 1.

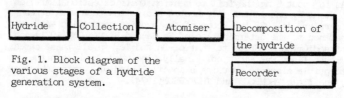

Fig. 1. Block diagram of the
various stages of a hydride
generation system.

3 - LEVELS OF ARSENIC AND SELENIUM IN WATERS

According to BOWEN'S(1966) classification, As and Se are elements of unknown biological function that are regarded as toxic because they accumulate in the human organism. In fact, many of numerous elements found in very poor concentration in living beings have toxic effects at concentrations that are slightly higher than normal.

The arsenic levels in rivers and lakes vary, but generally   10μg/l APHA et al.(1976) of As(III) may represent, i.e. 8 % of the total arsenic FLOWLER et al.(1979). The concentration in ground-water depend on the content of the bedrock and may be as high as 1.8 to 3.4 mg As/l (BENNETT(1981). In conditions of low aeration, As(III) may represent 25 to 50 % of total As in ground-waters FLOWLER et al.(1979). Endemic chronic arsenism has been observed in some areas with naturally high levels of As in drinking water (Taiwan, Chile, Argentina etc). Bottled mineral water in some cases may contain 20 times more As than ordinary drinking water. Soil and surface water may be polluted wit As via atmospheric sedimentation, from agricultural or industrial wastes and ashes, and from fertilizersand detergents manufactured from As-containing phosphate rock FISHBEIN(1976).

TABLE II . Natural waters quality criteria

	mg / l			
	As		Se	
	(*)	(**)	(*)	(**)
Argentinien (CATALAN,1981)	--	0.1	--	0
Canada (Degrémont, 1979)	0.01	0.5	--	0.1
Spain (1982)	--	0.05	--	0.02
Sweden (Degrémont, 1979)	--	--	--	--
Switzerland(Degrémont,1979)	--	--	--	--
URSS (CATALAN, 1981)	--	0.05[a]	--	0.001
USA (1980)	--	0.05	--	0.01
EEC (1980)	--	0.05	--	0.01
FAO (1973)[b]	--	0.2	--	0.05
FAO (1973)	0.1	2.0	0.02	0.02

(*) Guide levels   (**) Maximum Admisible Concentration
(a) As(III) and As(V)
(b) Cattle drinking water   (c) Sprinkler irrigation

The selenium levels in ground and surface waters range from 0.1 to 400 μg/l, and even more, depending on geological characteristics of area, water, pH, iron content etc. Drinking water usualy contains   10 μg Se/l, EPHA et al. (1976). TABLE II summarises some natural waters quality criteria.

4 - HYDRIDE GENERATION SYSTEM

Reaction chambers have include large test-tubes, erlenmeyers flasks, wash-bottles or pear-shaped flasks. The hydride evolved has been transferred in two principal ways: either the hydride is conveyed directly into the atomization cell as it is generated, or some form of storage is used before the transfer to the atomizer. In the former case the system acts as a continuous--flow system; in the latter case all of the hydride formed is introduced in a burst into atomizer, and a "transient" sharp signal is observed GOULDEN and BROOKSBANK(1974) called this a "batch operation".

4.1 - Hydride generation

Initially, metal-acid reactions were used to generate hydrides for quantitative analysis, the most widely employed being

$$A^{m+} + 2H^{\cdot} \rightleftharpoons AH_n + H_2 \quad (\text{excess})$$

$$NaBH_4 + 3OH_2 + HCl \rightleftharpoons H_3BO_3 + NaCl + 4H_2 \quad (\text{excess})$$

where A is the element to be determined and "m" may or may not be equal to "n". In order to avoid the problems associated with metal-acid reactions, in 1972 BRAMAN et al.(1972) used the reaction with sodium borohydride and SCHMIDT and ROYER (1973) were the first to report the generation of hydrides by reduction with sodium borohydride. Sodium borohydride was found to be superior to zinc with regard to reduction yields and contamination of the blank. Ever, since, it has been the reductant of choice VERLINDEN et al. (1981) and BALUJA-SANTOS and GONZALEZ-PORTAL (1986).

4.2 - Collection of the hydride

Most of the early procedures included some form of collection of the generated hydride.

4.2.1 - The Balloon method

The collector described in 1971 FERNANDEZ and MANNING (1971) has since been modified for automatic systems for MANNING (1971). Its main drawback is the frequency with which the balloon must be changed; the usefull life of balloon has been varously estimated as 50 MANNING(1971) or as few a 15-20 CHU et al.(1972).

4.2.2 - Syphon in liquid nitrogen.

This was the first collecting system employed. Various workers WATLING and WATLING(1980) consider that the use of syphon in liquid nitrogen is a specialized technique that is too slow and complex routine work.

4.2.3 - Other collector systems.

250 ml glass POLLOCK and WEST(1973) or plastic SIEMER and HAGEMANN (1975) flasks have been used as an alternative to balloons. Investigations ^{75}Se show that the recovery of H_2Se generated with sodium borohydride is subject to large and variable losses, due mainly to adsorption on the apparatus used, but also to incompleteness of the reaction involved JARON and FIJALKOWSKI(1985).

4.2.4 - Absorben solutions

MADSEN(1971) put forward a method in which the hydride (i.e. arsine) is generated in a 125 ml conical flask and collected in a 10 ml test tube containing 5 ml of a 0.01 M solution of $AgNO_3$. The hydride is absorbe over 15 minutes and the absorbent solution is then aspirated to the arsine flame of an AAS apparatus

4.3 - Atomization techniques

The first application of hydride generation usually involved atomization in an argon-hydrogen (entrained air) flame.The excess of hydrogen generated along with the hydride often perturbe the flame, changing its composition and thus also causing a change in the absorption of the flame,VERLINDEN(1981)

4.3.1 - Tubes heated by flames

This is the most widely used method HOLAK(1969) and THOMPSON and THOMERSON(1974) were the first to employ a special open tube having the adventages of not requiring a collector and of removing virtually all background absorption by the flame. The tube is fixed to the head of the burner.

4.3.2 - Electrically heated tubes

CHU, et al.(1972) used an electrically heated tube as an alternative to the flame. The carrier gas used was argon. The great adventage of these atomizers is the ease with which temperature can be controlled so sa to provide and optimal temperature for each hydride. TABLE III summarises the physico-chemical properties and decomposition of hydrides of As and Se PIERCE et al. (1976), VERLINDEN et al.(1981).

TABLE III. Physical-chemical properties

	Hydrire	Boiling point (ºC)	Thermal decomposition(ºC)	Stability
As	H_3As	-55	950-975	H_2O_2 and HNO_3 destroyer
Se	H_2Se	-42	850-950	

5 - LIGHT SOURCE

The most commonly used light source is the hollow cathode lamp (HCL) but it is not advisable for the determination of As or Se, for which the signal-to-noise ratio in single-beam apparatus is rather small. For these elements FERNANDEZ(1973) has obtained good sensitivities and detection limits using an electrode-less discharge lamp (EDL) in conjuntion with sodium borohydride, a balloon collector and an argon-hydrogen flame (fluorescence atomic spectroscopy and ICP/MIP techniques naturally need no light source.

6 - AUTOMATIC SYSTEMS

Various articles on the design and application of atomatic and semi-automatic systems have been published VIJAN and WOOD(1976). Two groups may be distinguihed, those in which the reagent is under pressure. The formar group are the more highly automated because they involve less manipulation but the latter group have the adventage of allowing very strict control of hydride generation conditions ,BROEKAERT and LEIS(1980), ANDERSON, et al. (1986).

7 - INTERFERENCES

The interferences affecting HG-AAS may be classified in two groups:

7.1 - Factor causing interferences during hydride generation

7.1.1 - Oxidation state

The redox state of the element to be determined has a considerable influence on the absorption signal. The arsenic can exist in hydride in(III) or (V) states. The signal produced by Se(VI), for example, are negligible compared with those of Se(IV). McDANIEL et al.(1976) investigated with radiotracers the reduction efficiency of several hydride-generation methods and found the Zn-HCl systems very inefficient 88 % of the Se present remaining un reacted.

7.1.2 - Concentration of acid

When sodium borohydride is used as hydride generation agent, the absorption signal of As remain constant when acid concentration is increased FERNANDEZ(1973) and is critical in the determination of selenium. Hydrochloric acid (0.4-6M) has usually been the acid of choice, although sulphuric acid is equally suitable VERLINDEN (1980).

7.1.3 - Organic bonds

Organic matter should be totally oxidized and the element to be analysed should be present in a single, appropriate oxidation state.

7.2 - Factors causing interferences during atomization

7.2.1 - Interferences of other hydride forming

VERLINDEN and DEELSTRA (1979) studied the efects of other hydride-forming elements on the determination of selenium:

Bi $\backsim$ As $\backsim$ Te $<$ Ge $<$ Se $<$ Sn

7.2.2 - Interferences from other elements

Almost all papers on hydride generation AAS deal to some extent with posible interferences SMITH (1975) made the first systematic study on the influence of 1 mg of each of 48 cations on the determination of 1 µg of As and 2 µg of Se, sodium borohydride was used as the reductant, with an argon-hydrogen(entrained air) flame CASTILLO et al(1989)study 27 different ions.

7.3 - Elimination of interferences

For the determination of As by HG-AAS using an argon-hydrogen flame, KIRKBRIGHT and TADDIE(1978) have developed a method for the elimination of interferences from Cu,Ni,Pt and Pd by means of thiosemicarbazide, EDTA or 1,10-phenanthroline. The precipitation with lanthanum has been used to remove interference from Cu in the determination of As, Sn, Se and Te, BEDRAD and KERBYSON(1976)

8 - SPECIATION

The determination of the particular form of an element in its natural environment, speciation, is the most fascinating and rapidly developing area of atomic spectroscopic application to have emerged in recent years.
The element that receives considerable attention in speciation is As. URASA and FEREDE (1987) described a method whereby As(III) and As(V) detected by d.c.P/AES after Ion-chromatographic separation. A detailed study and evaluation of As speciation including adventages and drawbacks of the cited technique with a review of recent progess in arsenic hydride methodology was presented in an excellent paper wel documented, arsenate, arsenite, monomethyl-, monoethyl-arsonate and dimethylarsinate have been speciated in tap, lake, river, rain and seawaters using HG with cold trap and volatilised sequentially into a quartz furnace. A study of the interferences from 22 cations on the determination of organic species and total inorganic As species ranged from 3.2 to 5.4 % and the practical detection limits were 2.0-3.8 ng/l, Van CLEUVENBERGEN, et al.(1988).
It is now widely recognosed that the physico-chemical form, or species, of an element markedly affects properties such as toxicity, biological uptake and environmental fate. Arsenic, a potentially toxic element, shows a

TABLE IV. The sensitivities of spectroscopy methods

	Sensitivities (µg / 1)	
	As	Se
ET-AAS	0.3	1.0
AAS	630	230
HG-AAS	0.1	0.01
FAS	100	40
HG-FAS	0.1	0.06
ICP	40	30
HG-ICP	0.8	0.8
MIP[a]	--	3.2
HG-MIP	0.35	1.25
APAN[b]	0.2	5

(a) Sensitivities in aqueous medium

APAN : Atmospheric pressure"active nitrogen" D'SILVA et al, Appl. Spectrosc., 1980, 34, 578.

remarkable graduation in its toxicity. For example, different forms of As vary in toxicity greatly BRANCH, et al.(1989), from arsenite, the most toxic, to arsenobetaina, in the order shown below:

Arsenite $>$ Arsenato $>$ Monomethylarsonic Acid $>$

Dimethylarsinic Acid $>$ Arsenobetaina

In 1977 the non-toxic organo-As compounds arsenobetaina was isolated EDMONDS et al.(1977). This compounds has now been identified in a wide variety of marine biota,LAWRENCE, et al.(1986). Several reviews have already appeared describing the many applications of the techniques.

The speciation of Se(IV) and Se(VI) in water samples was described by PETTINE, et al.(1986) and APTE and HOWARD(1986) using hydride generation quartz furnace AAS, Selenite in driking water is highly toxic, whereas selenate is not,SCHROEDER, et al.(1970)

9 - APPLICATIONS

As usual in AAS the analysis of waters is one of the most straighforwar applications because the sample is already in a liquid form and matrix

TABLE V. Some application for As and Se in waters by atomic spectrometry

Technique	Material examined	Elements determined (References)
ETAAS	Waters	As(IVANOV,et al.,1978)
	Drinking waters	As(Sb)(KWOK,1973)* As,Se'DUJMOVIV,1976) As,Se(Hg)(KOCK,et al.,1976)
	Reference water	As(FANG,et al.,1985)***
d.c.P/ES	waters	As(Sb)(MIYAZAKI, et al. 1977)
HG-AAS	Waters	As,Se(SCHMIDT;ROYER,1973), As,Se(CORBIN; BARNARD,1976) As(YAMAMOTO,et al. 1976) As(FUKAMUCHI,1978) As[x](NAKASHIMA,1978) As,Se(REICHERT;GRUBER, 1978) Se[x](NAKASHIMA, 1979), As,Se(REICHERT; GRUBER,1979), Se(CUTTER,1983),As(HE,1985) Se(QIN, 1987)
	Natural waters	As[x](NAKASHIMA,1979), As(YANAGI, 1980) Se(SINEMUS;MAIER, 1985)
	River water	As,Se(NARASAKI,1985),As,Se(NARASAKI,1988)[xx]
	Surface water	As,Se(PIERCE,et al. 1976)**
	Ground water	As(SUBRAMANIAN, et al.1984)
	Biotic materials	Se(PIWONKA,et al.,1985)
HG-ETAAS	Waters	As(SHAIKH;TALLMAN, 1977)
HG-ICP	Waters	Se(PYEN;FISHMAN,1978),As,Se(Sb,Bi,Te)(THOMPSON,et al.1981)[z] As,Se(Sb)(PYEN; BROWNER,1988)[xxx],As,Se(Sb)(PYEN;BROWNER, 1988)[a]
HG-FAS	Natural waters	As,Se(Hg) (D'ULIVO et al.1985)[b]

(*) Automatic (xx)Semi-automatic   (***) Zeeman Effect
(x) Concentred by flotation   (z) Simultaneous
(xxx) Simultaneous HG and ICP   (a) Simultaneous Flow Injection, ICP
(b) Simultaneous No-dispersive

TABLE VI. Speciation of As and Se in waters by atomic spectrometry

Technique	Material examined	Elements determined (References)
ETAAS[a]	Water	As*(PUTTEMANS,1982),As**(TSALEV,et al.1987)
ETAAS[b]		As*(GRABINSK, 1981)
d.c.P/ES[b]	Aqueous	As*(URASA;FEREDE, 1987)
ICP/ES[c]	Water	As,Se(TALMI;NORWEL,1975)
AAS[d]	Drinking water	As***(BUSHEE, 1984)
	Aqueous	As (BRANCH, et al. 1989)
HG-AAS	Waters	Se(ROBBERECHT; vanGRIECKEN, 1982) As*(YU;LIU,1982) Se[x](YU;LIU, 1982) Se(IV)[xx](LIU,et al.1983), As,Se(Sb,Te)(YU; LIU,1983), As(LAWRENCE et al.,1986)
	Natural waters	As(ANDREAE, 1977), Se(CUTTER, 1978) As(ANDERSON, 1986),Se(APTE;HOWARD,1986)
	Lake waters	As,Se(Sb,Bi,Te)**[**](SINEMUS,et al.,1981)
	Ground waters	As(III)(SUBRAMANIAN, et al.1984)
(b)		Se(RODEN;TALLMAN, 1982)
(c)	Aqueous	As,Se(Sb,Sn) (VIEN;FRY, 1988)
HG-ETAAS[a]	Natural waters	As(SHAIKH,TALLMAN,1978),As(BROVKO, 1987)
	Differents water	As(van CLUEVENBERGEN, 1988)
HG-FAS[e]	River water (Wastewater)	As*,Se[xx] (HAN;WANG, 1985)

(a) Graphite tube   (b) Separation by Ion-Chromatography
(c) Separation by Gas Chromatography   (d) High Performace Liquid Chro-
(*) As(III) and As(V).   matography.
(**) Inorganic species   (***) Anions
(**) Species of valence state   (x) Total element
(xx) Species of Se(IV) and (VI)   (e) No-dispersive

interferences are not usually severe. Higher power of detection can thus be achieved for element arsenic, selenium etc. TABLE IV lists the sensitivi- ties of spectroscopic methods with and without hydride generation system. TABLES V and VI summarizes the main applications for waters.

10 - REFERENBES

AGGETT,J;ASPELL, A.C. "The determination of As(III) and total As by atomic absorption spectrometry". Analyst, 101, 1976, 341.

Annual Report on Analytical Atomic Spectroscopy. Reviewing . The Ro- yal Chemical Society. 1971-1984. London.

ANDERSON,R.K; THOMPSON,M.;CULBARD,E. "Selective reduction of As spe- cies by continuos HG.II. Validation of methods for application to natural waters". Analyst, 111, 1986, 1153.

ANDREAE,M.O. "Determination of arsenic species in natural waters" Anal. Chem., 49, 1977, 820.

APHA, AWWA, WPCF " Standard Methods for the Examination of Water and Watewater " 14th Edition, Washington(EUA),pp 159, 237 and 282, 1976.

APTE,S.C.;HOWARD,A.G. "Determination of dissolved inorganic Se(IV) and Se(VI) in natural water by HG-AAS" J. Anal. At. Spectrom.,1,1986,379.

BALUJA-SANTOS,C.; GONZALEZ-PORTAL,A. "Application de la Spectrophoto-
metrie Atomique a l'Analyse du Vins.Systeme de Generation d'Hydrures"
IVO, 3, 1986, 347-372.

BOWEN, H.J.M. "Trace elements in Biochemistry". Ed. Academic Press Inc
London, 1966.

BRANCH,S.; BANCROFT,K.C.C.;EBDON,L.;O'NEILL,P. "The determination of
As species by coupled High-performance Liquid Chromatography-Atomic
Spectrometry". Anal. Proc., 26, 1989, 73.

BROEKAERT, J.A.C.;LEIS, F. "Application of two different ICP hydride
techniques to the use determination of As" Fresenius Z. Anal. Chem.
300, 1980, 22.

BUSHEE,D.S.;KRULL,I.S.;DEMKO,P.R.;SMITH,S.B."Trace analysis and speciation for As anions by HPLC hydride generation ICP/ES" J. Liq.
Chromatogr.,7, 1984, 861.

CALDWLL,J.S.;LISHK,R.J.;McFARREN,E.F. "Evaluation of a low cost As and
Se determination at ppb levels". J. Am. Wat. Wks. Ass., 65, 1973,731.

CAMPBELL, A.D."Critical evaluation of Analytical methods for the determination of trace elements in various matrixes. Part I. Determination of Se in Biological materials and water". Pure Appl. Chem.,
56,1984, 645.

CASTILLO,J.R.;MIR,J.M;GOMEZ,M.T. "A comparative study by volatile covalent hydride AAS of interferences for As(III)/As(V) and Se(IV) with
direct atomization in an air-acetylene flame and in a silica tube".
Microchem. J., 39, 1989, 213.

CATALAN-LAFUENTE, J. "Química del Agua" 2ª Ed. Editorial Blume, Madrid
1981, pp352

Van CLUEVENBERGEN, R.;VanMOLL,W.E.;ADAMS,F.C. "As speciation in water
by hydride cold trapping-quartz furnace atomic absorption Spectrometry: An evaluation". J. Anal. At. Spectrom., 3, 1988, 169.

CUTTER, G.A. "Species determination of selenium in natural waters"
Anal. Chim. Acta, 98, 1978, 59.

CRESSER,M.S.;EBDON,L.C.;McLEOD,C.W.;BURRIDGE,J.C." Atomic Spectro
metry Update-Environmental Analysis". J. Anal. At. Spectrom., 1,
1986, 1R-28R.

CRESSER, M.S.;EBDON, L.C.;DEAN, J.R. "Atomic Spectrometry Update-
Environmental Analysis" J. Anal. At. Spectrom., 3, 1987, 1R-44R.

Ibid., Idem., J. Anal. At. Spectrom., 4, 1989, 1R-45R.

CHU,R.C;BARRON,G.P.;BAUMGARNER,P.A.W. "Arsenic determination at sub-
microgram levels by AsH3 evolution and flameless atomic absorption
technique". Anal. Chem., 44, 1972, 1476.

DEGREMONT " Manual Técnico del Agua" 4ª Edición. Bilbao. Ed. Degré-
mont, 1979, pp1121-1146.

EBDON,L.C.;CRESSER,M.S;McLEOD,C.W. "Atomic Spectrometry Update-Environmental Analysis". J. Anal. At. Spectrom., 2, 1987, 1R-42R.

EEC,Off. J. Eur. Comm., nº 229, 1980, 11-29.

FAO-UNESCO "Calidad del Agua para la Agricultura" FAO, Roma (Ita-
lia) 1977, pp 71-73.

EPA., "Water Quality Criteria Documents: Availability". Federal Re-
gister., 45, 1980, pp 79325-79339.

FERNANDEZ, F.J. "Atomic absorption of gaseous hydrides utilizing sodium borohydride reduction". At. Absorpt. Newsl., 12, 1973, 93.

FISHBEIN, L." Environmental metallic carcinogens: an overview of exposure levels" J. Toxicol. Environ. Health., 2, 1976, 77.

FLOWLER,B.A.;ISHINISHI,N.;TSUCHIYA,K.;VAHTER,M. "Arsenic" in "Hamd-
book of the Toxicology of Metals. Editors FRIBERG,L;NORDBERG,G.F.;
VOUK,V.B., Amsterdam, Elsevier/North-Holland, 1979, pp293.

GODDEN,R.G.;THOMERSON, D.R. "Generation of Covalent Hydrides in AAS"
    Analyst., 105, 1980, 1137.

HOLAK, W. " Gas-sampling technique for As determination by AAS"
    Anal. Chem., 41, 1969, 1712.

NAKASHIMA, S. " Flotation separation and AAS determination of Se(IV)
    in water". Anal. Chem., 51, 1979, 654.

NARASAKI, H. "Semi-automated determination of As and Se in river wa-
    ter by HG-AAS using a gas collection device". J. Anal. At. Spectrom.
    3, 1988, 517.

ROBBINS,W.B.;CARUSO,J.A. "Development of Hydride Generation Methods
    for Atomic Spectroscopic Analysis". Anal. Chem., 51, 1979, 889A.

VERLINDEN,M;DEELSTRA,H.;ADRIAENSSENS,E. "Determination of selenium
    by AAS. A reviews". Talanta., 28, 1981, 637.

WHO. " International Standards for Drinking Water". 2th Ed Genève
    WHO, 1971.

YU,M.;LIU,G.;JIN,Q. "Determination of trace As,Sb,Se and Te in various
    oxidation states in water by HG and AAS after enrichment and sepa-
    ration with thiol cotton". Talanta, 30, 1983, 265.

Chapter 38

AN EFFECTIVE SINGLE WELL INJECTION/DETECTION TECHNIQUE FOR THE EVALUATION OF THE LOCAL SCALE DISPERSIVITY OF AQUIFERS

J P L Ferreira (National Laboratory of Civil Engineering, Lisbon, Portugal)

ABSTRACT

This paper deals with a subject which is fundamental for the development of groundwater pollution studies: tracer experiments for the evaluation of the longitudinal dispersivity of porous media. An effective single-well injection/detection technique, developed in FERREIRA 1986 and 1987, and its results are presented and compared with those obtained with the two-well technique. The tracer experiments were carried out in Rio Maior (Central Portugal), with sodium chloride solutions, using two fully penetrating piezometers, either in separate or together, and a pumping well.

KEY WORDS: groundwater studies; tracer experiments; Rio Maior, Portugal

1. DESCRIPTION OF THE SINGLE WELL INJECTION/DETECTION TECHNIQUE

The fixed equipment used in the tracer experiments consisted of a groundwater pumping station, designated by B5 and of two piezometers, each 1 1/2" in size, denominated by PC7/2 and PC8. The piezometers are located radially with reference to the well, at distances of 15 m for the PC7/2 and 40 m for the PC8, Fig. 1. Well B5, fully penetrating into the aquifer, is 77 m deep. At 30 m depth a pump was installed, which discharges 80 m^3/h under the mean pressure of 0.88 MPa. The two piezometers fully penetrating into the aquifer were constructed in PVC. PC7/2 is 70.8 m deep, its screened zone is 39 m long and begins at 30 m depth. PC8 is 78.6 m deep, its screened zone is 53 m long and begins at 25 m depth.

Using this equipment, six experiments with sodium chloride solutions were conducted. The objectives of those experiments were as follows: (1) To obtain NaCl breakthrough concentration curves over time in order to evaluate the longitudinal dispersivity of the Rio Maior white-sand aquifer and (2) to analyse the scale effect on the experimental results.

The experience gathered with Experiments 3 and 6 for a local scale (experimental distance of 0.5 m) will be hereinafter described. The results of these experiments will be compared with those of Experiment 1, which was developed for a global scale (experimental distance of 25 m).

The experiments were performed by injecting 1 m^3 of tracing solution, with the following NaCl concentrations: Experiment 1 - 50.35 g/l and Experiments 3 and 6 - 0.89 g/l.

To carry out Experiments 3 and 6, a new injection/detection technique, that only requires a single observation well, was developed. The piezometer used was PC7/2.

This new method consists of the following steps, Fig. 2: (1) Injection of 1 m^3 of fresh water from the aquifer. (2) Injection of 1 m^3 of tracing solution, straight afterwards step 1, (3) Injection of 1 m^3 of fresh water from the aquifer, straight afterwards step 2.

Step (1) will only be necessary to avoid secondary effects, if the piezometer was not previously clean, or if the aquifer water is unclean as a result of previous tests.

Fig. 3 shows the plotting as given directly by the recorder, for Experiment 3. The concentration peak recorded was 371.31 mg/l, and the difference between the groundwater initial concentration and the peak concentration was 188.79 mg/l.

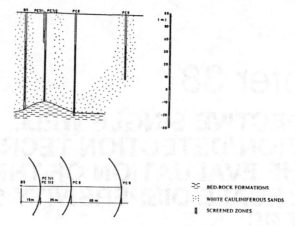

Figure 1: Pumping well B5 and surrounding piezometers

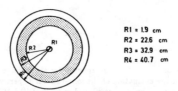

R1 = 1.9 cm
R2 = 22.6 cm
R3 = 32.9 cm
R4 = 40.7 cm

Figure 2: Slug rings obtained in Experiment 3

The determination of the distance travelled by the slug ring after the 2nd fresh-water injection is not accurate. In fact, the calculation of the radii shown in Fig. 2 should be based on the values of the total porosity of the aquifer white-sands, on the values of the effective porosity, or else on an intermediate value. This calculation becomes still more difficult on account of the coarse sand filter (2 to 3 mm grain size) that envelops the piezometer, whereas the grain size of the aquifer sands is 0.5 to 1 mm. The thickness of the filter is likely to be small, its existence, however, complicates the analysis of the results. Assuming a total porosity of 31% for the white-sand aquifer and an effective screened zone in the piezometer 18.0 m long, the following calculations were performed (Fig. 2): $R1 = 1.9$ cm (radius of the piezometer); $R2 = 22.6$ cm; $R3 = 32.9$ cm; $R4 = 40.7$ cm; Mean radius of the pollution slug = 27.8 cm.

The concentration peak is therefore expected to travel a total distance of aprox. 0.50 m onwards and backwards with reference to the piezometer (0.25 m in each opposite direction).

A second experiment similar to Experiment 3 , designated by Experiment 6, was carried out two month later to confirm the results previously obtained. The difference of concentrations recorded in Experiment 6 was similar to that of Experiment 3: 186.1 mg/l and 188.8 mg/l respectively. The shapes of the two curves were also similar in both experiments. For Experiment 6 the injection of 1 m³ of tracing solution took 28 min to be absorbed by the piezometer, whereas the 1 m³ of clean water injected afterwards took 20 min. The total injection time was 48 min. The peak concentration was registred after 5 h of experiment.

The single well injection/detection technique has the following advantages: (1) The technique is easy to perform. (2) The fixed equipment required for the experiments only consists of one observation well (the existence of a pumping well in the vicinity only diminishes the time necessary for the experiment). (3) The time of experiment is fairly reduced. (4) The experiment is independent of the aquifer natural flow direction. (5) The sodium chloride concentration in the injected solution is very low, and so the tracer (NaCl) can be considered an ideal tracer. (6) The curves obtained from the experiments are fairly perfect, allowing good post-processing. The reduction of the concentration curve peak referred to the injection concentration is very low (about 4.7 times).

As drawbacks the following can be pointed out: (1) The fact that filters envelop the piezometers leads to distortion of the information recorded. (2) The scale of the experiments is local and thus the area of the aquifer covered by the experiment is too reduced. (3) The experimental results may be affected by variations of the positioning of probes in the piezometers.

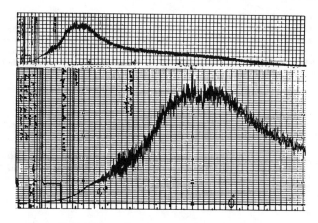

Figure 3: Breakthrough curve recorded during Experiment 3

2. COMPARISION OF THE RESULTS WITH THOSE OBTAINED WITH THE TWO-WELL TECHNIQUE

To finalize let us compare the results obtained with the method of moments applied to the breakthrough curves over time of Experiments 1 and 6, i.e. respectively with experimental distances of 25 m and 0.5 m. Fig. 4 shows the dimensionless concentration curves of the tracer experiments carried out in Rio Maior and of a mathematical model.

The global scale longitudinal dispersivity of the white-sand aquifer of Rio Maior (experimental distance equal to 25 m), was evaluated with the results of Experiment 1, being equal to the following value: $\alpha_L = 0.13$ m.

In Experiment 6 the analysis is less obvious than in Experiment 1, owing to the difficulty of the definition of the experimental distance. The most correct and practical analysis, as it seems, consists in the calculation of the concentration curve moments by using an experimental distance corresponding to twice the return distance of the slug towards the piezometer, after the injections of the tracer and of the second fresh water (respectively steps 2 and 3) are over. This means to consider that the shape of the breakthrough curve is originated both during the injection and during the return of the slug towards the piezometer. The total dispersivity α_{L_c}, i. e. corresponding to the onward and backward of the slug ($\alpha_{L_c} = \alpha_L/2$), obtained with the method of moments for Experiment 6 has the following value: $\alpha_{L_c} = 0.03$ m. This value was considered representative of the local dispersivity of the white-sand aquifer of Rio Maior, for an experimental distance of 0.5 m.

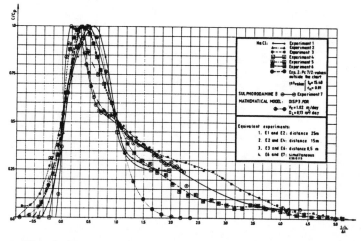

Figure 4: Dimensionless concentration curves of the tracer experiments and of a mathematical model

The values mentioned above for the dispersivities, i.e. $\alpha_{L_c} = 0.03$ m for the local scale and $\alpha_L = 0.13$ m for the global scale, are in accordance with the bibliographical analysis developed for the study. The scales of the tracer experiments described in the literature were found to affect considerably the values of the longitudinal dispersions of the porous media. Experiments in the local scale, namely laboratory experiments, have quantified longitudinal dispersions in two or three orders of magnitude below those observed in real aquifers.

REFERENCES

Ferreira, J. P. Lobo (1986)   *A Dispersão de Poluentes em Águas Subterrâneas.* Tese de Especialista, Laboratório Nacional de Engenharia Civil, Lisboa.

Ferreira, J. P. Lobo (1987)   *A comparative analysis of mathematical mass transport models and tracer experiments for groundwater pollution studies.* Doktor-Ingenieur Dissertation, Technische Universität Berlin, Berlin. (Memória N. 724, Laboratório Nacional de Engenharia Civil, Lisboa.)